Soil Fertility Advances and Nutrient Management

Soil Fertility Advances
and Nutrient Management

Edited by **Lester Bane**

R CALLISTO
REFERENCE

New York

Published by Callisto Reference,
106 Park Avenue, Suite 200,
New York, NY 10016, USA
www.callistoreference.com

Soil Fertility Advances and Nutrient Management
Edited by Lester Bane

International Standard Book Number: 978-1-63239-564-1 (Hardback)

Printed in the United States of America.

Contents

Preface

The advances in the field of soil fertility are described in this book along with information regarding nutrient management. It compiles contributions of various soil fertility experts and researchers. It extensively discusses the topic of soil mapping and soil fertility testing, describing spatial heterogeneity in soil nutrients within natural and managed ecosystems and latest soil testing methods and information on response of soil fertility indicators to agricultural practices. It further discusses, the theme of organic and inorganic amendments for increasing soil fertility, describing fertilizing materials that provide important amounts of essential nutrients for plants. The book also covers topics on integrated nutrient management planning. Herein, case studies describing successful application of this approach expansively across both geographically large as well as remote areas, to increase the production of staple crops and forages have been presented from Central Europe, South America and Africa.

Significant researches are present in this book. Intensive efforts have been employed by authors to make this book an outstanding discourse. This book contains the enlightening chapters which have been written on the basis of significant researches done by the experts.

Finally, I would also like to thank all the members involved in this book for being a team and meeting all the deadlines for the submission of their respective works. I would also like to thank my friends and family for being supportive in my efforts.

Editor

Part 1

Soil Mapping and Soil Fertility Testing

1

Soil Phosphorus Tests and Transformation Analysis to Quantify Plant Availability: A Review

Z.M. Zheng[1,2] and T.Q. Zhang[2]
[1]School of Environmental Sciences,
University of Guelph, Guelph, Ontario,
[2]Greenhouse and Processing Crops Research Centre,
Agriculture and Agri-Food Canada, Harrow, Ontario,
Canada

1. Introduction

Phosphorus (P) is regarded as the most important soil nutrient after nitrogen (N) for plant growth and development as it plays key roles in plant metabolism, structure and energy transformation. It has also been recognized as a potential pollutant in waters (Anderson, 1980). The P dynamics in soils and cycling in agro-ecosystems are of increased interest due to its contribution to the current environmental, agronomic and economic issues (Sharpley and Tunney, 2000).

Soil P tests involve extraction of P from soils with chemical or ion-sink extractants followed by a quantification of P in the extracting solution. Soil test is expected to determine the amount of P that can contribute to crop growth or water contamination. From the standpoint of availability to plants, soil P can be divided into functional pools of differing bioavailability (Tiessen et al., 1982). The information on soil P transformation between those pools is useful to predict P bioavailability as well as the risk of P transfer from soil to surface waters. However, soil P transformation has received less attention attributable to the difficulties associated with separation of inorganic P (P_i) and organic P (P_o) fractions and compositional identification of soil P_o pools. Such investigation is currently possible with an improved sequential fractionation procedure and adoption of advanced techniques such as nuclear magnetic resonance spectroscopy (NMR) and synchrotron-based techniques like X-ray absorption near-edge structure (XANES).

In this chapter, soil P tests with chemical and ion-sink methods and the influences of farming practices on P status and P transformation in soils were summarized. The emphasis was on description of the interrelationships between soil P pools of differing bioavailability. The analytical methods to assess P transformation in soils including advanced techniques such as path analysis, modeling and synchrotron-based techniques were described briefly. The limitations of methodologies of soil P tests and P transformation analyses were discussed and findings from our studies were integrated in this context.

2. Soil tests for available P

2.1 Conventional chemical extractions

Available P is the amount of P in soils that can be extracted or mined by plant roots and utilized by plant for its growth and development. It is a quantitative (extensive) parameter and is influenced by the prevailing soil conditions at a particular time and the plant's ability to extract P from soil solutions (Raven and Hossner, 1993; Holford, 1997). However, most people often use available P synonymously with P availability, an intensive parameter that does not reflect the amount or concentration of available P (White and Beckett, 1964).

Availability of P for plant utilization is not a function of its concentration in soil, but rather on the rate of its release from soil surface into soil solution (Abdu, 2006). Available P is composed of soil solution P and is replenished by P that enters the solution by desorption or dissolution of P_i associated with the soil solid phase or by mineralization of P_o (Hedley et al., 1982). The measurement of available P therefore needs to consider both the amount and rate of release of P from the solid phase. Very few appropriate methods have been developed. Isotopic dilution (^{32}P) techniques theoretically permit researchers to quantify the processes of P_o mineralization, dissolution of insoluble minerals and desorption of aggregate P, and could be likely used for this purpose. But errors involved in the measurement of change rates make it difficult to extrapolate the continuing release (or isotopic dilution) rates to a temporal scale corresponding to cropping seasons and growth cycles under field conditions (Tran et al., 1988; Sharpley et al., 1994). Thus, some limitations have to be overcome to give results that have practical application.

The most widely used soil P tests are chemical extractions that use chemical reagents to extract available P from soils. Water was probably the first extractant used to measure P in soils. The small amounts of soil P extracted by water and difficulties related to chemical analysis limit the use of water as an extractant. Bray and Kurtz (1945) used a combination of HCl and NH_4F to remove easily acid soluble P forms, largely Al- and Fe-phosphates. In 1953, Mehlich introduced a combination of HCl and H_2SO_4 acids (Mehlich 1) to extract P from soils in the north-central region of the U.S. In the early 1980s, Mehlich modified his initial soil test and developed a multi-element extractant (Mehlich 3) which is suitable for removing P and other elements in acid and neutral soils (Mehlich, 1984). Olsen et al. (1954) introduced 0.5 M sodium bicarbonate ($NaHCO_3$) solution at a pH of 8.5 to extract P from calcareous, alkaline, and neutral soils. The routine soil P tests may not give insight into the level of plant available P as the chemical reagents may solubilize non-labile P. For instance, the acidic Bray and Mehlich I extractants can dissolve Al- and Fe-phosphates, while Olsen extractant removes dissolved and adsorbed P on calcium carbonate and Fe-oxide surfaces (Mallarino, 1997). Moreover, these chemical extractants are not applicable over all soil types, which underscore the use of them for soil P extractions (Myers et al., 2005). Bray and Mehlich-3 extractants were designed to extract P from non-calcareous soils, whereas the Olsen method was meant for non-acidic soils. Furthermore, those conventional soil P tests derived from mineral soils may not necessarily be applicable for organic soils, although some routine soil P tests are being adopted to make agronomic recommendations in muck soils (Castillo and Wright, 2008; Wright, 2009). This, however, is an "alternative-than-never" choice, at this moment without specific test for organic soils available.

2.2 Ion-sink extractions

The ideal soil P test should be able to extract P in a similar manner as plant roots (Abdu, 2006). Ion sink tests have been developed to simulate how plant roots extract available P from soils. Methodologies were detailed in Sibbesen (1978), Schoenau and Huang (1991), Chardon et al. (1996) and Myers et al. (2005).

The ion-sink concept used in P extraction includes ionic exchange resin membranes, resin bags and FeO coated filter papers or strips. The exchange membrane resin is employed with either the batch or miscible displacement techniques. The batch technique involves the use of a wide soil to solution ratio, which varies the P concentration in the solution and the quantity of desorbed P as the reaction proceeds. If there is inadequate mixing of solution with the ion exchanger, a limited rate of reaction may occur (Sparks, 1985). This may also lead to a change in surface chemistry of the colloids and break down of soil particles (Barrow and Shaw, 1977). With miscible displacement technique, there can be error in dilution that leads to error in interpretation by altering the P concentration of soil (Sparks, 1999), which is more pronounced in colloids having low ion absorbing power (Carski and Sparks, 1985). Also, dispersion of soil colloids may not be fully achieved (Sparks, 1999). Even though anion exchange resin extracts more P than the FeO-coated papers, the additional P extracted may not be plant available (Robinson and Sharpley, 1994). Soil particles can contaminate the FeO-coated papers during shaking, which can lead to error in estimating desorbable P (Uusitalo and Yli-Halla 1999). This can, however, be minimized by the use of $CaCl_2$ solution as the background electrolyte to minimize soil dispersion (Myers et al., 2005), but reaction with $CaCl_2$ may reduce the amount of P extracted (Koopmans et al., 2001).

In spite of the aforementioned disadvantages, ion-sink methods, especially the anion exchange membranes, are still regarded as the most effective method of plant available P extraction. In addition to their capability to extract P from a variety of soil types, regardless of soil properties (Sharpley ct al., 1994), the ion sink methods simulate plant roots and extract P from soil without alteration of the soil chemical and physical characteristics (Raven and Hossner, 1993). Further more, the resin membranes can be re-used several times without losing its extracting power (Schoenau and Huang, 1991). This property makes it relatively cheaper than the FeO-coated papers. The compatibility of resins with soil solution chemistry and pH can be achieved by charging the resins with either HCO_3^- or Cl^- (Agbenin and Raij, 2001). The use of HCO_3^--resin is more advocated than Cl^--resin, as plant roots accumulate bicarbonate in the rhizosphere, leading to an increase in rhizosphere pH in acid to neutral soils and a decrease in rhizosphere pH in calcareous soils (Sibbesen, 1978), while when Cl^--resin is used, the Cl^- accumulated in solution inhibits the exchange reaction (Myers et al., 2005).

2.3 Phosphorus determination

Analysis of extracted P is typically done by colorimetry, most notably the Murphy and Riley (1962) method. A specific ion reacts with the color developing reagents to form colorful complex (e.g. blue antimony phospho-molybdate), then light absorption by the formed complex is detected at a specific wavelength. Colorimetric procedures are sensitive, reproducible and enable to automated analysis. In addition, the methods can be accommodated to water samples, digested solutions and extracts (Pierzynski et al., 2008).

Inductively coupled plasma (ICP) spectrophotometry is also now commonly used for P determination, particularly in routine soil P tests offered by public and commercial laboratories. The use of ICP has increased as the use of multi-element soil extractants becomes more popular. Results from ICP are not always directly comparable to those from colorimetric analyses (Pierzynski et al., 2011), as ICP estimates the total P in a solution while the colorimetric procedures measure P that can react with the color developing reagents. Moreover, there are certain limitations that must be considered while evaluating data generated by ICP, such as the matrix effects, spectral interference, etc. (de Boer et al., 1998).

Nuclear magnetic resonance (NMR) is a physical phenomenon based upon the magnetic property of atomic nucleus. It is observed that magnetic nuclei, like ^{1}H, ^{13}C and ^{31}P, could absorb radio frequency when placed in a magnetic field with a specific strength, described as the resonance of the nucleus. Different atoms in a molecule resonate at different frequencies at a given field strength. This is a powerful method that allows researchers to determine the structure of chemical compounds. The use of solution ^{31}P NMR spectroscopy has allowed us to identify P forms in soils and residual materials, and confirm P forms estimated by commonly used chemical extractants, such as sequential fractionation schemes. This technique has enabled more accurate determination of organic forms of P in soils and residual materials (Zhang et al., 1999; Turner and Leytem, 2008). In addition, the use of synchrotron-based techniques (e.g. XANES) has provided insights into both P_i and P_o forms in soils and residual materials. Descriptions of these approaches were detailed by Beauchemin et al. (2003) and Shober et al. (2006). These analytical advances have been critical in gaining a more detailed understanding of soil P transformation and reaction products following land application of residual materials. This information has helped assess the fate, reactivity, behavior of specific forms of P and the environmental implications of land application of materials such as biosolids and animal manures (Pierzynski et al., 2008).

3. Soil P transformation

3.1 Soil P fractionation

From the standpoint of availability to plants, soil P is divided into three functional fractions each including both P_i and P_o forms: (1) readily labile P pools, (2) moderately labile P pools and (3) sparingly soluble P pools (Tiessen et al., 1982). Due to the technical difficulties involved in isolating specific P compounds from soils, the financial, practical and safety limitations of using P radioisotopes, and limited access to NMR spectroscopy and synchrotron-based techniques, most of field studies on P transformation in soils must rely on alternative methods of measuring changes in soil P fractions. One technique is to sequentially separate soil P into various P_i and P_o fractions based on their bioavailability and biological associations, by first removing labile P, then more stable forms. This sequential extraction method was originally presented by Chang and Jackson (1957) and was later modified by Petersen and Corey (1966) and Williams et al. (1967). Briefly, this soil P sequential procedure uses NH_4Cl to extract labile P_i, NH_4F to dissolve specifically Al-associated P_i followed by NaOH to extract Fe-bound P_i and by dithionite-citrate for reductant-soluble or 'occluded' P_i forms. The HCl dissolved Ca-bound P_i and the final residue is analyzed by Na_2CO_3 fusion for total P. However, the procedure presents many interpretational problems. Since P_i reprecipitates during the fluoride extraction, the separation of Al- and Fe-associated P_i is not reliable and the reductant-soluble or 'occluded' P_i is an ill-defined pool. Furthermore, the P_o fraction is ignored (Williams et al. 1967).

An improved P fractionation scheme was developed by Hedley et al. (1982). This sequential extraction aimed to quantify labile P_i, Ca-associated P_i, Fe- and Al- associated P_i, as well as labile and more stable forms of P_o. With this improved procedure, soil P is initially extracted with an anionic-exchange resin (resin-P), and then with $NaHCO_3$ ($NaHCO_3$-P_i/P_o). Resin-P and $NaHCO_3$-P are assumed to be the readily plant-available fractions and generally regarded as P that is sorbed on carbonates, sesquioxides or crystalline minerals (Mattingly, 1975). Moderately labile P, sorbed on amorphous Fe and Al minerals and the 'protected P' that is occluded or contained within aggregates, is then extracted with NaOH (NaOH-P_i/P_o). The sparingly soluble apatite-type P minerals are extracted with HCl (HCl-P), and the residual P is dissolved by a H_2O_2 (or $K_2S_2O_8$)-H_2SO_4 digestion (residual H_2SO_4-P). The H_2SO_4-P fraction, chemically stable and not readily available to plants, may consist of either P_o or P_i, or both. This approach is currently the only one to evaluate both available P_i and P_o in soils with moderate success (Tiessen and Moir, 1993), although modifications to the procedure are often study-specific, for example, the use of de-ionized water instead of anionic-exchange resins to extract readily labile P. It was proven useful for establishing the effects of long-term cropping practices on P_i and P_o fractions and transformation (O'Halloran et al., 1987a; Paniagua et al., 1995; Richards et al., 1995; Tran and N'dayegamiye, 1995; Zhang and Mackenzie, 1997a, 1997c; Zhang et al., 2004; Zheng et al. 2001, 2002, 2004a) and for assessing soil P status in field soils (Simard et al., 1995; Beauchemin and Simard, 2000; Zheng et al., 2004b). It was also valuable for quantifying changes in soil P fractions in short-term incubations (Hedley et al., 1982; Iyamuremye et al., 1996) and from greenhouse experiments (Ivarsson, 1990). In addition, it offers a useful index of the relative importance of P cycling by biological versus geochemical processes in soils at different stages of development (Cross and Schlesinger, 1995).

3.2 Conceptual model of soil P transformation

The P transformation in soils involves complex mineralogical, chemical and biological processes (Fig. 1). The P cycle in soil is a cohesive dynamic system under the influence of long-term chemical transformations and short-term changes due to plant uptake or cropping. The leaching of bases, the removal of carbonates and the increasing Fe and Al activity that accompany the development of soils cause a shift from primary to secondary P_i forms and also influence the stabilization of organic matter and its associated P_o (Walker and Syers, 1976). The abundance and activity of various P_i forms and the turnover of P_o in soils control the replenishment of labile solution P following plant uptake. Surface associated or amorphous P_i replenishes the labile P pool while more stable crystalline species act as a sink as well as long-term reservoir of P, depending on other soil properties such as pH (Murrman and Peech, 1969). In this way, hydroxide or acid extractable P_i (NaOH-P_i or HCl-P) may act as the quantity factor that buffers the more labile P forms. The P_o may perform a similar function through controlled mineralization-immobilization processes (McGill and Cole, 1981). Soil organic carbon has a major role of promoting processes involved in P transformation, through its contribution as energy source for microbial activity (Stevenson, 1986). Microbes are heavily involved in P transformation in three ways: (1) by decomposition of P_o compounds, with release of available P_i; (2) by immobilization available P into cellular material; and (3) by promoting the solubilization of fixed or insoluble mineral forms of P, such as through the production of chelating agents (Stevenson, 1986; Frossard et al., 2000).

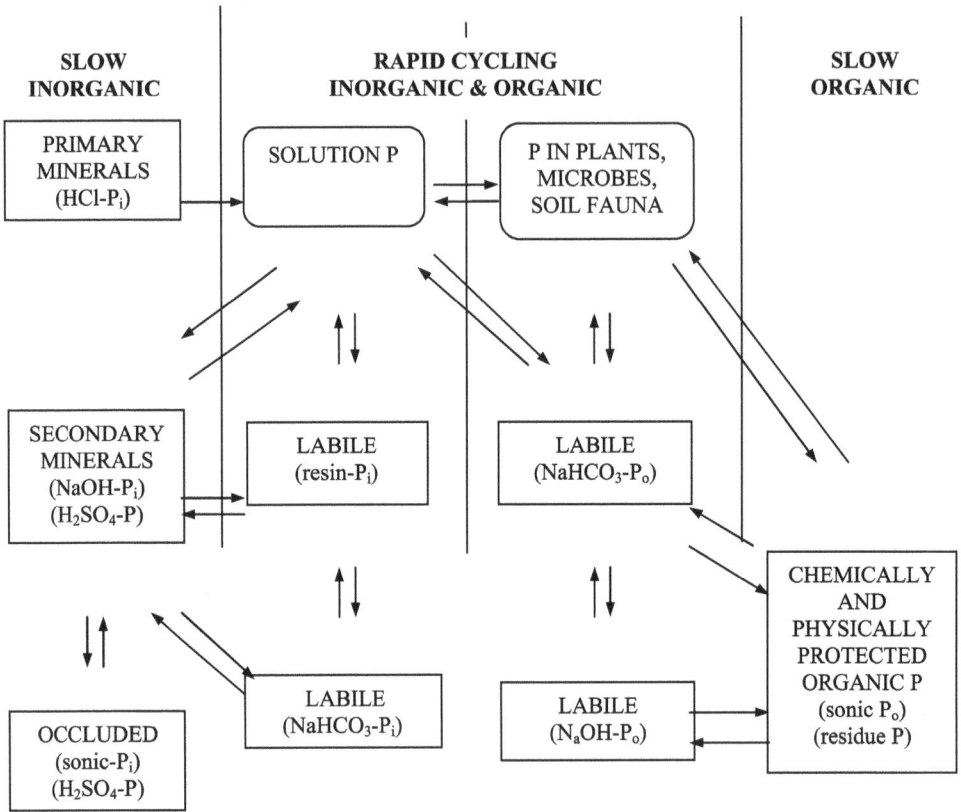

Fig. 1. A conceptual model of soil phosphorus transformation with its measurable components (Source: Tiessen et al., 1984)

3.3 Impacts of farming practices on soil P fractions and transformation

3.3.1 Effects of fertilization

Numerous studies have investigated the effects of fertilizer additions on P fractions and transformation. Generally, resin-P, $NaHCO_3$-P_i and $NaOH$-P_i reflect the difference between fertilizer P and crop P removal. Stable fractions are less affected by inorganic fertilizer P or manure application (O'Halloran, 1993; Richards et al., 1995; Tran and N'dayegamiye, 1995; Zhang and MacKenzie, 1997a, 1997c). The P_o fraction is either unaffected by excess inorganic fertilizer P (McKenzie et al., 1992a, 1992b; Zhang and MacKenzie, 1997a; Zheng et al., 2001) or increased with high rates (Schmidt et al., 1996), but it increases when inorganic fertilizer is combined with farmyard manure (O'Halloran, 1993; Tran and N'dayegamiye, 1995; Zhang and MacKenzie, 1997a). The P_o is a source of P to plants when fertilizer P is inadequate to meet crop P requirements (Zhang and MacKenzie, 1997a, Zheng et al., 2001, 2004b). In many cases, the readily labile P increases more when the two P nutrient sources are applied together than when only one is applied (Paniagua et al., 1995). Consequently, the effects of P fertilization on different P pools in soils depend on nutrient sources and rates of P applied as chemical fertilizer or manure (O'Halloran, 1993; Zhang and MacKenzie, 1997a, 1997b; Zheng et al., 2001, 2002).

Long-term organic residues or manure application increases microbial activity and potential mineralization of soil organic matter (N'dayegamiye and Angers, 1990). Consequently, it may induce transformation of soil P_o to P_i fractions and increase the available P in surface and subsurface soil horizons (Sharpley et al., 1984; Tran and N'dayegamiye, 1995, Zheng et al., 2002, 2004a). This mineralization of P_o during the growing season is very important for P availability to plants. Nevertheless, the excessive application of manure or of P-rich organic amendments plays a significant role in decreasing the P sorption capacity in soils (Sharpley et al., 1993; Simard et al., 1995; Beauchemin et al., 1996; Zheng et al., 2001). Decreased P sorption capacities might increase the risk of contamination of the receiving water bodies.

3.3.2 Effects of tillage

The impacts of tillage on P forms and availability have been the subject of many investigations. There was an increase in total and labile P contents in surface soil layer due to no-till operations (Cruse et al., 1983; Weill et al., 1990; Selles et al., 1997). This was attributed to the enhanced microbial activity and mineralization of soil P_o (Follett and Peterson, 1988) or to a lack of soil mixing with fertilizer P thus reducing P fixation by soil colloids (Sweeney, 1993; Selles et al., 1997, 1999). As a result, extractable P in surface soils is increased in no-tillage compared to conventional tillage. Soil P_o is generally increased near the bottom of plow layer by tillage due to the incorporation of crop residues (Dick, 1983, Weill et al., 1990). Nevertheless, O'Halloran et al. (1987a) did not observe a significant difference in the size of soil P fractions except for NaOH-P_o between stubble mulch, bare sod and no-till practices in a grassland soil. On the other hand, Sharpley et al. (1993) found that no-till plots tended to have higher soil moisture content and lower temperature, which resulted in more dissolved P in the runoff than conventionally tilled plots. Blevins et al. (1990) observed that movement of P was reduced using a system of chisel plowing compared to direct seeding or conventional plowing. The chisel plowing was associated with a reduced runoff following the spreading of fertilizers on the soil surface. Therefore, as with fertilization, tillage can alter distribution of P fractions and affect the transformation and transfer of P in soils.

3.3.3 Effects of cropping systems

Cropping systems also have major effects on changes in P fractions in soils. As cropping systems change from less intensive systems (i.e. crop-summer fallow) to more intensive cropping sequences (i.e. continuous wheat and wheat-wheat-fallow), soil P availability and transformation become less predictable due to larger P_i and P_o components from greater residue and litter which are maintained on soil surface (O'Halloran, 1987a, b). Wagar et al. (1986) found a buildup of P_o occurred with and without added P fertilizer when cropping systems changed from conventional winter wheat-fallow to more intensive rotations. Bowman and Halvorson (1997) observed that a change from wheat-fallow to continuous wheat cropping produced a significant increase of labile P fractions in 0-5 cm surface soils. Zheng et al. (2001) observed that forage-forage-barley rotation produced larger labile P fractions than barley monoculture in the 30-60 cm layer in a Labarre silty clay after 10 years of cultivation. Recently, Zhang et al. (2006) investigated the P status after 45 years of consistent cropping practices under three cropping systems of continuous corn, rotation corn and continuous bluegrass. Compared to the adjacent native soil in forest ecosystem, the study showed that continuous cropping without P fertilization decreased all P forms significantly except for water extractable P_o, with the largest decrease in labile P_i and moderately labile P_o; continuous cropping with fertilization resulted in comparable total P concentrations in the continuous corn and rotation corn systems and increased total P in the continuous bluegrass sod. The study also indicated that long-term cropping significantly enhanced the rate of moderately labile P_o mineralization, regardless of P fertilization, especially for continuous corn cropping system. However, compared to monoculture, crop rotations showed no effect on nutrient contents in the 0-20 cm soil layer in an experiment with soybean, corn and wheat in an Oxisol and in a Rhodic Ferralsol (DeMaria et al., 1999).

3.4 Soil texture effects on soil P fractions and transformation

Soil texture can be related to changes in P fractions and transformation. O'Halloran et al. (1985) found that up to 90% of the spatial variability in total P content of a Mollisol was explained by texture. A similar study showed that significant proportions of variability of all P fractions, except for H_2SO_4-P in a Brown Chernozemic loam, could be attributed to changes in sand content (O'Halloran et al., 1987b). Increasing silt plus clay content has been significantly correlated with larger soil resin-P, $NaHCO_3$-P_i/P_o and $NaOH$-P_i/P_o pools. HCl-P was positively correlated with sand content (O'Halloran et al., 1987b). There is evidence that P transformation in soils were closely linked to microbial activity and C dynamics, which affect P mineralization and immobilization (Hedley et al., 1982). For example, a loam soil supported higher microbial biomass than the sandy loam and sandy soils (Cooper and Warman, 1997). Huffman et al. (1996) indicated that soil texture had a greater effect on P transformation than did the combined effects of residue addition, residue placement and nutrient addition, because soil texture affected the labile P_i, liable P_o, and microbial P pools. Therefore, among those factors, the particle size of soil fractions tends to account for larger proportion of variability in soil P fractions. Those findings emphasize the importance of considering soil texture when planning studies of the effects of cropping practices on P fractions and transformation.

3.5 Profile-wise distribution of soil P fractions

Much effort has been devoted to investigate the vertical changes in forms and distribution of soil P as influenced by cropping practices. These changes in P forms were mainly limited to the top of 30 cm (O'Halloran, 1993; Sharpley et al., 1993; Bowman and Halvorson, 1997; Selles et al., 1997), or to depths of 65 cm (Reddy et al., 1980; Mozaffari and Sims, 1994; Richards et al., 1995), with a few studies extending to depths $\geq$ 90 cm (Chang et al., 1991;

Simard et al., 1995; Beauchemin et al., 1996; Zheng et al., 2001). These studies suggested that the impacts of cropping practices on soil P fractions extended deeper in soil profile than the depth distributed by primary tillage, and P profile-wise distribution was often complicated by P downward movement. The P_o movement in soils was greater than that of P_i (O'Halloran, 1993; Richards et al., 1995). This may be due to the fact that P_o was not retained as strongly as P_i by sorbing components of solid phase, or perhaps that P_o movement was driven by mineralization process. The P_o migrated to subsoil can be mineralized, since P_o was used by microorganisms as an energy source for their metabolism in reduced layers of poorly drained soils (Zheng et al., 2001).

The downward migration of P is associated with changes in soil P sorption capacity (Reddy et al., 1980). Simard et al. (1995) found that P sorption index (P_{si}) in agricultural soils was markedly decreased in all three horizons compared with forest soils from a watershed in Quebec. Accumulation of added P has lead to an increase in total labile P pool in the A horizon (0-30 cm), while resilient P pools were the major sinks for mobile P in the B (30-60 cm) and C (60-90 cm) horizons in agricultural soils (Simard et al., 1995; Beauchemin et al., 1996; Zheng et al., 2001). This finding was in line with previous studies suggesting that repeated additions of manure decreased P sorption capacity of soils and accelerated P downward movement (Reddy et al., 1980; Sharpley et al., 1993; Mozaffari and Sims, 1994).

4. Analysis of P transformation in soils

4.1 Path analysis of soil P transformation

Path analysis is a statistical technique used to examine interrelationships among variables that are often illustrated by a path diagram. It provides not only plausible explanations of observed correlations by constructing the cause-and-effect model, but also allows decomposition of observed correlations into direct and indirect effects (Johnson and Wichern, 1988). Path analysis was useful in linking changes in soil P fractions and clarifying concepts of P transformation in soils (Tiessen et al., 1984; Beck and Sanchez, 1994; Zhang and MacKenzie, 1997a; Zheng et al., 2002, 2004a). Using path analysis, Tiessen et al. (1984) found that in Mollisols, much of resin-P_i was derived from $NaHCO_3$-P_i and $NaOH$-P_i fractions, and in more weathered Ultisols, 80% of variability in labile P was accounted for by P_o forms. Beck and Sanchez (1994) showed that the $NaOH$-P_i fraction acted as a major sink for fertilizer P in soils and P_o was a major primary source of plant-available P in unfertilized soils of Peru. Zhang and MacKenzie (1997a), using the same approach for soils receiving manure and fertilizer P, indicated that P_o accumulated as $NaOH$-P_o through $NaHCO_3$-P_i. The $NaHCO_3$-P_i and $NaOH$-P_i were major sinks for added P. When mineral fertilizer was the only source of P, most $NaHCO_3$-P_i was directly supplied from fertilizer P through $NaOH$-P_i. Zheng et al. (2002, 2004a) investigated the P transformation in a Labarre silty clay as affected by nutrient sources and cropping systems in a long term study and showed that the roles of P_o pools were more important than P_i pools for P transformation and $NaHCO_3$-P_o was sensitive to P source and was likely acted as a transitory pool rather than as a sink or source of soil P (Fig. 2). Thus, path analysis can reflect the changes in P transformation depending on soil type, climatic conditions and cropping practices.

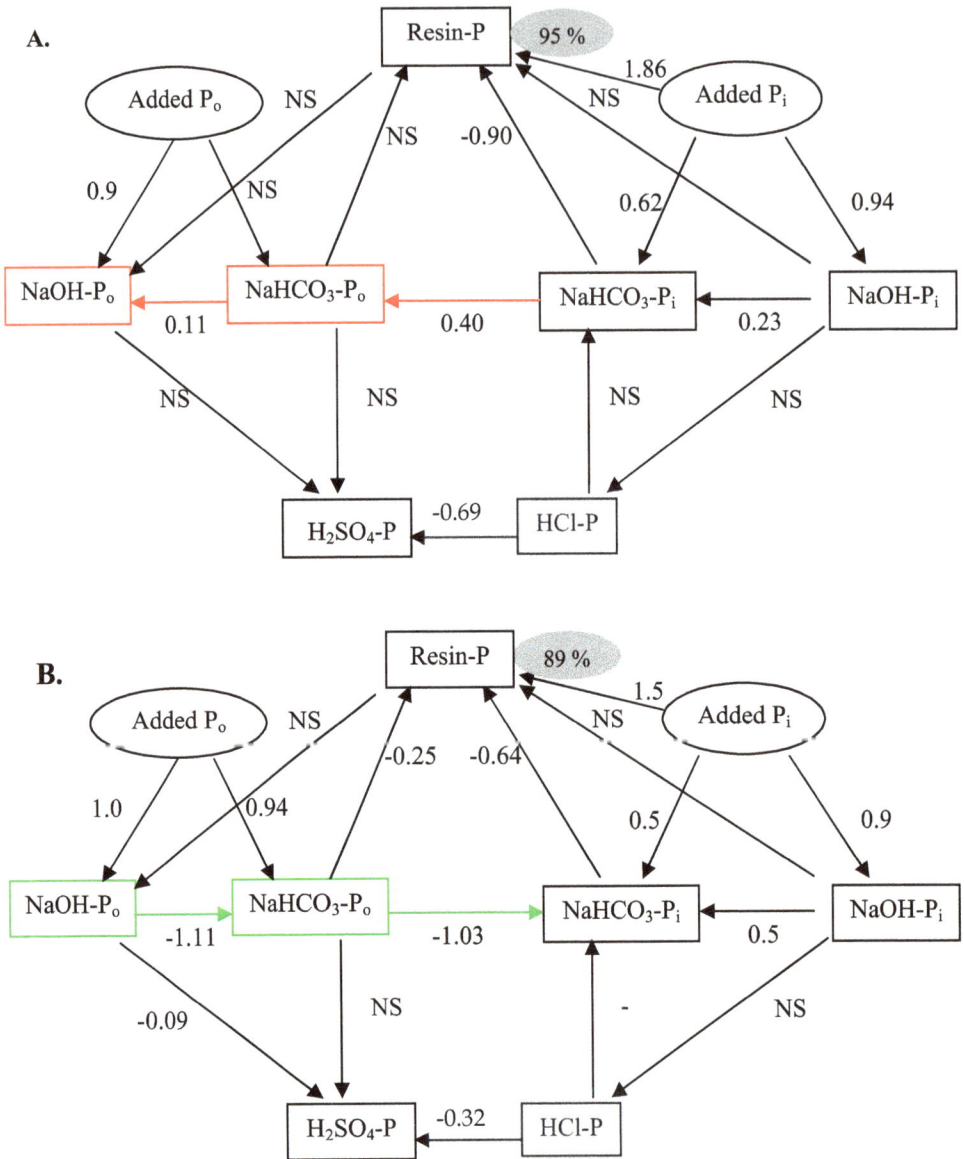

Fig. 2. Pathways of P transformation after 10 annual applications of mineral fertilizer (A) and liquid dairy manure (B) in the 0-15 cm soil of a Labarre silty clay under barley monoculture. The percentage value indicates the partial correlation between assed P_i and resin-P; numbers are path coefficients; NS, not significant at $P \leq 0.05$ (Source: Zheng et al., 2004a)

4.2 Modeling of soil P transformation

The P concentration in soil solution is in turn controlled by P transformation processes such as mineralization, precipitation and adsorption. The complexity of interactions among these processes has led to the use of both descriptive and mechanistic mathematical models to describe them. The interpretation of changes in P concentrations in soil solution has been modeled using a variety of approaches, such as adsorption model, sorption-transport model and multi-reaction model. Most of earlier simulations of P adsorption have been based on simple rate constants (Jones et al., 1984), such as Langmuir (Enfield et al., 1981), Freundlich (Shaviv and Shacher, 1989) and Elovich equations (Chien and Clayton, 1980). A few studies have used kinetic equations such as first- or second-order reaction rates to describe the P sorption in soils (Bowden et al., 1980; Beauchemin et al., 1996). Nevertheless, adsorption models did not fully describe the mechanisms of P transformation in soils, and hence were unlikely to be generally applicable. Van de Zee and Gjaltema (1992) proposed a sorption-transport model in which both sorption reversibility and precipitation irreversibility were taken into account, but reaction rates remained uncertain.

A mechanistic multi-reaction model was conceived to represent P concentrations under dynamic boundary conditions, and to explain temporal and spatial P distribution among water soluble and resin-, $NaHCO_3$-, $NaOH$-, and HCl-P fractions (Grant and Heaney, 1997; Grant et al., 2004). The model describes adsorption-desorption, precipitation-dissolution and ion pairing. The model can explain the temporal and spatial distribution of soluble and solid fractions under specified changes in boundary conditions in different soil types. However, the model requires more detailed, explicit information about soil chemical composition, which may not be readily available in routine fashion for many soils. Thus, where necessary, some assumptions must be made. To date, despite several P transformation models have been developed to describe certain processes, the majority of existing models need to be verified in heterogeneous soils under field conditions. Furthermore, it is unlikely that one model can kinetically represent all processes involved in P transformation.

5. Summary

Phosphorus is an essential element for plant growth and development, as it plays key roles in plant metabolism, structure and energy transformation. It is also a potential pollutant of water. The P dynamics in plant-soil-water systems is of increased interest due to its importance for environmental, agronomic and economic issues.

Soil P test involves P extraction from soils followed by a quantification of the nutrient in the extracting solution. The most widely used soil P tests are chemical extractions, such as Bray 1 & 2, Olsen, Mehlich I & III methods. Those chemical extractants are not applicable over all soil types, which is a limitation for soil P extraction. The ion-sink extractions, including ionic exchange resin membranes, resin bags, FeO coated filter papers or strips, simulate plant roots to extract P from soils without alteration of soil characteristics and have the advantage of extracting P from variety of soil type regardless of soil properties. The mainstay of P determination is the use of colorimetric procedures, most notably Murphy and Riley (1962). Inductively coupled plasma (ICP) spectrophotometry is becoming more popular for multi-element determination. The isotopic dilution (^{32}P) techniques, the ^{31}P NMR solution and the synchrotron-based techniques can provide insights into P chemistry and forms in soil and residual materials. Although these advanced analytical techniques are capable of gaining more detailed understanding of P dynamics in soils, unfortunately not all researchers have access to the required instrumentation.

Based on its availability to plants, soil P can be divided into functional groups of readily labile P pools, moderately labile P pools and sparingly soluble P pools, each includes both P_i and P_o forms. The P transformation in soils plays important roles in P bioavailability and mobility from soil to water, and was intensively affected by cropping practices, i.e. fertilization, tillage and cropping systems. It involves complex chemical, mineralogical and biological processes. The complexity of interactions among these processes has led to the use of both descriptive and mechanistic mathematical models to describe them. Path analysis is powerful in plausibly explaining interrelationships among P pools of differing bioavailability and clarifying changes in P fractions and transformation in soils. Mechanistic multi-reaction models that describe transformation processes of mineralization-mobilization, adsorption-desorption, precipitation-dissolution and ion pairing are available to interpret the temporal and spatial distribution of P fractions in soils. However, the models require more detailed, explicit information on soil chemical composition that are not routinely available for many soils. Furthermore, most mechanistic models need to be verified in heterogeneous soils under field conditions.

6. References

Abdu, N. 2006. Soil-phosphorus extraction methodologies: A review. Afr. J. Agric. Res. 1: 159-161.

Agbenin, J.O, B.V. Raij. 2001. Kinetics and energetics of phosphate release from tropical soils determined by mixed ion-exchange resin. Soil Sci. Soc. Am. J. 65:1108-1114.

Anderson, G. 1980. Assessing organic phosphorus in soils. p. 411-431. In F.E. Khasawneh et al. (eds.) The role of phosphorous in agriculture. ASA, Madison, WI.

Barrow, N.J., T.C. Shaw. 1977. Factors affecting the rate of phosphate extracted from soils by anion exchange resin. Geoderma. 18:309-323.

Beauchemin, S., R.R. Simard, and D. Cluis. 1996. Phosphorus sorption-desorption kinetics of soil under contrasting land uses. J. Environ. Qual. 25:1317-1325.

Beauchemin, S., and R.R. Simard. 2000. Phosphorus status of intensively cropped, neutral to calcareous soils of the St. Lawrence lowlands. Soil Sci. Soc. Am. J. 64:659-670.

Beauchemin, S., D. Hesterberg, J. Chou, M. Beauchemin, R.R. Simard, and D.E. Sayers. 2003. Speciation of phosphorus in phosphorus-enriched agricultural soils using x-ray absorption near-edge structure spectroscopy and chemical fractionation. J. Environ. Qual. 32:1809-1819.

Beck, M.A., and P.A. Sanchez. 1994. Soil phosphorus fraction dynamics during 18 years of cultivation on a Typic Paleudult. Soil Sci. 34:1424-1431.

Blevins, R.L., W.W. Frye, P.F. Baldwin, and S.D. Robertson. 1990. Tillage effects on sediment and soluble nutrient losses from a Maury silt loam. J. Environ. Qual. 19:683-686.

Bowden, J.W., S. Nagarajah, N.J. Barrow, A.M. Posner, and J.P. Quirk. 1980. Describing the adsorption of phosphate, citrate and selenite on a variable charge mineral surface. Austr. J. Soil Res. 18:49-60.

Bowman, R.A., and A.D. Halvorson. 1997. Crop rotation and tillage effects on phosphorus distribution in the Central Great Plains. Soil Sci. Soc. Am. J. 61:1418-1422.

Bray, R.H., and L.T. Kurtz. 1945. Determination of total, organic, and available forms of phosphorus in soils. Soil Sci. 59:39-45.

Carski, T.H, D.L. Sparks. 1985. A modified miscible displacement technique for investigating adsorption-desorption kinetics in soils. Soil Sci. Soc. Am. J. 49:1114-1116.

Castillo, M.S., and A.L. Wright. 2008. Soil phosphorus pools for Histosols under sugarcane and pasture in the Everglades, USA. Geoderma 145:130–135.

Chang, C., T.G. Sommerfeldt, and T. Entz. 1991. Soil chemistry after eleven applications of cattle feedlot manure. J. Environ. Qual. 20:475-480.

Chang, S.C., and M.L. Jackson. 1957. Fractionation of soil phosphorus. Soil Sci. 84:133-144.

Chardon, W.J, R.J. Menon, S.H. Chien. 1996. Iron oxide impregnated filter paper (Pi test): A review of its development and methodological research. Nutr. Cycling Agroecosyst. 46:41-51.

Chien, S.H., and W.R. Clayton. 1980. Application of Elovich equation to the kinetics of phosphate release and sorption in soils. Soil Sci. Soc. Am. J. 44:265-268.

Cooper, J.M., and P.R. Warman. 1997. Effects of three fertility amendments on soil dehydrogenase activity, organic C and pH. Can. J. Soil Sci. 77:28-283.

Cross, A.F., and W.H. Schlesinger. 1995. A literature review and evaluation of the Hedley fractionation: Applications to the biogeochemical cycle of soil phosphorus in the natural ecosystems. Geoderma 64:197-214.

Cruse, R.M., G.A. Yakle, T.C. Colvin, and D.R. Timmons. 1983. Tillage effects on corn and soybean production in farmer managed, university-monitored field plots. J. Soil Water Conserv. 38:512-514.

De Boer, J.L.M., U. Kohlmeyer, P.M. Breugem, and T. van der Velde-Koerts. 1998. Determination of total dissolved phosphorus in water samples by axial inductively coupled plasma emission spectrometry. J. Anal. Chem. 360:132-136.

De Maria, I.C., P.C. Nnabude, and O.M. de Castro. 1999. Long-term tillage and crop rotation effects on soil chemical properties of a Rhodic Ferralsol in southern Brazil. Soil Tillage Res. 51:71-79.

Dick, W.A. 1983. Organic carbon, nitrogen and phosphorus as influenced by tillage intensity. Soil Sci. Soc. Am. J. 47:102-107.

Enfield, C.G., T. Phan, D.M. Walters, and R. Ellis. 1981. Kinetic models for phosphate transport and transformations in calcareous soils: 1. Kinetics of transformations. Soil Sci. Soc. Am. J. 45: 1059-1070.

Follett, R.F., and G.A. Peterson. 1988. Surface soil nutrient distribution as affected by wheat-fallow tillage systems. Soil. Sci. Soc. Am. J. 52:141-147.

Frossard, E., L.M. Condron, A. Oberson, S. Sinaj, and J.C. Fardeau. 2000. Processes governing phosphorus availability in temperate soils. J. Environ. Qual. 29:15-23.

Grant, R.F., and D.J. Heaney. 1997. Inorganic phosphorus transformation and transport in soils: mathematical modeling in ecosystems. Soil Sci. Soc. Am. J. 61:752-764.

Grant, R.F., M. Amrani, D.J. Heaney, R. Wright, and M. Zhang. 2004. Mathematical modelling of phosphorus losses from land application of hog and cattle manure. J. Environ. Qual. 33:210-233.

Hedley, M.J., J.W.B. Stewart, and B.S. Chauhan. 1982. Changes in inorganic and organic phosphorus induced by cultivation practices and laboratory incubations. Soil Sci. Soc. Am. J. 46:970-976.

Holford, J.C.R. 1997. Soil phosphorus: its measurement, and its uptake by plants. Aust. J. Soil Res. 35: 227-239.

Huffman, S.A, C.V. Cole, and N.A. Scott. 1996. Soil texture and residue addition effects on soil phosphorus transformation. Soil Sci. Soc. Am. J. 60:1095-1101.

Ivarsson, K. 1990. The long-term soil fertility experiments in southern Sweden. IV. Changes in inorganic and organic soil phosphorus after a pot trial. Acta Agric. Scand. 40:205-215.

Jones, C.A., C.V. Cole, A.N. Sharpley, and J.R. Williams. 1984. Simplified soil and plant phosphorus model: I. Mobility of phosphorus. Soil Sci. Soc. Am. Proc. 38: 446-451.

Johnson, R.A., and D.W. Wichern. 1988. Multivariate linear regression models. p. 273-333. *In* Applied multivariate statistical analysis. 2nd ed. Prentice Hall, Englewood Cliffs, NJ.

Kamprath, E.J., and M.E. Watson. 1980. Conventional soil and tissue tests for assessing the phosphorus status of soils. p. 433-469. *In* F.H. Khasawneh et al. (eds.) The role of phosphorus in agriculture. ASA, CSSA, and SSSA, Madison, WI.

Iyamuremye, F., R.P. Dick, and J. Baham. 1996. Organic amendments and phosphorus dynamics: II. Distribution of soil phosphorus fractions. Soil Sci. 161:436-443.

Koopmans G.F., M.E. Van Der Zeeuw, P.F.M.A. Romkens, W.J. Chardon, O. Oenema 2001. Identification and characterization of phosphorus rich sandy soils. Neth. J. Agric. Sci. 49:369-384.

Mallarino, A.P. 1997. Interpretation of soil phosphorus tests for corn in soils with varying pH and calcium carbonate content. J. Prod. Agric. 10:163-167.

Mattingly, G.E.G. 1975. Labile phosphate in soils. Soil Sci. 119: 369-375.

McGill, W.B., and C.V. Cole. 1981. Comparative aspects of C, N, S and P cycling through soil organic matter during pedogenesis. Geoderma 26:267-286.

McKenzie, R.H., J.W.B. Stewart, J.F. Dormaar, and G.B. Schaalje. 1992a. Long-term crop rotation and fertilizer effects on phosphorus transformation: I. In a Chernozemic soil. Can. J. Soil Sci. 72:569-579.

McKenzie, R.H., J.W.B. Stewart, J.F. Dormaar, and G.B. Schaalje. 1992b. Long-term crop rotation and fertilizer effects on phosphorus transformation: II. In a Luvisolic soil. Can. J. Soil Sci. 72:581-589.

Mehlich, A. 1984. Mehlich no.3 extractant: A modification of Mehlich no.2 extractant. Commun. Soil Sci. Plant Anal. 15:1409-1416.

Mozaffari, M., and J.T. Sims. 1994. Phosphorus availability and sorption in an atlantic coastal plain watershed dominated by animal-based agriculture. Soil Sci. 157:97-107.

Murphy, J., and J.P. Riley. 1962. A modified single solution method for determination of phosphate in natural waters. Anal. Chim. Acta. 27:31-36.

Murrman, R.P., and M. Peech. 1969. Relative significance of labile and crystalline phosphates in soil. Soil Sci. 107:149-155.

Myers, R.G., A.N. Sharpley, S.J. Thien, G.M. Pierzynski. 2005. Ion-sink phosphorus extraction methods applied on 24 soils from the continental USA. Soil Sci. Soc. Am. J. 69: 511-521.

N'dayegamiye, A., and D.A. Angers. 1990. Effets de l'apport prolongé de fumier de bovins sur queques propriétés physiques et biologiques d'un loam limoneux Neubois sous culture de mäis. Can. J. Soil Sci. 70:259-262.

O'Halloran, I.P., R.G. Kachanoski, and J.W.B. Stewart. 1985. Spatial variability of soil phosphorus as influenced by soil texture and management. Can. J. Soil Sci. 65:475-487.

O'Halloran, I.P., J.W.B. Stewart, and E. De Jong. 1987a. Changes in P forms and availability as influenced by management practices. Plant Soil 100:113-126.

O'Halloran, I.P, J.W.B. Stewart, and R.G. Kachanoski. 1987b. Influence of texture and management practices on the forms and distribution of soil phosphorus. Can. J. Soil Sci. 67:147-163.

O'Halloran, I.P. 1993. Effect of tillage and fertilizer on the inorganic and organic phosphorus. Can. J. Soil Sci. 73:359-369.

Olsen, S.R., C.V. Cole, F.S. Watanabe, and L.A. Dean. 1954. Estimation of available phosphorus by extraction with sodium bicarbonate. USDA Circ. 939, U.S. Govt. Print. Office, Washington, D.C.

Paniagua, A., M.J. Mazzarino, D. Hass, L. Szott, and C. Fernandez. 1995. Soil phosphorus fractions under tropical agro-ecosystems on a Volcanic Soil. Austr. J. Soil Res. 33: 311-320.

Petersen, G.W., and R.B. Corey. 1966. A modified Chang and Jackson procedure for Routine fractionation of inorganic soil phosphates. Soil Sci. Soc. Am. Proc. 30: 563-565.

Pierzynski, G.M., A.N. Sharpley, and J.L. Kovar. 2008. Methods of phosphorus analyses for soils, sediments, residues and water: Introduction. p. 1-4. In J.L. Kovar and G.M. Pierzynski (eds.), Methods for phosphorus analysis for soils, sediments, residuals and waters. 2nd Edition. Southern Cooperative Series Bulletin No. 408.

Pierzynski, G.M., H.L. Zhang, A. Wolf, P. Kleinman, A. Mallarino, and D. Sullivan. 2011. Phosphorus determination in waters and extracts of soils and by-products: inductively coupled plasma spectrometry versus colorimetric procedures. http://www.sera17.ext.vt.edu/SERA_17_Publications.htm. Accessed on July 13, 2011.

Raven, K.P, and L.R. Hossner. 1993. Phosphate desorption quantity-intensity relationships in soil. Soil Sci. Soc. Am. J. 57:1505-1508.

Reddy, K.R., M.R. Overcash, R. Khaleel, and P.W. Westerman. 1980. Phosphorus adsorption-desorption characteristics of two soils utilized for disposal of animal waste. J. Environ. Qual. 9: 86-92.

Richards, J.E., T.E. Bates, and S.C. Sheppard. 1995. Change in the forms and distribution of soil phosphorus due to long-term corn production. Can. J. Soil Sci. 75:311-318.

Robinson, J.S., and A.N. Sharpley. 1994. Effect of organic phosphorus on the sink characteristics of iron oxide-impregnated filter paper. Soil Sci. Soc. Am. J. 58:758-761.

Schmidt, J.P., S.W. Buol, and E.J. Kamprath. 1996. Soil phosphorus dynamics during seventeen years of continuous cultivation: fractionation analysis. Soil Sci. Am. J. 60: 1168-1172.

Schoenau, J.J., and W.N. Huang. 1991. Anion-exchange membrane, water and sodium bicarbonate extractions as soil tests for phosphorus. Commun. Soil Sci. Plant Anal. 22: 465-492.

Selles, F., R.A. Kochann, J.E. Denardin, R.P. Zentner, and A. Faganello. 1997. Distribution of phosphorus fractions in a Brazilian Oxisol under different tillage systems. Soil & Tillage Res. 44:23-34.

Selles, F., B.G. McConkey, and C.A. Campbell. 1999. Distribution and forms of P under cultivator and zero-tillage for continuous- and fallow-wheat cropping systems in the semi-arid Canadian prairies. Soil Tillage Res. 51:47-59.

Sharpley, A.N., S.J. Smith, B.A. Stewart, and A.C. Mathers. 1984. Forms of phosphorus in soil receiving cattle feedlot waste. J. Environ. Qual. 13:211-215.

Sharpley, A.N., S.J. Smith, and W.R. Bain. 1993. Nitrogen and phosphorus fate from long-term poultry litter applications to Oklahoma soils. Soil Sci. Soc. Am. J. 57:1131-1137.

Sharpley, A.N., J.T. Sims, and G.M. Pierzynski. 1994. Innovative soil phosphorus availability indices: assessing inorganic phosphorus. p. 116-142. *In* Soil testing: prospects for improving nutrient recommendations, SSSA special publication 40. SSSA, Madison, WI.

Sharpley, A.N., and H. Tunney. 2000. Phosphorus research strategies to meet agricultural and environmental challenges of 21st century. J. Environ. Qual. 29:176-181.

Shaviv, A., and N. Shachar. 1989. A kinetic-mechanistic model of phosphorus sorption in calcareous soils. Soil Sci. 148:172-178.

Shober, A.L., D.L. Hesterberg, J.T. Sims, and S. Gardner. 2006. Characterization of phosphorus species in biosolids and manures using XANES spectroscopy. J. Environ. Qual. 35:1983-1993.

Sibbesen, E. 1978. An investigation of the anion-exchange resin method for soil phosphate extraction. Plant Soil 50:305-321.

Simard, R.R., D. Cluis, G. Gangbazo, and S. Beauchemin. 1995. Phosphorus status of forest and agricultural soils from a watershed of high animal density. J. Environ. Qual. 24:1010-1017.

Sparks. D.L. 1985. Kinetics of ionic reactions in clay minerals and soils. Adv. Agron. 38: 231-266.

Sparks, D.L. 1999. Kinetics and mechanisms of chemical reactions at the soil mineral/water interface. p. 135-191. *In* D.L. Sparks (ed.) Soil physical chemistry. 2nd edition. CRC Press, Boca Raton, FL.

Stevenson, F.J. 1986. The phosphorus cycle. p. 231-284. *In* Cycles of soil: carbon, nitrogen, phosphorus, sulfur, micronutrients. John Wiley & Sons, New York, NY.

Sweeney, D.W. 1993. Fertilizer placement and tillage effects on grain sorghum growth and nutrient uptake. Soil Sci. Soc. Am. J. 57:532-537.

Tiessen, H., J.W.B. Stewart, and J.R. Bettany. 1982. Cultivation effect on the amounts and concentration of carbon, nitrogen and phosphorus in grassland soils. Agron. J. 74:831-835.

Tiessen, H., J.W. Stewart, and C.V.Cole. 1984. Pathways of phosphorus transformation in soils of differing pedogenesis. Soil Sci. Soc. Am. J. 48:853-858.

Tiessen, H., and J.O. Moir. 1993. Characterization of available P by sequential extraction. p. 75-86. *In* M.R. Carter (ed.) Soil sampling and methods of analysis. Can. Soil Sci. Soc., Lewis Publ., Bica Raton, FL.

Tran, T.S, J.C. Fardeau, and M. Giroux. 1988. Effects of soil properties on plant-available P determined by the isotopic dilution P-32 method. Soil Sci. Soc. Am. J. 52:1382-1390.

Tran, T.S., and A. N'dayegamiye. 1995. Long-term effects of fertilizers and manure application on the forms and availability of soil phosphorus. Can. J. Soil Sci. 75:281-285.

Turner, B.L., and A.B. Leytem. 2008. Phosphorus speciation in soils and manures by solution ^{31}P NMR spectroscopy. p. 95-101. *In* J.L. Kovar and G.M. Pierzynski (eds.) Methods for phosphorus analysis for soils, sediments, residuals and waters. 2nd Edition. Southern Cooperative Series Bulletin No. 408.

Uusitalo, R, and M. Yli-Halla. 1999. Estimating errors associated with extracting phosphorus using iron oxide and resin methods. J. Environ. Qual. 28:1891-1897.

Van der Zee, S.E.A.T.M., and A. Gjaltema. 1992. Simulation of phosphate transport in soil columns. I. Model development. Geoderma 52:87-109.

Wagar, B.I., J.W.B. Stewart, and J.O. Moir. 1986. Changes with time in the form and availability of residual fertilizer phosphorus on Chernozemic soils. Can. J. Soil Sci. 66:105-119.

Walker, T.W., and J.K. Syers. 1976. The fate of phosphorus during pedogenesis. Geoderma 15:1-19.

Weill, A.N., G.R. Mehuys, and E. McKyes. 1990. Effect of tillage reduction and fertilizer type on soil properties during corn production. Soil & Tillage Res. 17:63-76.

Williams, J.D.H., J.K. Syers, and T.W. Walker. 1967. Fractionation of soil inorganic phosphate by a modification of Chang and Jackson's procedure. Soil Sc. Soc. Am. Proc. 31:736-739.

White, R.E., and P.H.T. Beckett. 1964. Studies on the phosphate potentials of soils. 1. The measurement of phosphate potential. Plant and Soil 20:1-16.

Wright, A.L. 2009. Phosphorus sequestration in soil aggregates after long-term tillage and cropping. Soil & Tillage Res. 103:406-411.

Zhang, T.Q., and A.F. MacKenzie. 1997a. Changes of soil phosphorus fractions under long-term corn monoculture. Soil Sci.Soc. Am. J. 61:485-493.

Zhang, T.Q., and A.F. MacKenzie. 1997b. Phosphorus in zero-tension soil solution as influenced by long-term fertilization of corn. Can. J. Soil Sci. 77:685-691.

Zhang, T.Q. and A.F. Mackenzie. 1997c. Changes of phosphorous fractions under continuous corn production in a temperate clay soil. Plant Soil. 192:133-139.

Zhang, T.Q., A.F. Mackemzie, and F. Saurlol. 1999. Nature of soil organic phosphorus as affected by long-term fertilization under continuous corn (*Zea mays* L.): A ^{31}P NMR study. Soil Sci. 164:662-670.

Zhang, T.Q., A.F. Mackenzie, B.C. Liang, and C.F. Drury. 2004. Soil test phosphorus and phosphorus fractions with long-term phosphorus addition and depletion. Soil Sci. Soc. Amer. J. 68:519-528.

Zhang, T.Q., C.S. Tan, C.F. Drury, and W.D. Reynolds. 2006. Long-term (≥43 years) fate of soil phosphorus as related to cropping systems and fertilization. World Congress of Soil Science. Philadelphia, USA. July 9-16.

Zheng, Z., R.R. Simard, J. Lafond, and L.E. Parent. 2001. Changes in phosphorus fractions of a Humic Gelysol as influenced by cropping systems and nutrient sources. Can. J. Soil Sci. 81:175-183.

Zheng, Z., R.R. Simard, J. Lafond, and L.E. Parent. 2002. Pathways of soil phosphorus transformations after 8 years of cultivation under contrasting cropping practices. Soil Sci. Soc. Am. J. 66:999-1007.

Zheng, Z., J.A. MacLoed, J.B. Sanderson, and J. Lafond. 2004a. Soil phosphorus dynamics during 10 annual applications of manure and mineral fertilizers: fractionation and path analyses. Soil Sci. 169:449-456.

Zheng, Z., J.A. MacLeod, and J. Lafond. 2004b. Phosphorus status of a Humic Cryaquept profile in frigid continental climate as influenced by cropping practices. Biol. Fertil. Soils 39:467-473.

2

Soil Fertility of Tropical Intensively Managed Forage System for Grazing Cattle in Brazil

Alberto C. de Campos Bernardi,
Patrícia P. A. Oliveira and Odo Primavesi
Embrapa Pecuária Sudeste, São Carlos – SP,
Brazil

1. Introduction

In Brazil the predominant beef and dairy cattle production systems are based mostly on grazing and rely on native and cultivated pastures, which are grazed by continuous stocking all year round and are the main source of animal feed. About 90% of the nutrients required by the ruminants are obtained directly through grazing and supplemental forage feeding is only utilized in intensive dairy production systems and feed-lot systems (Euclides et al., 2010).

Most of the Brazilian cattle are maintained on pastures grown on acidic and low fertility soils that do not receive any lime or fertilizer. This lack of lime and nutrient input in the establishment and the maintenance phases, and the inadequate management of grasses are the main causes of pasture degradation in Brazil (Cantarella et al., 2002). Estimates indicate that approximately 80% of the 50 million hectares of pastures are degraded or under degradation process in the Cerrado (Savannah) region. Pasture degradation is the main cause of low productivity in the Brazilian livestock sector. In these low-input forage systems, lack of regular nitrogen (N) supply plus declining soil P levels are expected to eventually necessitate a reduction in stocking rate and animal production (Macedo, 2002).

The average stocking rate of Brazilian pasture is less than 1.0 animal per ha per year (Carvalho, 2002) with an annual productivity of meat and of milk around 50 and 2,000 kg per hectare, respectively. On the other hand, using appropriate technology in cultivated pastures it is possible to improve animal production significantly. Well established pastures that are properly managed and fertilized are the main source of food for cattle and most practical as the least costly source of feeding (Camargo et al., 2002). In intensive cattle production, well managed pastures allow for increased rates of stocking and productivity (Corsi and Nussio, 1993; Primavesi et al, 1999; Lugão et al., 2003) based on replanting with better grass varieties, grazing rotation, forage availability in the dry season, regulation of stocking densities, genetic improvement of cattle herds and building up soil fertility by balanced fertilization supply and improving soil organic matter (SOM) content.

In the intensive cattle production system, pastures are mainly composed of *Brachiaria* and *Panicum* species (C4 grasses with high herbage dry matter yield potential). These grass forages are used in the summer season, and in the winter, feed is based on chopped sugarcane (corrected with crude protein or nitrogen) or silage (corn or sorghum), due the decrease in forage yield. This feeding system leads to increasing stocking rate up to 6 to 12 cow ha[-1] in pasture for about 200 days year[-1] without irrigation, with higher stocking rates possible when pastures are irrigated during the dry season (Corsi et al., 2001; Santos et al., 2003). Crop-pasture rotation and the tropical grass pasture management intensification increases animal production to more than 25,000 kg of milk per ha per year and 900 kg of liveweight gain per ha per year (Corsi et al., 2001).

Carvalho and Batello (2009) pointed out that as stocking rate is increased, individual animal performance decreases, while production/unit-area increases to some maximum and then declines as a result of concurrent process controlling plant production and utilization by the grazing animal. Increasing grazing intensity decreases plant solar energy capture because of the negative impact on leaf area index. The harvest efficiency is increased with increasing grazing intensity because the forage intake per unit area is increased. Conversely, with increasing the number of animals, the competition for forage decreases the individual animal intake, which diminishes the assimilation efficiency.

Of the controllable factors determining forage yield and quality, soil fertility including fertilizer application is one of the most important. On these tropical acid soils naturally poor in plant nutrients, soil liming and a balanced nutrient supply of nitrogen (N), phosphorus (P), potassium (K), calcium (Ca), magnesium (Mg), sulphur (S), boron (B), copper (Cu), molybdenum (Mo), manganese (Mn) and zinc (Zn) are therefore essential to ensure high yielding and high quality forage (Corsi and Nussio, 1993; Primavesi et al, 1999; Camargo et al., 2002). In order to intensify cattle production and reach high animal productivity, the main point to be considered is the correction of soil acidity and the balanced supply of mineral nutrients.

Maintenance of soil fertility depends on the nutrient recycling and inputs to the system. In extensive systems, natural rates of nutrient cycling may be sufficient, since it works with very low stocking rates, requiring plenty of waste returned to the soil, but it presupposes the need for large areas. In the case of intensive production, which is carried out in small areas, requiring high stocking rate, and thus higher biomass productivity of the grass, is essential to correct soil acidity with lime and fertilizer use.

However, the used rates of fertilizer in Brazilian pastures are still extremely low, around 5 to 6 kg per ha of NPK. In intensively managed pastures, the maximum doses of nutrients for technical and economical response are quite high. Research results point to economic doses around 800 kg per ha per year of N and K_2O in irrigated pastures, applied as 8 to 9 split applications, and 500 kg per ha per year of these nutrients in non-irrigated pastures (Martha Júnior et al., 2004; Sousa et al., 2004; Primavesi et al., 2004; Primavesi et al., 2003; Oliveira et al., 2003).

Intensive management also contributes to turn to a more profitable and competitive livestock production because it reduces the pasture land area, reduces the potential deforestation of natural forests, increases the possibility of environmental conservation, and release areas for another agricultural land use (Kaimowitz and Angelsen, 2008). Moreover, it contributes to increase carbon sequestration, lower energy loss by animals that otherwise would take long walks in search of food and water, and generates less methane per unit of product. So, intensification is a sustainable practice, since the recent development of Brazilian agriculture has been strongly based on productivity gains, and, to a minor extent, on land area expansion (Contini and Martha Jr., 2010).

This chapter provides information regarding the management of liming and fertilization of intensively managed pastures (based on soil analysis and requirement of the grass). Sources of information are scientific articles (mostly in Portuguese) and personal experience of experts.

2. Fertility of Brazilian soils

Brazilian tropical soils usually produce low yields due to the high Al saturation, low concentrations of most mineral nutrients that are essential for plant development, low organic matter content, leading to low CEC and high P fixation (Bernardi et al., 2002). A summary of the extent of soil-related limitations, both physical and chemical, in the acid infertile soils of the tropical Latin America region was given by Sanchez and Salinas (1981) and is presented in Table 1. Deficiency of N and P was shown to be the most severe limitation to crop growth. The list of major chemical constraints is completed by the toxicity of Al, deficiency of K, high P fixation, and low cation exchange capacity. Other physical hindrances are shown but they are of minor relevance.

Soil constraints	Tropical America		Acid soils	
	1.000.000 ha	% total	1.000.000 ha	% total
Nitrogen deficiency	1332	89	969	93
Phosphorus deficiency	1217	82	1002	96
Potassium deficiency	799	54	799	77
Calcium deficiency	732	49	732	70
Magnesium deficiency	731	49	739	70
Sulphur deficiency	756	51	745	71
Cu deficiency	310	21	310	30
Zn deficiency	741	50	645	62
High P fixation	788	53	672	64
Low CEC	620	41	577	55
Aluminum toxicity	756	51	756	72
Low water availability	626	42	583	56
High erosion risk	543	36	304	29
Flooding	306	20	123	12
Compaction	169	11	169	16
Laterization	126	8	81	8
Water stress (> 3 month)	634	42	299	29

Source: Adapted from Sanchez and Salinas (1981).

Table 1. Geographical extent of the major soil constraints to crop production in tropical America.

The most important function of organic matter in soil is as a reserve of nutrients required by plants (Craswell and Lefroy, 2001). Nevertheless, SOM also plays an extremely important role in tropical soils, since it affects soil properties such as electrical charge and nutrient supply (Sanchez, 1976). The main factor responsible for negative charges, and therefore the SOM which contributes 60 - 80% of total soil CEC (Raij, 1969). The organic matter content is affected by vegetation type, as well as the parent material from geological formations and increases with soil clay content and rainfall (Tognon et al., 1998).

SOM can be increased by addition of crop residues, cover crops, green manure crops, compost, animal manure, by reduced or no-tillage and by avoiding residue burning. Enhanced SOM increases soil aggregation, water holding capacity and P availability; reduces P fixation, toxicity of Al and Mn, leaching nutrients by adsorbing exchangeable Ca, Mg and K (Baligar and Fageria, 1997). SOM also provides a source of nutrients, as was shown by Pereira et al. (2000) who evaluated changes in chemical properties of a Xanthic Hapludox managed under pasture, using two rotational systems with *Brachiaria brizantha* and *Panicum maximum*. The organic material incorporated into the soil through vegetable and animal residues, influenced the chemical characteristics, increasing the levels of Ca, Mg, K, P, N, C, OM and pH, and decreasing the Al levels, indicating that SOM has a buffering effect and a complexing effect on Al.

3. Soil testing

Liming and fertilizer recommendations for pasture should be based mainly on soil analysis and expected yield. Potassium rates are recommended based on soil exchangeable K values. However P recommendation is based on two analytical methods: ion exchange resin-extractable P and Mehlich-1 P. Due to the differences between the analytical protocols used for P determination, there are differences in the interpretation levels as shown in Table 2 and 3. Soil analysis for micronutrients, extracted with hot water (B) and DTPA-TEA or Mehlich-1 (Fe, Cu, Mn, and Zn) is also used as criterion for fertilizer recommendation (Tables 4 and 5).

Fertility class	P resin			
	Forestry	Perennial	Annual	Horticultural
		mg kg^{-1}		
Very low	0 - 2	0 - 5	0 - 6	0 - 10
Low	3 - 5	6 - 12	7 - 15	11 - 25
Medium	6 - 8	13 - 30	16 - 40	26 - 60
High	9 - 16	31 - 60	41 - 80	61 - 120
Very high	>16	>60	>80	>120

Source: adapted from Raij et al. (1996).

Table 2. Soil fertility classes and limits for interpretation of soil-P availability by the P resin method

Clay content	Soil fertility class				
	Very low	Low	Medium	Adequate	High
%			mg kg⁻¹		
60-100	≤ 2.7	2.8 – 5.4	5.5 – 8.0	8.1 – 12.0	> 12.0
35-60	≤ 4.0	4.1 – 8.0	8.1 – 12.0	12.1 – 18.0	> 18.0
15-35	≤ 6.6	6.7 – 12.0	12.1 – 20.0	20.1 – 30.0	> 30.0
0-15	≤ 10.0	10.1 – 20.0	20.1 – 30.0	30.1 – 45.0	> 45.0

Source: adapted from Alvarez et al. (1999).

Table 3. Soil fertility classes and limits for interpretation of soil-P availability by the Mehlich-1 P method, considering soil clay content.

Soil fertility class	B (hot water)	Cu	Fe	Mn	Zn
			DTPA-extractable		
			mg kg⁻¹		
Low	0 - 0.2	0 - 0,2	0 - 4	0 - 1,2	0 - 0,5
Medium	0,21 - 0,6	0,3 - 0,8	5 - 12	1,3 - 5,0	0,6 - 1,2
High	>0,6	>0,8	>12	>5,0	>1,2

Source: adapted from Raij et al. (1996).

Table 4. Soil fertility classes and limits for interpretation of micronutrient availability by the hot water (B) and DTPA (Cu, Fe, Mn and Zn) methods.

Micronutrient	Soil fertility class				
	Very low	Low	Medium	Adequate	High
			mg kg⁻¹		
Cu	≤ 0.3	0.4 -0.7	0.8 – 1.2	1.3 – 1.8	> 1.8
Fe	≤ 8	9 – 18	19 – 30	31 – 45	> 45
Mn	≤ 2	3 – 5	6 –8	9 –12	> 12
Zn	≤ 0.4	0.5 – 0.9	1.0 – 1.5	1.6 – 2.2	> 2.2

Source: adapted from Alvarez et al. (1999).

Table 5. Soil fertility classes and limits for interpretation of micronutrient (Cu, Fe, Mn and Zn) availability by the Mehlich-1 extraction method

4. Foliar diagnosis

Leaves are the first and the principal part grazed by animals (Moraes and Palhano, 2002) so their chemical composition reveals their nutritional value. The principle of foliar diagnosis is based on comparing nutrient concentrations in leaves with standard values. Crops are considered to integrate factors such as presence and availability of soil nutrients, weather variables, and crop management. So plant tissue analyses are the best reflection of what the plant has taken up. Traditionally, whole plant shoots are sampled for forage nutritional diagnosis (Monteiro, 2004). The range of levels considered adequate for forages in Brazil by Werner et al. (1996) are shown in Table 6.

Usually, forage value is based on its protein content however tissue mineral composition also plays an important role in animal nutrition. Gerdes et al. (2000) mentioned that the low nutritional value of tropical forage, often mentioned in literature, is associated with reduced crude protein and minerals, the high fiber content and low dry matter digestibility.

Forage	N	P	K	Ca	Mg	S
	g kg⁻¹					
Panicum maximum cv Colonião	15 - 25	1.0 - 3.0	15 - 30	3 - 8	1.5 - 5.0	1.0 - 3.0
Pennisetum purpureum cv Napier	15 - 25	1.0 - 3.0	15 - 30	3 - 8	1.5 - 4.0	1.0 - 3.0
Cynodon dactylon cv. Coast-cross	15 - 25	1.5 - 3.0	15 - 30	3 - 8	2.0 - 4.0	1.0 - 3.0
Cynodon spp cv Tifton	20 - 26	1.5 - 3.0	15 - 30	3 - 8	1.5 - 4.0	1.5 - 3.0
Brachiaria brizantha	13 - 20	0.8 - 3.0	12 - 30	3 - 6	1.5 - 4.0	0.8 - 2.5
Andropogon gaianus	12 - 25	1.1 - 3.0	12 - 25	3 - 6	1.5 - 4.0	0.8 - 2.5
Brachiaria decumbens	12 - 20	0.8 - 3.0	12 - 25	2 - 6	1.5 - 4.0	0.8 - 2.5
Medicago sativa	34 - 56	2.5 - 5.0	20 - 35	10 - 25	3.0 - 8.0	2.0 - 4.0

Source: Adapted from Werner et al. (1996).

Table 6. Adequate shoot macronutrient concentrations for tropical forages and alfalfa.

5. Pasture fertilization

The pasture productivity is determined by many factors as species, climatic and soil conditions and management practices. Research in tropical and subtropical regions has highlighted the need to supply the pasture system with macro- and micro-nutrients, as well as soil amendments, since fertilization is one of the factors that most contribute to increase forage dry matter (DM) productivity and quality (Primavesi et al., 1999; Cantarella et al., 2002).

Pastures fertilization consists of two phases: fertilization during the establishment period, which aims to provide nutrients for the development of fresh established pasture and correcting deficiencies in the soil nutrient supply; and maintenance fertilization, which aims to provide or restore nutrients extracted or lost during grazing. Table 7 based on Macedo (2004) summarizes the N, P and K recommendations for pastures from three sources: Werner et al. (1997), Cantarutti et al. (1999), and Vilela et al. (2002). These actual lime and fertilizer recommendations for forage in the published literature are adequate for semi-intensive forage management systems, but do not capture the whole production potential of the tropical forage.

Bernardi et al (2008) carried out a survey with 232 farmers and rural extension workers who adopted the intensive managed forage system. Questions about the adoption of technical soil conservation and fertility, leaf and soil analysis, irrigation, use of fertilizers and limestone and costs of these techniques on production were made. The results pointed to an usual use of soil analysis, lime, micronutrients and soil conservation practices by 96%, 97%,

78% and 99% of interviewees, respectively. Regardless of the soil test, most used a N, P_2O_5 and K_2O formulation (NPK) for pasture fertilization of 400 kg ha-1 of 8-28-16 at planting or seeding, and 700 kg ha-1 of 20-5-20 at topdressing. However, 93.8% of the producers do not perform leaf analysis. The approximate relationship between fertilizer use and animal production for this group of interviewees was 1 ton fertilizer to 1000 liters milk and 1 ton fertilizer to 195 kg meat.

Recommendation	Grass-based forage phases			
	Establishment		Maintenance	
	N (kg ha-1)	Criteria	N (kg ha-1)	Criteria
Werner et al. (1996)	40	20-40 days after germination	40 to 80 50	Nutritional need each grazing cycles
Cantarutti et al. (1999)	0 to 150	60% of soil coverage by pasture	50 100-150 200-300	Technological level: Extensive Medium Intensive
Vilela et al. (2002)	40 to 50	75% of soil coverage by pasture	40	Extensive
	P_2O_5 (kg ha-1)	Criteria	P_2O_5 (kg ha-1)	Criteria
Werner et al. (1996)	20 to 150	Nutritional need	20 to 50	Nutritional need (annual application)
Cantarutti et al. (1999)	15 to 120	Technological level	15 to 60	Technological level (annual application)
Vilela et al. (2002)	20 to 180	Nutritional need	20	Nutritional need (biannual application)
	K_2O (kg ha-1)	Criteria	K_2O (kg ha-1)	Criteria
Werner et al. (1996)	20 to 80	Nutritional need	20 to 60	Nutritional need (annual application)
Cantarutti et al. (1999)	20 to 60	Technological level	40 to 200	Technological level (annual application)
Vilela et al. (2002)	20 to 60	Nutritional need	50	$K < 30$ mg dm-3

Source: adapted from Macedo (2004).

Table 7. Nitrogen, P and K fertilizer recommendations (levels and fertilization criteria) for establishment and maintenance of grass-based forage for grazing cattle in tropical pastures of Brazil.

6. Nitrogen

In general, N is the nutrient that is the most limiting for plant growth and affects the productivity of pastures (Jarvis et al., 1995). Many authors pointed out that an increment in N fertilization increases grass dry matter yield. Vicente-Chandler et al. (1959) found a positive response to the application of up to 1,800 kg N per ha per year. Growth response of 40 to 70 kg DM kg^{-1} of applied N can be expected at fertilizer rates of as high as 400 to 600 kg N ha^{-1} during the growing season (Vicente-Chandler et al., 1974; Minson et al., 1993).

Other authors, however, demonstrated that higher responses are obtained with doses of 300 to 400 kg N per ha per year (Werner et al. 1977; Olsen, 1972, Gomes et al., 1987). According to Cantarella et al. (2002) N use efficiency (expressed in kg of dry matter produced per kg fertilizer N applied), decreases with increasing dose of N applied.

Additions of N through the mineralization of SOM, atmospheric (wet and dry) deposition and biological-fixing activities in soils are uncertain and are generally inadequate to sustain high pasture productivity (Martha Jr. et al., 2004).

Positive results of N fertilization on dry mass production of species of *Brachiaria* were achieved by Bonfim-da-Silva and Monteiro (2006), Primavesi et al. (2006), Rodrigues et al. (2006), and Benett et al. (2008), for *Cynodon dactylon* cv. coast-cross by Corrêa et al. (2007), for *P. maximum* by Lugão et al. (2003), Volpe et al. (2008) and for *Pennisetum* by Martha Jr. et al. (2004).

From a forage quality standpoint, N fertilization increased whole plant crude protein concentration (Alvim et al. 1998) but effects on other forage quality variables were less consistent. Results on the effect of N on *in vitro* dry matter digestibility (IVDMD) are conflicting (Monson and Burton, 1982; Gomide and Costa, 1984; Cáceres et al., 1989).

Urea has been the most used N-source in Brazil (ANDA, 2008), due to lower cost per unit of N. But N use efficiency of urea may be reduced because of losses from agricultural system by volatilization of ammonia to atmosphere. This is one of the main factors responsible for the low efficiency of urea, and may reach extreme values, with losses close to 80% of N applied (Lara-Cabezas et al., 1997, Cantarella et al., 1999; Martha Junior et al., 2004). Such results have been reported even in acid soils, since the liming increases soil pH and favors volatilization. Mulch present in no-tillage or pasture systems may also increase the amount of N lost by volatilization, especially when urea is applied on soil surface. Anjos and Tedesco (1976) compared the losses by N volatilization from N sources and reported values of 30% for urea and less than 1% for ammonium sulfate. In urea fertilized pastures, the volatilization of ammonia (NH_3) into the atmosphere is the most significant N loss (Whitehead, 1995). These losses are enhanced when the urea is applied in the topdressing and at the end of the rainy season (Martha Júnior et al., 2004, Primavesi et al., 2004), specifically with greatest losses (60%) when applied on wet soil (field capacity) followed by no rain or irrigation, and lowest losses (1%) when used on dry soil followed by around 10 mm rain or irrigation (Primavesi et al. 2001). These losses after N fertilization reduce pasture growth and consequently, the stocking rate and weight gain (Whitehead, 1995). In a Marandu grass (*Brachaiaria brizantha*) pasture in Brazil, Oliveira et al. (2007) found gaseous losses of N fertilized with urea ranging from 14 to 38%, but the losses were much lower when urea was incorporated in the soil by a no-till double-disc soil cultivator.

Ammonia losses from N fertilizer can be reduced through the use of sources less susceptible to volatilization (nitrate-based fertilizers), or by soil incorporation of the urea (a process hindered by direct fertilization) (Primavesi et al., 2003). Slow urea liberation has also be observed with the addition of acids, salts of K, Ca and Mg, and by the choice of specific urea's grain size distribution (Allaire and Parent 2004). The urea-N losses can also be reduced using zeolites as additives in the fertilizers to control the retention and release of NH_4^+ (Bernardi et al., 2009).

Teitzel et al. (1991) associated the use of N fertilizers in intensively managed tropical pastures with positive economic responses. Results from Euclides et al. (2007) showed that fertilizing *P. maximum* with N had significant responses on animal yield as weight gain and production per hectare.

Results from Primavesi et al. (2004) and Primavesi et al. (2006) indicated that the N fertilization doses of 500 kg ha^{-1} year^{-1} of N can be reduced in approximately 10% in subsequent years until stabilizing in the sixth or seventh year with maintenance dose, due the increased SOM levels.

Efficient pasture use in intensive production systems depends on the balanced mineral nutrition of the forage plant (Hopkins et al., 1994). Nevertheless physiological processes of plants are affected specially by high doses of N fertilizer. Data presented by Primavesi et al. (2001 and 2003) suggest that the maximum level of N in tropical forage grasses is 24 g kg^{-1}, from which starts the accumulation of nitrate in forage losses occur more intense and nitrate leaching. The forage protein content should be at least 70 g kg^{-1} or higher to stimulate an animal intake and digestibility.

When N fertilizers are supplied to grasses, there may be increasing in levels of provided nutrient, but there may be side effects of this application, resulting in increases or reductions in levels of other nutrients. Primavesi et al. (2005) determined the uptake of cations and anions of coastcross grass fertilized with N from 0 to 1,000 kg ha^{-1} (split in 5 applications during the rainy season). High doses of N fertilizer as urea or ammonium nitrate applied on coastcross grass favored absorption of cations and anions, although increasing rates of N caused higher K$^+$ uptake in relation to other cations and in Cl$^-$ among the anions. Batista and Monteiro (2010) evaluated changes in K, Ca and Mg concentrations of *B. brizantha*, *cv.* Marandu (Marandu palisadegrass) due N and S fertilizer inputs. N fertilization influenced Ca and Mg concentrations as well as the proportions of K, Ca and Mg in the above-ground part of Marandu palisadegrass.

7. Phosphorus

Acid tropical soils normally contain a limited P reserve and often have a high sorption capacity (Novais and Smith, 1999). According to Sanchez (1976), there are two main processes responsible for P fixation in acid soils: (i) precipitation by exchangeable Al and; (ii) adsorption on the surface of sesquioxides. Phosphorus fixation tends to be high in acid soils where the Fe and Al–oxyhydroxides are ubiquitous. The reversibility of P sorption is important since desorption often is a limiting step in the uptake of P by crops. Hence, P is considered to be the most limiting nutrient in uncultivated tropical soils and frequently found only as a trace (below 1 mg per kg of soil).

Phosphorus plays an important role in plant metabolism as cell energy transfer, respiration and photosynthesis. The response to fertilization depends, among other factors, the availability of P in the soil, the availability of other nutrients such as N and K, the species and climatic conditions (Sousa et al., 2004). Phosphorus deficiencies limit pasture establishment and growth (Corrêa and Reichardt, 1995; Corrêa and Haag, 1993). Souza and Lobato (2004) and Macedo (2005) also verified is that P is the most limiting nutrient related to pasture establishment and pasture sustainability. Results from Werner (1986), Corrêa and

Haag (1993), Hoffmann et al. (1995) and Belarmino et al. (2003) pointed out that P fertilization significantly increased root and tiller growth. Once P is available in sufficient supply, N availability drives pasture production. Cultivars of *P. maximum* generally show high response to P fertilization (Gheri et al., 2000).

8. Potassium

Potassium is the cation in higher concentration in forage plants and has relevant physiological and metabolic functions such as enzymes activation, photosynthesis, photo-assimilates translocation, stabilization of internal pH, stomatal function, turgor-related processes, N absorption and protein synthesis. The addition of K increases its levels in plant tissue and reduces the Ca and Mg levels in equivalent quantities (Mattos et al., 2002).

Providing an adequate supply of K is important for plant production and is essential to maintain high quality and profitable yields. In order to determine the best time and way of supplying of a nutrient source, their dynamics in soil and role on plant metabolism should also be considered (Benites et al., 2010). Band application of K in the furrow or topdressing application on soil surface is possible due its uptake by mass flux and its high mobility within the plant (Benites et al., 2010). Proper management of K fertilizer in relation to doses and application methods (banding, broadcast and split applications) can minimize losses, avoid depletion of soil K, increase the soil available K pool for a beneficial residual effect of infrequent K fertilization and increase crop yields per unit of K applied to soil (Vilela et al., 2002). The most appropriate time and manner of application of K and of any other nutrient are determined according to plant requirement and element dynamics in soil. The strategy for K fertilization must be accomplished in two steps: first corrective fertilization and then maintenance fertilization.

9. Liming

Soil acidity is one of the most yield-limiting factors for crop production, since is a complex of numerous factors involving nutrient/element deficiencies and toxicities, low activity of beneficial microorganisms and reduced plant root growth that limit nutrient and water uptake (Fageria and Baligar, 2003). High amounts of Al, and sometimes Mn, and the low contents of Ca, Mg, and other nutrients frequently account for the low productivity of crops grown on the acid soils. High concentrations of Al inhibit root development and tend to limit absorption of other nutrients, especially of Ca and Mg since their uptake is directly related to root growth and plant development (Lathwell and Grove, 1986).

The clay fraction of Oxisols and Ultisols s usually dominated by sesquioxides, gibbisite, kaolinite and intergrade minerals. These compounds have low intrinsic amounts of negative charges and, therefore, most of the CEC of these soils depends on organic matter (see below) and depends on the soil solution pH. As a consequence, such soils exhibit a strong relationship between charge and pH. In some cases the soils may show net positive charge at low pH, which affects the availability of some nutrients (Sanchez, 1976). Cation exchange capacity is responsible for the equilibrium of ions in the solid/liquid interface in soils. So the usually low values of CEC combined with low pH lead to leaching of K, Ca, and Mg. Low concentrations of K, Ca and Mg, and the low CEC associated with high Al contents are serious fertility constraints in acid tropical soils. Evaluation of these parameters in subsurface layers (below 0.2 m) should be undertaken.

Liming is a low-cost and effective way to neutralize soil acidity and to improve crop yields. Liming reduces Al and Mn toxicity, improves P, Ca and Mg availability, increases CEC, promotes N_2 fixation, and improves soil structure. Overall, liming improves soil capacity to supply needed nutrients and the ability of plants to absorb nutrients and water due to better root growth and activities of beneficial microorganisms. Also an increase in exchangeable bases and pH can stimulate decomposition and mineralization of organic matter by creating a more favorable environment for microbial populations (Sanchez, 1976; Havlin et al., 1999; Fageria and Baligar, 2008). The quantity of lime required depends on the soil type, quality of liming material, costs and crop species or cultivars (Fageria and Baligar, 2008).

Information on appropriate liming rates for tropical forages grown on acid soils in Brazil requires further studies, because the forage response to this practice have been differentiated (Cruz et al., 1994; Oliveira et al., 2003; Paulino et al., 2006), probably due to differences in soil properties and the variability of tolerance to soil acidity of tropical forage grasses. Macedo (2005) believes that the controversy lies in the Cerrado, since clays in these Oxisols affect their response to liming in a way that is quite different from other regions of Brazil. Besides, forages commonly used in this region have high tolerance to soil acidity. Limestone rates are calculated to raise soil base saturation as a percentage of the cation exchange capacity (CEC) of the soil at pH 7.0, to levels which vary with forage species. Werner et al. (1996) recommended a soil base saturation (BS) of 70% for planting and 60% for pasture maintenance, for *Pennisetum purpureum, P. maximum, Cynodon dactylon, Digitaria decumbens, Hyparrhenia rufa* and *Chloris gayana* pastures; 60% of BS for the establishment period and 50% of BS for the maintenance of *B. brizantha, Andropogon gayanus* e *Cynodon plectostachyus* pastures; and 40% of BS for the establishment and maintenance of *Brachiaria decumbens, Brachiaria humidicola, Melinis minutiflora, Paspalum notatum* and *Setaria anceps* pastures. *B. decumbens* pasture, which had the adequate basis saturation of 40%, after four years of intensive use of N and K fertilizer was observed high depletion in Ca and Mg levels in soil that could lead to plants death (Primavesi et al., 2004 and 2008).

Nevertheless, some intensive managed pastures appear to require more lime, with 80% of basis saturation being the optimum level. This apparent controversy can be explained by the fact that commonly used N, P and S fertilizers are acid-forming. The acidifying effect varies with the forms of these elements in the specific fertilizer used. For example, urea, ammonium nitrate and ammonium sulphate are soil-acidifying and require respectively 1.8, 1.8 and 5.4 kg of lime per kg of fertilizer to neutralize the produced soil acidity (Havlin et al., 1999).

Disking for lime incorporation in soil cultivated with pastures is a controversial practice (Arruda et al., 1987; Soares Filho et al., 1992; Luz et al., 1998). A slight increase in forage yield (130 kg ha⁻¹) was observed when lime was incorporated into the soil of a degraded *P. maximum* pasture by disking (Luz et al., 1998). On the other hand, disking decreased the dry aboveground matter of *B. decumbens* pasture (Soares Filho et al., 1992). In the recovery of degraded pastures, Primavesi et al. (2004 e 2008) showed that lime can be applied superficially especially when high N dose are used.

10. Sulphur

Sulphur (S) is an important macronutrient for plant metabolism and growth, being a component of essential amino acids (methionine, cysteine) and other organic compounds. The extraction of sulfur by forage plants may be around 50 kg ha^{-1} yr^{-1}, considering yields of 20 t ha^{-1} year^{-1} of dry matter and S concentration in shoots of 2.5 g kg^{-1} (Werner et al., 1996). Since plants have a lower demand for S than N, it has been neglected in Brazilian pasture fertilization (Monteiro et al., 2004). In Brazil, as a result of the constant use of concentrated NPK fertilizers, besides some edaphic and climatic factors, S became a limiting nutrient for plant development. Moreover in intensive forage management system low response to N fertilizer may be associated to low levels of S in the soil (Cunha et al. 2001; Mattos and Monteiro, 2003, Oliveira et al. 2005; Bomfim and Monteiro-Silva, 2006).

Monteiro et al. (2004) suggested that S fertilization of pasture grasses should be recommended when these forages are well fertilized with N. Stevens (1985) emphasized that both N and S supply are directly related and they must be in plant tissues in adequate proportions and amounts for the optimal synthesis of protein. The N:S ratio is an important nutritional status index since it remains constant at different stages of grasses development (Vitti & Novaes, 1986). According with Scott (1983) to ensure proper development to forage plants, the optimum N:S ratio must be around 16.5:1. The uses of ammonium sulphate or simple superphosphate or gypsum are adequate sulphur sources.

11. Micronutrients

Micronutrients play important roles in plant metabolism, acting as a constituent of organic compounds or as regulators of the functioning of enzyme systems. Micronutrients, also known as trace minerals, which chiefly include boron (B), molybdenum (Mo), copper (Cu), zinc (Zn), manganese (Mn) and iron (Fe), are required in extremely small quantities by crops and cattle. Review articles related to micronutrients on tropical forages prepared by Gupta et al. (2001) and Monteiro et al. (2004) provide information about forage yield responses to these minerals.

Soil acidity is one of the primary factors affecting the availability of micronutrients to crops (Sanchez, 1976). With the exception of Mo, the plant availability of other micronutrients, e.g., Zn, Mn, B and Fe decreases with liming (Gupta et al., 2001; Monteiro et al., 2004).

Micronutrients needs are assessed by soil testing and eventually supplied by fertilizers. In more intensive agricultural systems, such as intensively managed grassland with high productivity and high stocking rates, the tendency is a greater need for micronutrients. So an adequate supply is important to avoid a reduction in forage production. Positive response of grasses to zinc addition has been reported in Cerrado low fertility soil and for intensive managed pasture system with high N fertilization (Monteiro et al., 2004).

12. "Adubapasto" software

As presented in this chapter, the criteria for lime and fertilization recommendations for intensively managed pastures are not organized in a specific publication. Thus, there was a clear need to gather, organize, and make available this existing information to producers and agricultural extension agents. So, Embrapa Pecuária Sudeste made free, online software

Adubapasto 1.0 with remote access software by Web service. The structure of this software is based on: 1) architecture of the environment: CLIENT / SERVER, 2) Server Operating System: LINUX, 3) Web Server: APACHE, 4) Application Server: Zope/Plone; 5) Database Server: FIREBIRD and 6) Language development: PYTHON / JAVA SCRIPT.

Software is available at the site: http://www.cppse.embrapa.br/adubapasto. Based on the results of soil analysis, characteristics of a property and cattle schedule (stocking rates, number of days in pasture, supplemental feeds, etc), algorithms for lime and fertilizer recommendation were established, based on results of studies published in scientific and technical literature and experience of experts in soil fertility, fertilizer use, plant nutrition, animal nutrition and forage production.

The calculation routines include recommendations for liming, gypsum, N, P and K fertilizer for pasture establishment and maintenance periods, depending on the forage species, cattle management and stocking rate.

As a result, the software generates reports of recommendations for correction and fertilization, stocking rate expected and achieved. It is also possible to assess the historical evolution of soil fertility, since the data is stored in the database. This software operates as a management tool for agricultural technicians, extension workers, producers and researchers who can organize their information in the database.

13. Final remarks

The chapter showed that intensive pasture management practices can make cattle more profitable and competitive, and also conserve soil and water, thereby reducing the potential for deforestation and increasing the possibility of environmental preservation. So, High pasture productivity, leading to improve and high cattle production (milk and meat) is a sustainable practice that can meet societal demands for development without environmental degradation, better quality of life and improved resource availability with the opportunity to progressively combat social inequalities in all sectors and especially in the agricultural sector.

Soil chemical analysis is an important tool to know the soil fertility and make appropriate lime and fertilizer recommendations. There are differences between analytical protocols used for soil testing in Brazil, therefore, attention should be paid to the results to avoid any ambiguity in the interpretation and recommended doses.

Pastures need N to accumulate carbon, and also K, Ca, Mg, P, S and micronutrients for high yield and quality and profitability of cattle. Nutrient deficiencies and some soil chemical constraints can be avoided by regular monitoring of soil fertility. Careful attention to stocking rates to prevent overgrazing is also important.

The practices of soil amendment and fertilizer depend on the production system that farmer adopts and will always want to obtain satisfactory economic return with low environmental impacts. Table 8 summarizes the suggestions for liming and fertilization of intensive managed pasture systems on tropical acid soils in Brazil. Soil testing must be carried out every year.

Liming	Increase basis saturation to 70 or 80% (Ca: 55 to 60% of CEC*; Mg: 15 to 20% of CEC)
Nitrogen	Doses established as a function of forage species, animal stocking rate and soil organic matter: 40 to 50 kg ha^{-1} of N per AU** considering 3 to 7 AU** per ha. Reducing doses in 10% each year, from 6 or 7th year just N fertilizer for replace the exportation. Foliar diagnosis for evaluation keeping N in shoots approximately 24 – 25 g kg^{-1}.
Phosphorus	Begin with 10 mg dm^{-3} and increase until 30 mg dm^{-3}
Potassium	Begin with 3% of CEC* and increase until 6% of CEC*
Sulphur	60 - 90 kg ha^{-1}
Micronutrients	B = 0.5 to 1.0 kg ha^{-1} Cu = 1.0 to 2.0 kg ha^{-1} Zn = 2.0 to 4.0 kg ha^{-1}, or FTE*** = 30 to 40 kg ha^{-1}

*CEC = cation exchange capacity; **A.U. = animal unit = 450 kg of live weight; ***FTE = fritted trace elements (composition: Ca = 7.1%; S = 5.7%; B = 1.8%; Cu = 0.8%; Mn = 2.0%; Mo = 0.1% and Zn = 9.0%).

Table 8. Suggestions for liming and fertilization of intensive managed pasture systems in Brazil.

14. Acknowledgments

The authors wish to thank IPI (International Potash Institute) for the financial support to soil fertility studies of pasture systems.

15. References

Allaire, S.E.; Parent, L.E. 2004. Physical properties of granular organic-based fertilizers. Part 1 Static properties. Biosystems Engineering, 87: 79-87.

Alvarez V., V.H.; Novais, R.F.; Barros, N.F.; Cantarutti, R.B. & Lopes, A.S. 1999. Interpretação dos resultados das análises de solo. In: Riberio, A.C.; Guimarães, P.T.G. & Alvarez V., V.H. Recomendações para o uso de corretivos e fertilizantes em Minas Gerais (5a Aproximação). Viçosa, CFSEMG, p.25-32

Alvim, M.J.; Xavier, D.F.; Botrel, M.A. et al. 1998. Resposta do Coastcross (Cynodon dactylon) a diferentes doses de nitrogênio e freqüências de corte. Revista Brasileira de Zootecnia, 27: 833-840.

Anda. Associação Nacional para Difusão de Adubos. 2008. Anuário estatístico do setor de fertilizantes. São Paulo.

Anjos, J.T.; Tedesco, M. J. 1976. Volatilização de amônia proveniente de dois fertilizantes nitrogenados aplicados em solos cultivados. Científica, 4:49-55.

Arruda, N.G.; Cantarutti, R.B.; Moreira, E.M. 1987. Tratamentos físico-mecânicos e fertilização na recuperação de pastagens de Brachiaria decumbens em solos de Tabuleiro. Pasturas Tropicales, 9: 36-39.

Baligar, V.C.; Fageria, N.K., 1997. Nutrient use efficiency in acid soils: nutrient management and plant use efficiency. In: Moniz, A.C.; Furlani, A.M.C.; Schaeffert, R.E.; Fageria,

N.K.; Rosolem, C.A.; Cantarella, H. (eds.). Plant-soil interactions at low pH. Campinas, SP, Viçosa, MG: Brazilian Soil Sci. Soc., p.75-96.

Batista, K.; Monteiro, F.A. 2010. Variações nos teores de potássio, cálcio e magnésio em capim-marandu adubado com doses de nitrogênio e de enxofre. Revista Brasileira de Ciência do Solo, 34: 151-161.

Belarmino, M.C.J.; Pinto, J.C.; Rocha, G.P. et al. 2003. Altura de perfilho e rendimento de matéria seca de capim Tanzânia em função de diferentes doses de superfosfato simples e sulfato de amônio. Ciência Agrotecnologia, 27: 879-885.

Benett, C. G. S.; Buzetti, S.; Silva, K. S.; Bergamaschine, A. F.; Fabrício, J. A. 2008. Produtividade e composição bromatológica do capim marandu a fontes e doses de nitrogênio. Ciência e Agrotecnologia, 32: 1629-1636.

Benites, V.M; Polidoro, J.C.; Carvalho, M.C.S.; Resende, A.V.; Bernardi, A.C.C. Álvares, F.A. O potássio, o cálcio e o magnésio na agricultura brasileira. In: Prochnow, L.I. Boas práticas para uso eficiente de fertilizantes. Piracicaba: IPNI. 2010.

Bernardi A.C.C., Machado, P.L.O.A.; Silva, C.A. 2002. Fertilidade do solo e demanda por nutrientes no Brasil. In: Manzatto; C.M., Freitas Júnior, E.; Peres, J.R.R. Uso agrícola dos solos brasileiros. Rio de Janeiro: Embrapa Solos, p 61-77.

Bernardi, A.C.C.; Monte, M.B.M. 2009. O uso de zeólitas na agricultura. p. 493-508. In: Lapido-Loureiro, F.E.V., Melamed, R. Figueiredo Neto, J. eds. Fertilizantes, agroindústria e sustentabilidade. CETEM/MCT, Rio de Janeiro, RJ, Brazil.

Bomfim-Silva, E. M.; Monteiro, F. A. 2006. Nitrogênio e enxofre em características produtivas do capim-braquiária proveniente de área de pastagem em degradação. Revista Brasileira de Zootecnia, 35: 1289-1297.

Cáceres, O.; Santana, H.; Delgado, R. 1989. Influencia de la fertilización nitrogenada sobre el valor nutritivo y rendimiento de nutrimentos. Pastos y Forrages, 12: 189-195.

Camargo, A.C., Novo, A.L., Novaes, N.J., Esteves, S.N., Manzano, A.; Machado, R. 2002. Produção de leite a pasto. In: 'Simpósio sobre o manejo da pastagem', 18., Piracicaba. Anais... Piracicaba: Fealq. p. 285-319.

Cantarella, H.; Correa, L.A.; Primavesi, O.; Primavesi, A.C. 2002. Fertilidade do solo em sistemas intensivos de manejo de pastagens. p.99-131. In: Anais do Simpósio sobre Manejo de Pastagens. 2002. FEALQ, Piracicaba, SP, Brazil.

Carvalho, P.C.F. 2002. Country Pasture/Forage Resources Profile: Brazil, FAO, Rome, <http://www.fao.org/waicent/faoinfo/agricult/agp/agpc/doc/counprof/brazil.htm>

Carvalho, P.C.F; Batello, C. 2009. Access to land, livestock production and ecosystem conservation in the Brazilian Campos biome: the natural grasslands dilemma. Livestock Science, 120: 158-162.

Contini, E.; Martha Jr., G.B. 2010. Brazilian agriculture, its productivity and change. Bertebos Conference on "Food security and the futures of farms: 2020 and toward 2050". Falkenberg: Royal Swedish Academy of Agriculture and Forestry, August 29-31,

Correa, J.C.; Reichardt, K. 1995. Efeito do tempo de uso das pastagens sobre as propriedades de um latossolo amarelo da Amazônia Central. Pesquisa Agropecuária Brasileira, 30:107-114,

Corrêa, L.A.; Cantarella, H.; Primavesi, A.C.; Primavesi, O.; Freitas, A.R.; Silva, A.G. 2007. Efeito de fontes e doses de nitrogênio na produção e qualidade da forragem de capim-coastcross. Revista Brasileira de Zootecnia, 36: 763-772.

Corsi, M.; Martha Jr., G.B.; Nascimento Jr., D.; Balsalobre, M.A.A. 2001. Impact of grazing management on productivity of tropical grasslands. In: International Grassland CongresS, 19., São Pedro, 2001. Proceedings. São Pedro, SBZ, p.801-805.

Corsi, M.; Nussio, L.G. 1993. Manejo do capim elefante: correção e adubação do solo. In: 'Simpósio sobre o manejo da pastagem', 10., Piracicaba. Anais... Piracicaba: Fealq, pp.87-115.

Craswell, E.T., Lefroy, R.D.B. 2001. The role and function of organic matter nin tropical soils. In: Martius, C. Tiessen, H., Vlek, P.L.G. (Eds.). managing organic matter in tropical soils: scope and limitations. Kluwer, Dordrecht, The Netherlands, p.7-18.

Cruz, M.C.P.; Ferreira, M.E.; Lucheta, S. 1994. Efeito da calagem sobre a produção de matéria seca de três gramíneas forrageiras. Pesquisa Agropecuária Brasileira, 29: 303-312.

Cunha, M. K.; Siewerdt, L.; Silveira Jr., P; Siewerdt, F. 2001. Doses de nitrogênio e enxofre na produção e qualidade a forragem de campo natural de planossolo no Rio Grande do Sul. Revista Brasileira de Zootecnia, 30: 651-658,

Euclides, V.P.B. Valle, C.B.; Macedo, M.C.M.; Almeida, R.G., Montagner, D.B., Barbosa, R.A. 2010. Brazilian scientific progress in pasture research during the first decade of XXI century. Revista Brasileira de Zootecnia, 39: 151-168.

Euclides, V.P.B.; Costa, F.P.; Macedo, M.C.M. 2007. Eficiência biológica e econômica de pasto de capim-tanzânia adubado com nitrogênio no final do verão. Pesquisa Agropecuária Brasileira, 42: 1345-1355.

Fageria, N.K. And Baligar, V.C. 2003. Fertility management of tropical acid soil for sustainable crop production. In: Rengel, Z., ed. Handbook of soil acidity. New York, Marcel Dekker, p.359-385.

Fageria, N.K. and Baligar, V.C. 2008. Ameliorating soil acidity of tropical Oxisols by liming for sustainable crop production. Advances Agronomy, 99:345-431.

Gerdes, L.; Werner, J.C.; Colozza, M.T.; Possenti, R.A.; Schammass, E.A. 2000. Avaliação de características de valor nutritivo das gramíneas forrageiras marandu, setária e tanzânia nas estações do ano. Revista Brasileira de Zootecnia, 29: 955-963,

Gheri, E.O.; Cruz, M.C.P.; Ferreira, M.E. et al. 2000. Nível crítico de fósforo no solo para *Panicum maximum* Jacq. cv. Tanzânia. Pesquisa Agropecuária Brasileira, 35: 1809-1816.

Gomes, J.F.; Siewerdt, L.; Silveira Jr., P. 1987. Avaliação da produtividade e economicidade do feno de capim pangola (Digitaria decumbens Stent) fertilizado com nitrogênio. Revista da Sociedade Brasileira de Zootecnia, 16: 491-499.

Gomide, J.A.; Costa, G.G. 1984. Adubação nitrogenada e consorciação de capim-colonião e capim-jaraguá. III. Efeitos de níveis de nitrogênio sobre a composição mineral e digestibilidade da matéria seca das gramíneas. Revista da Sociedade Brasileira de Zootecnia, 13: 215-224.

Gupta, U.C.; Monteiro, F.A.; Werner, J.C. 2001. Micronutrients in grassland production. In: INTERNATIONAL GRASSLAND CONGRESS, 19., São Pedro, 2001. Proceedings. São Pedro, SBZ, p.149-156.

Guss, A.; Gomide, J.A.; Novais, R.F. 1990.Exigência de fósforo para o estabelecimento de quatro espécies de *Brachiaria* em solos com características físico-químicas distintas. R. Bras. Zootec, 19: 278-289.

Havlin, J.; Beaton, J.D.; Tisdale, S.L. & Nelson, W.L. 1999. Soil fertility and fertilizers: an introduction nutrient management. Upper Saddle River: Prentice Hall. 499p.

Hopkins, A.; Adamson, A.H. and BOWLING, P.J. 1994. Response of permanent and reseeded grassland to fertilizer nitrogen: 2. Effects on concentrations of Ca, Mg, K, Na, S, P, Mn, Zn, Cu, Co and Mo in herbage at a range of sites. Grass Forage Science, 49:9-20.

Jarvis, S.C.; Scholefield, D.; Pain, B. 1995. Nitrogen cycling in grazing systems. p.381-419. In: Bacon, P.E. ed. Nitrogen fertilization in the environment. Marcel Dekker, New York, EUA.

Kaimowitz, D.; Angelsen, A. 2008. Will livestock intensification help save Latin America´s forests? Journal of Sustainable Forestry, 27: 6-24.

Lara-Cabezas, W.A.R; Korndörfer, G.H.; Motta, A.S. 1997. Volatilização de N-NH3 na cultura de milho: II. Avaliação de fontes sólidas e fluidas em sistema de plantio direto e convencional. Revista Brasileira Ciência do Solo 21:489-496.

Lathwell, D.J.; Grove, T.L., 1986. Soil-plant relationship in the tropics. Annu. Rev. Ecol. Syst., 17:1-16.

Lugão, S.M.B.; Rodrigues, L.R.A.; Abrahão, J.J.S.; Malheiros, E.B.; Morais, A. 2003. Acúmulo de forragem e eficiência de utilização do nitrogênio em pastagens de *Panicum maximum* Jacq. (acesso BRA-006998) adubadas com nitrogênio. Acta Scientiarum. Animal Sciences, 25: 371-379.

Luz, P.H.C.; Braga, G.J.; Herling, V.R.; Lima, C.G.L. 1998. Efeitos de tipos, doses e incorporação de calcário sobre as características agronômicas do *Panicum maximum* Jacq. cv. Tobiatã. In: REUNIÃO DA SOCIEDADE BRASILEIRA DE ZOOTECNIA, 35., Botucatu, 1998. Anais. Brasília: SBZ, p.248-250.

Macedo, M.C.M. 2002. Degradação,renovação e recuperação de pastagens cultivadas:ênfase sobre a região dos Cerrados. In: Simpósio Sobre Manejo Estratégico da Pastagem, 2002, Viçosa. Anais do Simpósio Sobre Manejo Estratégico da Pastagem. Viçosa, MG : Departamento de Zootecnia, UFV, p.85-108.

Macedo, M.C.M. 2004.Analise comparativa de recomendações de adubação em pastagens. In: 21° Simposio sobre Manejo da Pastagem, 2004, Piracicaba, SP. Fertilidade do Solo para Pastagens Produtivas. Piracicaba, SP: FEALQ, v. 1. p. 317-355.

Macedo, M.C.M. 2005. Pastagens no ecossistema Cerrados: evolução das pesquisas para o desenvolvimento sustentável. In: REUNIÃO ANUAL DA SOCIEDADE BRASILEIRA DE ZOOTECNIA, 42., 2005, Goiânia. Anais... Goiânia: Sociedade Brasileira de Zootecnia, p.56-84.

Martha Jr., G.B.; Corsi, M.; Trivelin, P.C.O. 2004. Nitrogen recovery and loss in a fertilized elephantgrass pasture. Grass and Forage Science, 59: 80-90.

Martha Junior, G.B.; Corsi, M.; Trivelin, P.C.O.; Alves, M.C. 2004a. Nitrogen recovery and loss in a fertilized elephant grass pasture. Grass and Forage Science 59:80-90.

Martha Júnior, G.B.; Vilela, L.; Barioni, L.G.; Sousa, D.M.G. 2004b. Manejo da adubação nitrogenada em pastagens. In: Pedreira, C.G.S.; Moura, J.C.; Faria, V.P. Fertilidade do solos para pastagens produtivas. Piracicaba: FEALQ, p. 101-138.

Mattos, W. T.; Monteiro, F. A. Produção e nutrição do capim braquiária em função de doses de nitrogênio e enxofre. Boletim de Indústria Animal, v.60, p.1-10, 2003.

Mattos, W. T.; Santos, A. R.; Almeida, A. A. S.; Carreiro, B. D. C.; Monteiro, F. A. 2002. Aspectos produtivos e diagnose nutricional do capim-Tanzânia submetido a doses de potássio. Magistra, 14: 37-44.

Minson D.J., Cowan T. And Havilah E. 1993. Summer pastures and crops. Tropical Grasslands, 27, 131–149.

Monson, W.G.; Burton, G.W. 1982. Harvest frequency and fertilizer effects on yield, quality and persistence of eight bermuda grasses. Agronomy Journal, 74: 371-374.

Monteiro, F.A. 2004. Concentração e distribuição de nutrientes em gramíneas e leguminosas forrageiras. In: Simpósio Sobre Manejo Estratégico Das Pastagem, 2., Viçosa, 2004. Anais. Viçosa: DZO,UFV, p.71-107.

Monteiro, F.A.; Colozza, M.T.; Werner, J.C. 2004. Enxofre e micronutrientes em pastagens. In: Simpósio Sobre Manejo Da Pastagem, 21., Piracicaba, 2004. Anais. Piracicaba: FEALQ,. p.279-301.

Moraes, A.; Palhano, A.L. 2002. Fisiologia de produção de plantas forrageiras. In: Wachowicz, C.M; Carvalho, R.I.N. (Ed.). Fisiologia vegetal - Produção e pós-colheita. Curitiba: Champagnat, p. 249-271.

Mott, G.O. 1981. Measuring forage quantity and quality in grazing trials. In: Southern Pasture And Forage Crop Improvement Conference, 37., 1981, Nashville. Proceedings... Nashville: United States Department of Agriculture, p.3-9.

Novais, R.F.; Smyth, T.J., 1999. Fósforo em solo e planta em condições tropicais. Viçosa, MG: Universidade Federal de Viçosa. UFV, Departamento de Solos. DPS. 399p.

Oliveira, P. P. A.; Trivelin, P. C. O.; Oliveira, W. S.; Corsi, M. 2005. Fertilização com N e S na recuperação de pastagem de Brachiaria brizantha cv. Marandu em um neossolo quartzarênico. Revista Brasileira de Zootecnia, 34: 1121-1129.

Oliveira, P.P.A.; Boaretto, A.E.; Trivelin, P.C.O. et al. 2003. Liming and fertilization to restore degraded Brachiaria decumbens pastures grown on an entisol. Scientia Agrícola, 60: 125-131.

Oliveira, P.P.A.; Trivelin, P.C.O.; Oliveira, W.S. 2003. Eficiência de fertilização nitrogenada com uréia (15N) em Brachiaria brizantha cv. Marandu associada ao parcelamento de superfosfato simples e cloreto de potássio. Revista Brasileira de Ciência do Solo, 27: 613-620.

Oliveira, P.P.A.; Trivelin, P.C.O.; Oliveira, W.S. 2007. Balanço do nitrogênio (15N) da uréia nos componentes de uma pastagem de capim-marandu sob recuperação em diferentes épocas de calagem. Revista Brasileira de Zootecnia 36: 1982-1989.

Olsen, F.J. 1972. Effect of large application of nitrogen fertilizer on the productivity and protein content of four tropical grasses in Uganda. Tropical Agricultural, 49: 251-260.

Paulino, V.T.; Costa, N.L.; Rodrigues, A.N.A. et al. 2006. Resposta de Panicum maximum cv Massai à níveis de calagem. In: Reunião Anual Da Sociedade Brasileira De Zootecnia, 43., 2006, João Pessoa.. Anais... João Pessoa: Sociedade Brasileira de Zootecnia (CD-ROM).

Pereira, W.L.M.; Veloso, C.A.C.; Gama, J.R.N.F., 2000. Propriedades químicas de um Latossolo Amarelo cultivado com pastagens na Amazônia Oriental. Scientia Agricola, 57:531-537.

Primavesi, A.C.; Primavesi, O.; Corrêa, L.A. Cantarella, H.; Silva, A.G.; Freitas, A.R.; Vivaldi, L.J. 2004. Adubação nitrogenada em capim-coastcross: efeitos na extração de nutrientes e recuperação aparente do nitrogênio. Revista Brasileira de Zootecnia, 33: 68-78.

Primavesi, A.C.; Primavesi, O.; Corrêa, L.A.; Cantarella, H. and Silva, A.G. 2005. Absorção de cátions e ânions pelo capim-coastcross adubado com ureia e nitrato de amônio. Pesq. Agropec. Bras., 40: 247-253.

Primavesi, A.C.; Primavesi, O.; Corrêa, L.A.; Silva, A.G.; Cantarella, H. 2006. Nutrientes na fitomassa de capim-marandu em função de fontes e doses de nitrogênio. Ciência e Agrotecnologia, 30: 562-568.

Primavesi, O.; Corrêa, L. A.; Primavesi, A. C.; Cantarella, H.; Silva, A. G. Adubação com uréia em pastagem de Brachiaria brizantha sob manejo rotacionado: eficiência e perdas. São Carlos: Embrapa Pecuária Sudeste, (nov) 2003. 6p. (Embrapa Pecuária Sudeste, Comunicado Técnico, 41). Available in: http://www.cppse.embrapa.br/080servicos/070publicacaogratuita/comunicadote cnico/ComuTecnico41.pdf

Primavesi, O.; Corrêa, L. A.; Primavesi, A.C.; Cantarella, H.; Armelin, M. J. A.; Silva, A. G.; Freitas, A. R. de. Adubação com uréia em pastagem de Cynodon dactylon cv. Coastcross: eficiência e perdas. São Carlos: Embrapa Pecuária Sudeste, (jun) 2001. 42p. (Embrapa Pecuária Sudeste, Circular Técnica, 30). Available in: http://www.cppse.embrapa.br/080servicos/070publicacaogratuita/circular-tecnica/Circular30.pdf

Primavesi, O.; Primavesi, A.C.; Camargo, A.C. 1999. Conhecimento e controle, no uso de corretivos e fertilizantes, para manejo sustentável de sistemas intensivos de produção de leite de bovinos a pasto. Revista de Agricultura 74:249-266.

Primavesi, O.; Primavesi, A.C.; Corrêa, L.A.; Armelin, M.J.A.; Freitas, A.R. Calagem em pastagem de Brachiaria decumbens recuperada com adubação nitrogenada em cobertura. São Carlos: Embrapa Pecuária Sudeste, (dez) 2004. 32p. (Embrapa Pecuária Sudeste, Circular Técnica, 37). Available in: http://www.cppse.embrapa.br/080servicos/070publicacaogratuita/circular-tecnica/Circular37.pdf

Primavesi, O.; Primavesi, A.C.; Correa, L.A.; Silva, A.G.; Cantarella, Heitor. Lixiviação de nitrato em pastagem de coastcross adubada com nitrogênio. Revista Brasileira de Zootecnia, 35: 683-690, 2006.

Raij, B. van, 1969. Capacidade de troca de frações orgânicas e minerais dos solos. Bragantia, 28:85-112.

Raij, B. Van; Cantarella, H.; Quaggio, J.A.; Furlani, A.M.C. Recomendações de adubação e calagem para o Estado de São Paulo. Campinas, Instituto Agronômico/Fundação IAC, 1996. 285p. (Boletim Técnico, 100).

Rodrigues, R.C.; Alves, A.C.; Brennecke, K.; Plese, L.P.M.; Luz, P.H.C. 2006. Densidade populacional de perfilhos, produção de massa seca e área foliar do capimxaraés cultivado sob doses de nitrogênio e potássio. Boletim de Indústria Animal, 63: 27-33.

Sanchez, P. A.; Salinas, J. G. 1981. Low-input technology for managing Oxisols and Ultisols in tropical America. Advances in Agronomy, 34: 280-407.

Sanchez, P.A., 1976. Properties and management of soil in tropics. New York: John Wiley. 619p.

Santos, F.A.P.; Martinez, J.C.; Voltolini, T.V.; Nussio, C.M.B. 2003. Associação de plantas forrageiras de clima temperado e tropical em sistemas de produção animal de regiões sub-tropicais. In: Simpósio Sobre Manejo Da Pastagem; Produção Animal

Em Pastagem, Situação Atual E Perspectivas, 20., Piracicaba, 2003. Anais. Piracicaba: FEALQ, p.215.

Scott, N.M.; Watson, M.E.; Caldwell, K.S. 1983. Response of grassland to the application of sulphur at two sites in Northeast Scotland. Journal of the Food and Agriculture, 34: 357-361.

Soares Filho, C.V.; Monteiro, F.A.; Corsi, M. 1992. Recuperação de pastagens degradadas: 1. Efeito de diferentes tratamentos de fertilização e manejo. Pasturas Tropicales, 14: 2-6.

Sousa, D.M.G.; Lobato, E.; Rein, T.A. 2004. Adubação com fósforo. In: Sousa, D.M.G.; Lobato, E. (Eds.) Cerrado correção do solo e adubação. 2.ed. Brasília: EMBRAPA, p.147-168.

Sousa, D.M.G.; Martha Júnior, G.B.; Vilela, L. 2004. Manejo da adubação fosfatada em pastagens. In: Pedreira, C.G.S.; Moura, J.C.; Faria, V.P. Fertilidade do solos para pastagens produtivas. Piracicaba: FEALQ, p. 155-215.

Stevens, R.J. 1985. Evaluation of the sulphur status of some grasses for silage in Northern Ireland. Journal of Agricultural Science, 105: 581-585.

Teitzel J.K., Gilbert M.A. And Cowan R.T. 1991. Sustaining productive pastures in the tropics. 6. Nitrogen fertilized grass pastures. Tropical Grasslands, 25, 111–118.

Tognon, A.A., Demattê, J.L.I.; Demattê, J.A.M. 1998. Teor e distribuição da matéria orgânica em latosolos das regiões da floresta amazônica e dos cerrados do Brasil Central. Scientia Agricola, 55(3): 343-354.

Vicente-Chandler J., Abruña F., Caro-Costas R., Figarella J., Silva S. And Pearson R.W. 1974. Intensive grassland management in the humid tropics of Puerto Rico. Bulletin 23, University of Puerto Rico, Rio Piedras, Puerto Rico: University of Puerto.

Vicente-Chandler, J.; Silva, S.; Figarella, J. 1959. The effect of nitrogen fertilization and frequency of cutting on the yield and composition of three tropical grasses. Agronomy Journal, 51: 202-206.

Vilela, L.; Soares, W.V.; Sousa, D.M.G.; Macedo, M.C.M. 2002. Calagem E Adubação Para Pastagens. Sousa, D.M.G.; Lobato, E. Cerrado: correção do solo e adubação. Planaltina, Embrapa Cerrados. p.367-382.

Vilela, L.; Sousa, D. M. G.; Silva, J. E. 2002. Adubação Potássica. In: Souza. D. M. G.; Lobato, E. Cerrado: correção do solo e adubação. Planaltina: Embrapa Cerrados, p.169-183.

Vitti, G.C.; Novaes, N.J. 1986. Adubação Com Enxofre. In: Simpósio Sobre Calagem E Adubação De Pastagens, 1., Nova Odessa. Anais. Piracicaba: POTAFOS, 1986. p.191-231.

Volpe, E.; Marchetti, M.E.; Macedo, M.C.M. et al. 2008. Acúmulo de forragem e características do solo e da planta no estabelecimento de capim-massai com diferentes níveis de saturação por bases, fósforo e nitrogênio. Revista Brasileira de Zootecnia, 37: 228-237.

Werner, J.C.; Paulino, V.T.; Cantarella, H.; Andrade, N.O.; Quaggio, J.A. 1996. Forrageiras. In: Raij, B. Van; Cantarella, H.; Quaggio, J.A.; Furlani, A.M.C. Recomendações de adubação e calagem para o Estado de São Paulo. Campinas, Instituto Agronômico/Fundação IAC, p.261-273 (Boletim Técnico, 100).

Werner, J.C.; Pedreira, J.V.S.; Caiele, E.L. 1977. Estudo de parcelamento e níveis de adubação nitrogenada com capim-pangola (Digitaria decumbens Stent). Boletim da Indústria Animal, 24: 147-151.

Whitehead, D.C. 1995. Volatilization of ammonia. p.152-179. In: Whitehead, D.C. ed. Grassland nitrogen. CAB International, Wallingford, England.

3

Natural Forests of Oak in NW Spain: Soil Fertility and Main Edaphic Properties

I.J. Díaz-Maroto[1], P. Vila-Lameiro[1], O. Vizoso-Arribe[1],
E. Alañón[2] and M.C. Díaz-Maroto[2]
[1]Santiago de Compostela University
[2]Castilla-La Mancha University
Spain

1. Introduction

The species of *Quercus* genus dominates a large part of the temperate forests of the Northern hemisphere and in subtropical transition areas. At present, about 400 species are recognized, most of which occur in Mexico and extend all over North America, Europe and most of Asia (Díaz-Maroto et al., 2005). In the Northwest Spain, the climax vegetation that currently covers the largest area is the broad-leaved forest characterized by various oak species (Buide et al., 1998). According to different studies (palynological and dendrochronological studies, principally) these forests were established in the NW Iberian Peninsula between five and seven thousand years ago, after the last glaciation in Quaternary (Guitián, 1995). Historical factors, site conditions, and requirements of the species give rise to different forest types with various floristic compositions and structures (Peterken & Game, 1984). All the forests belong to *Querco-Fagetea* class (Atlantic Province) with the maximum possible degree of evolution, and they would remain in the current state if environmental conditions did not change (Rivas-Martínez, 1987; Rivas-Martínez et al., 2001).

The forests correspond to the following phytosociological associations (Rivas-Martínez, 1987; Díaz & Fernández, 1994; Rivas-Martínez et al., 2001): 1) *Myrtillo-Quercetum roboris* P. Silva, Rozeira & Fontes 1950; 2) *Rusco aculeati-Quercetum roboris* Br.-Bl., P. Silva &Rozeira 1956; 3) *Blechno spicanti-Quercetum roboris* Tx. & Oberdorfer 1958; 4) *Linario triornithophorae-Quercetum petraeae* (Rivas-Martínez, Izco & Costa ex. F. Navarro 1974) F. Prieto & Vázquez 1987; 5) *Luzulo henriquesii-Quercetum petraeae* (F. Prieto & Vázquez 1987) Díaz & F. Prieto 1994; 6) *Linario triornithophorae-Quercetum pyrenaicae* Rivas-Martínez et al., 1984; 7) *Holco molli-Quercetum pyrenaicae* Br.-Bl., P. Silva & Rozeira 1956; 8) *Genisto falcatae-Quercetum pyrenaicae* Rivas-Martínez in Penas & Díaz 1985.

According to EU Directive 92/43/EEC, these forests are habitats of importance to conservation and, consequently, they will be part of Natura Network 2000. For this, their distribution, floristic diversity and main edaphic properties knowledge is essential (Díaz-Maroto et al., 2005, 2006). The surface area covered by oak forest in Northwest Spain has gradually decreased over the historical period, until ours days, with observable forest fragmentation and decrease in species number, similar to other temperate zones (Graae &

Sunde, 2000; Acar et al., 2004). The main reasons for such a decrease were: 1) Land clearing for establishment of crops and pastures; 2) Timber and firewood extraction; 3) Forest fires; 4) Unfortunate silvicultural treatments and, 5) Massive reforestation with pine and eucalyptus, more recently, more and less, since beginning of last century (Guitián, 1995).

Recently, these forests have increased substantially (Direction General of Nature Conservation [DGCONA], 2001, 2003). Native hardwood forests account for approximately 27% (375,922 ha) of the total woodland area of Galicia. *Quercus robur* L. stands cover an area of 187,789 ha, representing almost 14% of total woodland area, and *Quercus pyrenaica* Willd. stands cover an area of 100,504 ha, corresponding to more than 7% of the woodlands. In Asturias, data for genus *Quercus* suggest that these cover 76,871 ha, 7.25% of total woodland area. The future perspectives for these forests are very positive, as data from the first of four Forest National Inventories shows an important increase, nearly of 20% surface area occupied by them in the third Forest National Inventory of start of XXI century.

Soils in study area tend to be acidic, with pH ranging from 6.00 to 4.00, approximately. The high precipitation rate of near to 1500 mm per year, particularly the intensity of rainfall (storms with more than 200 mm rainfall are common) during a continuous period within the annual cycle (Díaz-Maroto & Vila-Lameiro, 2005) is the decisive factor in this territory. In addition, drainage conditions that favour water infiltration into the soil, fast runoff flow, and a temperate temperature regime (Carballeira et al., 1983), allow for soil acidification. Therefore, acidity in soils should not imply low forest site quality, which is based on productivity, growth rates, diversity, overstory vegetation,..., (Díaz-Maroto et al., 2005).

Slightly acidic soils are even more likely to sustain varied forest vegetation than alkaline soils (Hardtle et al., 2005). Soil fertility, expressed as the macronutrient richness, is more relevant than soil pH and physical characteristics (Thomas & Buttner, 1998; Hagen-Thorn et al., 2006). The soils under oak forests that grow on acidic rocks are oligotrophic or mesotrophic, similar to soils under pine and eucalyptus forests (Díaz-Maroto & Vila-Lameiro, 2005). These soils show a high carbon (C)/nitrogen (N) ratio, a very low saturation percentage, and low concentration of exchangeable cations, in mg kg^{-1} [(potassium (K), calcium (Ca) and magnesium (Mg)]. However, phosphorus (P) concentration, in mg kg^{-1}, is high. Always, if these data are compared with the data obtained by other authors in similar soils under oak forests, in general, but, in different areas.

This study aims to contribute to the knowledge of soils under *Quercus* spp., in general, and, *Q. robur, Q. pyrenaica* and *Q. petraea*, in particular, to Northwest Iberian Peninsula. Particularly, the aim of present work is to analyze and characterize the edaphic habitat of oak forests in the study area and, also describe their main properties, mainly plant chemistry and nutrition related to soil fertility.

2. Material and methods

2.1 Study area

The study area is located in the Northwest Iberian Peninsula (latitude: 42-44 °N and longitude: 6-9 °W). For *Q. robur* and *Q. pyrenaica,* the area covers the whole of Galicia, while for *Q. petraea* (Matt.) Liebl., the area also includes sites of Autonomous Communities of Galicia and Asturias and, the León province (Fig. 1).

This area has a complicated topography, an average altitude higher than 550 meters, and slopes steeper than 20% in half of the surface area. It is a territory with a rich and different lithological composition (Díaz-Maroto et al., 2005, 2006).

The climate is quite varied, but follows a general oceanic pattern with annual rainfall from 600 to more than 3000 mm (Carballeira et al., 1983). The study area was defined and the sampling zones were chosen based on the Forest Map of Spain (Ruiz de la Torre, 1991).

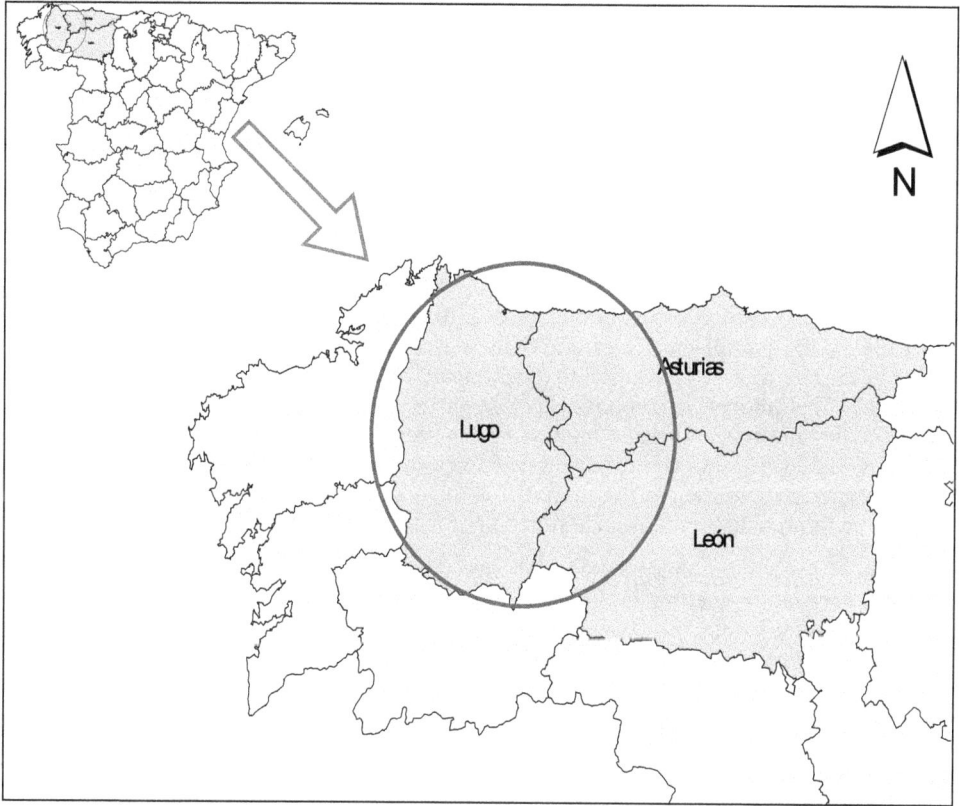

Fig. 1. Location of study area in the Northwest Iberian Peninsula

2.2 Experimental design and analysis techniques used

The number of sampling points was different for each species, but in all cases, we studied the physiographic features (altitude, slope, and orientation), general soil characteristics (parent material, depth, and description of horizons), soil profile, and physicochemical properties (Food and Agriculture Organization [FAO], 1977). To enable description of the soil profiles, a pit approximately 2 m long and 80 cm wide was dug at each site. The horizons were differentiated according to color and/or texture. One sample of each horizon (about 2 kg) was collected, and the following variables were analyzed (FAO, 1977; Klute, 1986): Munsell color, moisture (by oven drying to constant weight at 105 °C), pH (potentiometrically in water at 1:2.5 soil–solution ratio), total nitrogen (N) (semi-micro-

Kjeldahl method), carbon (Walkley and Black method), organic matter (OM) (multiplying the percent of C by a factor of 1.724), C/N ratio, exchangeable cations [by extraction with 1 N ammonium acetate at pH 7 and quantified by atomic absorption (Ca and Mg) and by emission (K) using a Perkin-Elmer 603 spectrophotometer, Waltham, Mass.] and available P by the Bray and Kurtz method (Bray-1 P).

At each sampling point, soil different parameters were estimated. The parameter total value in the whole soil profile was determined by using a standard weighted mean for Sand (S), Silt (Si), and Clay (Cl). For the rest of studied parameters, the weighted mean was considered by applying the method of Russell & Moore (1968). In the case of surface values, the information from the upper 20 cm of the soil was considered, except when more than one horizon was present at this depth. In that case, it was calculated as a weighted mean (Díaz-Maroto et al., 2005).

2.3 Statistical analysis

Univariate statistics were calculated (Walpole et al., 1999) and the following values were presented for the measured soil parameters: lower limit (LL), parameter minimum value; lower threshold (LT) or 10th percentile; mean (M); upper threshold (UT) or 90th percentile; upper limit (UL), parameter maximum value. Based on these values, the optimal edaphic habitat of the oak forests, in relation to a parameter, can be defined as the interval between UT and LT. This interval is composed by 80% of the sampling points, excluding the areas where the parameter showed the highest and the lowest values, marginal habitats (Díaz-Maroto et al., 2005). The optimal habitat defines the most suitable soil conditions for a particular species, whereas, in the marginal is the condition where some parameter is not adequate so the area has questionable suitability for oak growth (Hardtle et al., 2005).

Normality of the soil parameters was evaluated (Walpole et al., 1999), the result allowed a bivariable correlation analysis by Pearson (normal) or by Spearman's rank (Rozados et al., 2000; Sas Institute Inc. [SAS], 2004) to be applied. When soil parameters were significantly ($p<0.05$) correlated, possible linear relationships between the parameters was studied.

3. Edaphic habitat of oak forests in NW Spain

3.1 Physiographic characteristics of the stands

Most of the oak stands are located in zones with a steep slope or at the thalwegs, due to historical anthropogenic activities in these forests, at low and medium altitude with variable orientations, but shady sites are dominated in Q. robur and Q. petraea, while sunny sites in Q. pyrenaica stands (Díaz-Maroto et al., 2005; Díaz-Maroto & Vila-Lameiro, 2006).

Some stands of Q. robur are located in continental influence zones that are relatively distant from their suitable habitat. In these cases, this species is very easily hybridized, with Q. petraea and Q. pyrenaica (Díaz-Maroto et al., 2006), thus it is complicated to distinguish between these three species in such zones (Rivas-Martínez et al., 2001).

3.2 Parent rock and soil type

The geological material is heterogeneous with rocks from Precambrian and Paleozoic ages, tertiary and quaternary sediments (Ruiz de la Torre, 1991). Siliceous substrates were present in most of the stands, with dominance of granite and schist (Díaz-Maroto & Vila-Lameiro, 2005). Soil depth is elevated, and the presence of loose, fresh, deep soils is especially suitable for *Quercus* spp., in general, and above all for *Q. robur*, because these species has an extensive root system that penetrates very deep into soil (Göransson et al., 2006).

The soils in *Q. robur* stands are mainly Dystric Cambisols (70%), with smaller proportions of the following soil types: Humic Umbrisols (10.26%), Humic Regosols (10.26%), Dystric Regosols (7.70%), and Gleyic Cambisols (2.55%). Dystric Cambisols occurs on all substrate types with an A/Bw/C profile. The Bw horizon is the most characteristic for brownification process, and it can greater than 100 cm (Van de Moortel et al., 1998). The *Q. pyrenaica* stands are mainly established on Umbric Regosols (54%) and, to a lesser extent, on Humic Umbrisols (25.64%), Dystric Cambisols (15.38%), Umbric Leptosols (5.12%), and Gleyic Cambisols (2.56%). Finally, in the study area, *Q. petraea* occurs on less well developed soils than the other oaks, *Q. robur* and *Q. pyrenaica*. More than 75% of the stands occur on Umbric Regosols with an "A" horizon that is usually no deeper than 50 cm.

3.3 Chemical and nutritional properties

3.3.1 *Quercus robur* soils

The main properties of the soils under *Q. robur* forests in NW Spain were characterized. Eleven edaphic parameters (pH, OM, N, C/N ratio, P, K, Ca, Mg, S, Si and Cl) were estimated in 39 soil samples (Table 1).

The parameters that reflect the chemical and biological soil properties (pH, OM, N, and C/N ratio) show average values similar to the values reported as adequate by Castroviejo (1988) and Rozados et al. (2000) for these formations, but, the values obtained for OM are an exception because they are slightly higher than optimal, its average value is equal to 8.64 ± 5.19. According to the acidity scale suggested by Wilde (1946), most soils are to strongly acidic (4.7 ≤ pH < 5.5). Some of them present a value lower than 4 (extremely acidic) or higher than 6 (moderately acidic).

However, the studied soils show satisfactory humification, which is revealed by high values of the C/N ratio (Mansson & Falkengren-Grerup 2003; Hagen-Thorn et al. 2004). The pH of the upper horizons is the most relevant pH in the biological processes (Nornberg et al., 1993). This is because it depends on the residues of the dominant species in the vegetal cover and because the first roots of young plants are developed in the upper layer. The leaching due to rainfall is very important, because it can move dissolved salts down through the soil and cause an overall increase in pH at depth. Therefore, the analysis of the data obtained in the topsoil, the upper 20 cm of the soil, would mean a low mineralization rate due to the relatively high OM content, the low pH, and the C/N ratio near 15 (Nornberg et al., 1993; Neirynck et al., 2000; Landgraf et al., 2006) (Table 1).

Nutrient concentration is considered in the low or medium category (for instance, the exchangeable K concentration is only 73 mg kg⁻¹) as compared with other forests in the study area, except the Bray-P concentration, which is considerably higher (21.8 mg kg⁻¹ versus 1–6 mg kg⁻¹) (Castroviejo, 1988; Rozados et al, 2000).

The physical characteristics values are close to those reported by Castroviejo (1988) (Table 1). These data can be influenced by the steep slope on which the stands are located. Moreover, slight erosion occurs despite the vegetation cover and the soil protection by the extensive root system of *Q. robur*. This erosion leads to a gradual loss of the fine fractions, which are dragged downhill. The particle-size distribution of the soil determines the physical properties and influences the soil acidity (Van de Moortel et al., 1998).

Parameter	Average ± SD	CV %	Max-Min
pH	4.85 ± 0.46	9	6.15–3.92
OM (g kg^{-1})	8.64 ± 5.19	60	23.31–1.04
N (g kg^{-1})	0.307 ± 0.178	58	0.793–0.04
C/N	14.6 ± 4.5	31	29.6–6.9
P (mg kg^{-1})	21.8 ± 28.9	133	117.2–0.4
K (mg kg^{-1})	73 ± 40	55	231–9
Ca (mg kg^{-1})	120 ± 216	180	1297–3
Mg (mg kg^{-1})	29 ± 21	72	85–0
S (g kg^{-1})	66 ± 21	31	89–14
Si (g kg^{-1})	23 ± 20	92	85–7
Cl (g kg^{-1})	11 ± 5	45	27–1

Table 1. Average, standard deviation (±SD), coefficient of variation (CV), and the range of values (Max-Min) of soil parameters (n=39) in *Q. robur* forests in NW Spain

3.3.2 *Quercus petraea* soils

Soils under natural forests of *Q. petraea* in northwest Spain were characterized in an inventory of 52 stands in which the chemical properties and nutrient contents were determined by analyzing ten edaphic parameters (Table 2). Siliceous substrates (mainly slates) of high pH predominated in all stands. Umbric Regosols occurred in almost 80% of the stands, indicating that the soils were less well developed than those under *Q. robur*. The C/N ratio was particularly high and the OM content was less than 10 mg kg^{-1}.

In general, the soils are very acidic, with pH values that present little variation. The high rainfall in a large part of study area accentuates the acid character of the soils.

Although the mean value of the C/N ratio was close to 18, the low pH values do not allow adequate humification conditions (Mansson & Falkengren-Grerup, 2003). The rate of mineralization may be considered as slow, as analysis of the data corresponding to the topsoil shows that the mean content of OM is 10.09, the pH is 4.59 and the C/N ratio 17.92.

The concentrations of all macronutrients, except phosphorus, were higher than in the soils under pedunculate oak, and those of phosphorus and calcium were the most variable. The contents of base cations, fine earth (FE) and total gravel (GRA) were high in the optimal habitat (West of Galicia, East of Asturias and NW León).

Parameter	Average ± SD	CV %	Max-Min
pH	4.72 ± 0.36	8	5.65–4.23
OM (g kg⁻¹)	7.82 ± 4.03	52	19.82–1.82
N (g kg⁻¹)	0.250 ± 0.100	43	0.540–0.070
C/N	17.9 ± 3.8	21	25.1–8.9
P (mg kg⁻¹)	9.5 ± 7.5	79	24.8–0.9
K (mg kg⁻¹)	90.0 ± 52.9	59	275–26
Ca (mg kg⁻¹)	203 ± 276	136	1135–13
Mg (mg kg⁻¹)	46 ± 42	93	165–4
FE (g kg⁻¹)	44 ± 11	26	67–18
GRA (g kg⁻¹)	56 ± 12	20	82–33

Table 2. Average, standard deviation (±SD), coefficient of variation (CV), and the range of values (Max-Min) of soil parameters (n=52) in natural forests of $Q.$ $petraea$ in NW Spain

3.3.3 Quercus pyrenaica soils

The soils where natural stands of "rebollo" oak ($Quercus$ $pyrenaica$ Willd.) occur in NW Iberian Peninsula were characterized in an inventory of 40 different stands; the main properties of the soils were also determined by analyzing ten different edaphic parameters (Table 3).

The substrates are principally siliceous, with schist and quartzite predominating; the dominant textural composition is sandy-loam with small pockets of clay-loam and sandy textures. Soils are mainly the Umbric Regosol-type (about 60% of soils), although other soil types like Cambisols are found in areas with a soil depth greater than 80 cm and, normally located in thalwegs.

The values of the parameters that reflect the chemical properties of the soil, i.e., the pH, organic matter, nitrogen, and C/N ratio, are similar to those reported as suitable for $Quercus$ spp. forests, in general, to the study area. However, the concentrations of macronutrients, except for phosphorous, were higher than in soils under $Q.$ $robur$ (Table 3) (Díaz-Maroto et al., 2006).

Parameter	Average ± SD	CV %	Max-Min
pH	5.19 ± 0.29	6	6.13–4.30
OM (g kg⁻¹)	4.10 ± 2.69	66	11.30–1.01
N (g kg⁻¹)	0.151 ± 0.118	78	0.555–0.024
C/N	18.4 ± 4.3	27	31.9–10.1
P (mg kg⁻¹)	9.4 ± 13.3	140	71.6–0.8
K (mg kg⁻¹)	101.4 ± 61.9	59	281–22
Ca (mg kg⁻¹)	154 ± 223	144	935–0.1
Mg (mg kg⁻¹)	61 ± 62	102	237–4
FE (g kg⁻¹)	59 ± 16	27	88–28
GRA (g kg⁻¹)	41 ± 16	39	72–12

Table 3. Average, standard deviation (±SD), coefficient of variation (CV), and the range of values (Max-Min) of soil parameters (n=40) in natural forests of $Q.$ $pyrenaica$ in NW Spain

3.4 Characterization of edaphic habitat of *Quercus robur* forests

The optimal habitat of *Q. robur* forests, which was considered statistically to fall between the 10th and 90th percentiles (80% of plots ≈ sampling points), allows determination of the most suitable ecological range, and may be considered as the minimum potential area of the species in the study area. Marginal habitats are defined as those between the limits of the optimal habitat and the absolute extremes; they include 20% of plots and provide an estimation of the species adjustment to site conditions according to parameters values that remain outside the optimal habitat, since not all of them have the same level of significance as biotope descriptors (Fig. 2) (Díaz-Maroto et al., 2005).

In *Q. robur* stands, 80% of soils are Cambisols, so we calculated the optimal and marginal edaphic habitats for this soil-type only (Fig.2).

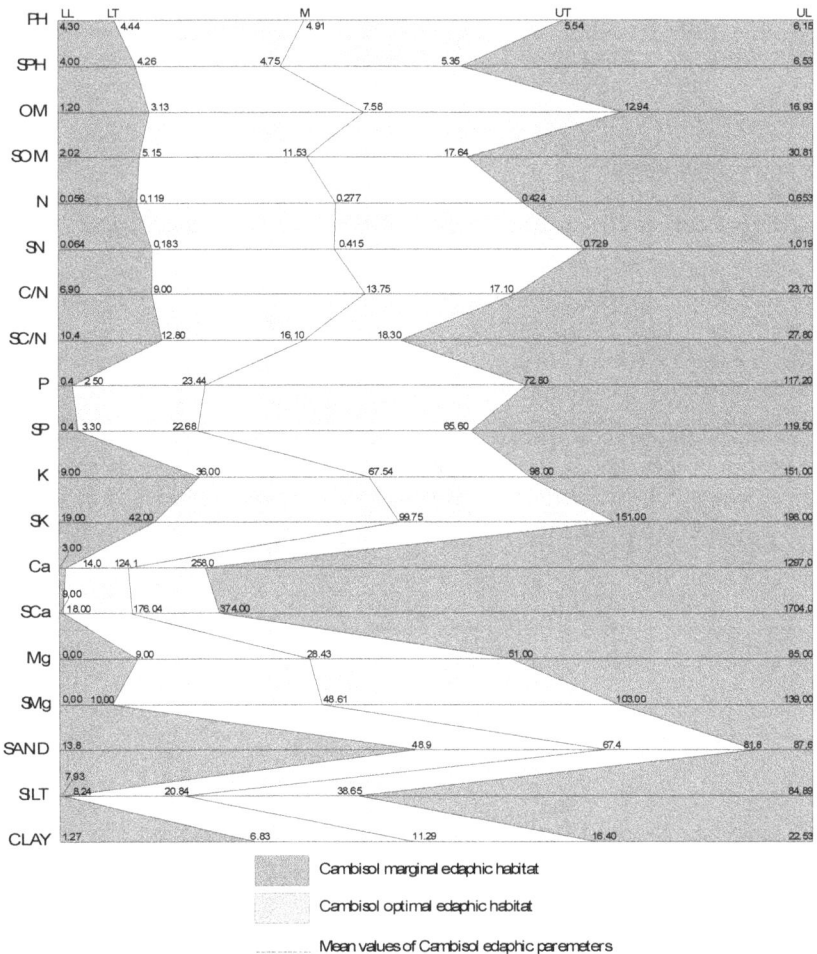

Fig. 2. Optimal and marginal edaphic habitats in Cambisols soils (*Q. robur* forests in Northwest Iberian Peninsula)

The wideness of the upper marginal intervals (upper threshold and limit) for Ca and SCa in the Cambisols edaphic habitat (Fig. 2), suggests that these parameters are not good indicators of the upper limit of optimal habitat (Díaz-Maroto et al., 2005, 2006):

- The soils located within the optimal edaphic habitat (Fig. 2) are mainly strongly acidic, with an average pH value equal to 4.91 for the whole profile, and equal to 4.75 for the topsoil. Within the marginal habitat, there are moderately acidic soils, with pH higher than 5.5 (Wilde, 1946). OM content is variable, with a minimum that does not reach 2 mg kg^{-1} for the whole profile and corresponds to the lower threshold of the edaphic habitat (Fig. 2), although the average is 7.58 mg kg^{-1} for the whole profile and 11.53 mg kg^{-1} for the topsoil, which characterizes a mull-moder humus (Camps et al., 2004).
- Optimal edaphic habitat shows low content of exchangeable cations (Fig. 2). However, the P has a considerably high value, with an average of 24 mg kg^{-1} for the whole profile.
- The soils have an elevated percentage of sand with an optimal habitat that ranges between 49-82 mg kg^{-1} and an average value of 67 mg kg^{-1}. Conversely, the clay proportion reaches a maximum of nearly 23 mg kg^{-1}, with an average of 11 mg kg^{-1}, because the substrates overlie parent material that releases a greater proportion of sand than clay during weathering (Castroviejo, 1988).

3.5 Characterization of edaphic habitat of *Quercus pyrenaica* forests

The same methodology was followed to distinguish the optimal and marginal edaphic habitats of *Quercus pyrenaica* forests in NW Spain. In this case, as about 60% of soils are Umbric Regosols. In the Fig. 3, we only calculated the optimal and marginal edaphic habitats to this soil-type. Also, it can be seen that certain edaphic parameters are not suitable for fixing the upper value of the optimal habitat, given the wide range of upper marginal values: OM, total and surface N, total and surface Mn and surface Ni as, to a lesser extent, total and surface P. The most notable characteristics were as follows (Fig. 3):

- The pH values of the optimal habitat ranged between 4.89 and 5.38, corresponding to strongly to moderately acidic soils (Wilde, 1946). In marginal habitats soil pH was less than 4.30 or greater than 6.13 (Fig. 3).
- The OM content was low and for optimal habitats ranged between 1.39-6.06 mg kg^{-1} for the total profile and between 1.89-7.72 mg kg^{-1} for the upper 20 cm (Fig. 3). The optimal values of the C/N varied widely and indicate adequate humification (Mansson & Falkengren-Grerup, 2003) and a normal rate of mineralization.
- The optimal habitat showed higher contents of exchangeable base cations than in the soils under *Quercus robur* within the study area. However, the concentration of P was lower, probably because the stands correspond to mature forests that had not undergone changes in land use (Díaz-Maroto et al., 2005).
- Wide suitable ranges of fine earth and total gravel contents were also apparent.

3.6 Characterization of edaphic habitat of *Quercus petraea* forests

Finally, the same methodology was followed to distinguish the optimal and marginal edaphic habitats of *Q. petraea* forests in NW Spain. In this species, more than 75% of the stands also occur on Umbric Regosols. Then, we analized the optimal and marginal edaphic habitats to this soil-type (Fig. 4). Parameters related to Ca concentration are not suitable for determining the edaphic habitat of sessile oak in northwest Spain, given the wide range of upper marginal values (Fig. 4):

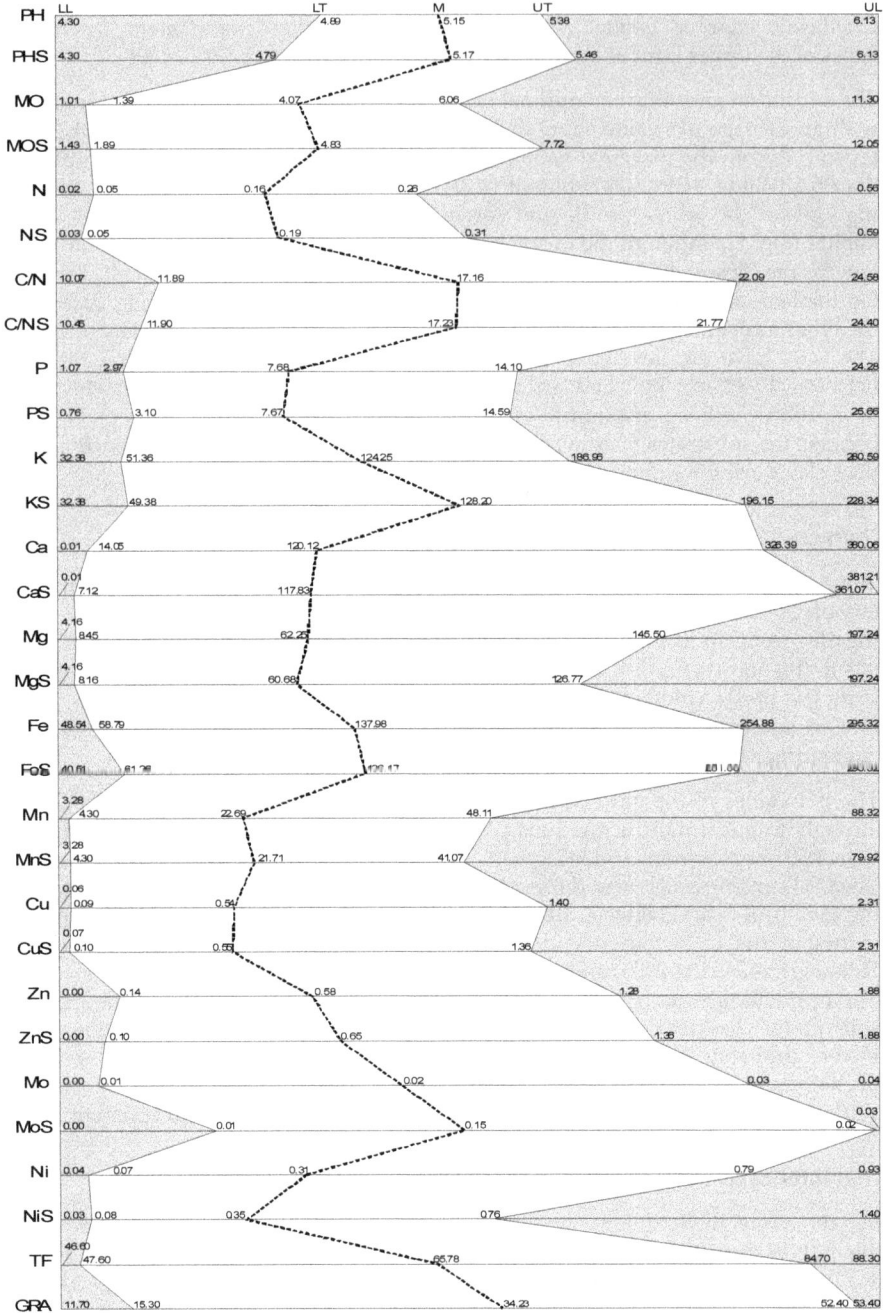

Fig. 3. Optimal and marginal edaphic habitats in Umbric Regosols (*Q. pyrenaica* forests in Northwest Iberian Peninsula)

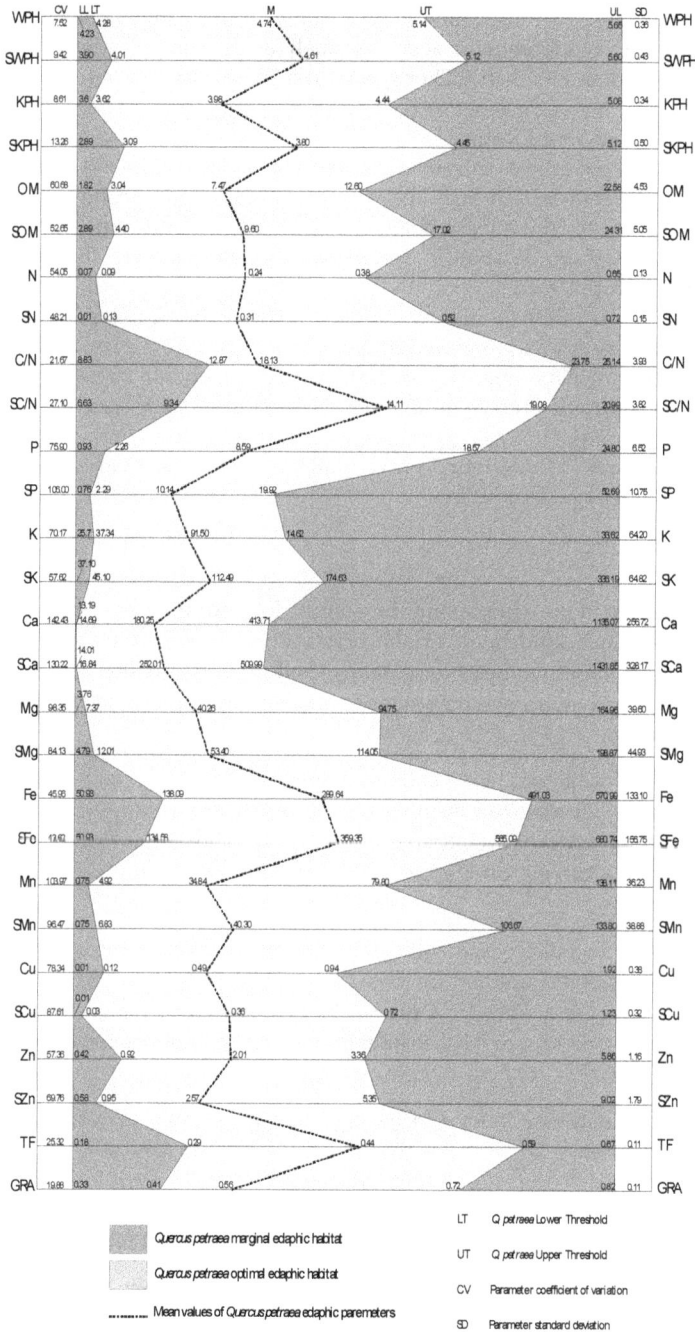

Fig. 4. Optimal and marginal edaphic habitats in Umbric Regosols (*Q. petraea* forests in Northwest Iberian Peninsula)

- The total pH in H_2O (WPH) values of the optimal habitat ranged between 4.28 and 5.14, corresponding to moderately or even extremely acidic soils (Wilde, 1946). In marginal habitats soil pH was less than 4.23 or greater than 5.65 (Fig. 4). These values varied little and, principally, the surface pH in H_2O (SWPH), partly due to a high rainfall that does not allow adequate humification.
- The OM matter content was intermediate (Hagen-Thorn et al., 2004) and for the optimal habitat ranged between 3.04 and 12.60 mg kg^{-1} for the total profile, and between 4.40 and 17.02 mg kg^{-1} for the topsoil (Fig. 4), giving rise to moder type humus with a slow rate of mineralization, in spite of suitable C/N ratio of more than 18 (Mansson & Falkengren-Grerup, 2003).
- The optimal habitat showed higher contents of base cations, except P, than in *Q. robur* soils (Díaz-Maroto & Vila-Lameiro, 2005).
- The concentration of the micronutrient Mn varied widely (4.92-79.80 mg kg^{-1}). Copper was the micronutrient present at lowest concentrations, with an average value of less than 0.5 mg kg^{-1} (Fig. 4).
- The total fine earth (FE) and total gravel (GRA) contents also varied widely, 29-59 mg kg^{-1} and 41-72 mg kg^{-1}, respectively.

4. Conclusions

Most of the studied soils correspond to stands located in areas with a complicated topography, a medium altitude and variable orientation. The substrates are mainly siliceous, with sandy-loam or loamy textures dominating. Soil depth is greater than 100 cm in more than half of the stands, although some stands can also develop on rocky substrates, mainly granite and schist.

Within the study area, *Quercus robur* is located on more developed soils than the other two species, principally, Dystric cambisol, with an A/Bw/C profile. *Q. pyrenaica* occurs predominately on Regosols and *Q. petraea* is present on poorly developed soils, with more than 75% of stands occurring on Umbric Regosols with an "A" horizon that is usually no deeper than 50 cm.

The chemical parameters of *Q. robur* soils (pH, OM, N and C/N ratio), show average values similar to those considered optimal for this species, except for the OM, which had higher values. They are strongly acidic ($4.7 \leq$ pH < 5.5), however, they have a good humification. The macronutrient content can be considered low or medium, except for P, since soils develop on substrates that are poor in base cations.

Organic matter, pH and clay tended to be significantly ($p<0.05$) correlated with other soil parameters, and with each other. A positive linear relationship ($p<0.001$) was noted between pH and Bray-P concentration, suggesting that many oak forests are located on abandoned agricultural land. In these stands, the high P content increases with the value of pH. Clay content shows significant differences for surface and total OM and nitrogen.

The analysis of edaphic habitat by soil type does not differ greatly from the results of the stands considered as a whole. *Q. robur* forests show a tendency to occur on the best soils, cambisol type, and not on poorer soils, where the occurrence of this species is often due to anthropogenic pressure.

In *Quercus pyrenaica* soils, the pH values were quite homogeneous, giving rise to very acidic soils, although these were not conducive conditions for humification, as verified by the C/N ratios and the presence of moder or mull type humus. Mineralization occurs at a normal rate and the chemical-nutritive characteristics are suitable for "rebollo" oak forests. Although these forests are developed on parent rock, which are poor in base cations, all studied macronutrients, except P, were present at higher concentrations than in soils under *Q. robur*, probably because the stands correspond to mature forests that had not undergone changes in land use.

The statistical analyses reveal that Ca and Mg contribute to the greater number of significant relationships. The influence of the physical properties (fine earth and gravel) has a smaller importance.

Finally, in *Q. petraea* soils, the pH values were also quite homogeneous, giving rise to very acidic soils, although, in this case, this did not allow adequate conditions for humification too, giving rise to moder type humus. Mineralization occurs slowly and the biological-chemical characteristics are suitable for sessile oak forests. Although these soils are developed on substrates that are poor in base cations, all macronutrients except P were also present at higher concentrations than in soils under *Q. robur*, probably because no changes in soil use have taken place, and also because of the loamy textures of the substrates.

Analysis of the edaphic habitat reveals that most highly developed forest soils, within the study area, correspond mainly with *Quercus robur*, and even *Q. pyrenaica*. The optimal habitat in *Q. petraea* stands shows a wide range of values for base cations, pH and physical parameters (FE and GRA).

5. References

Acar, C.; Acar, H. & Altun, L. (2004). The diversity of ground cover species in rocky, roadside and forest habitats in Trabzon (North-Eastern Turkey). *Biologia*, Vol. 59, No.4, (July 2004), pp. 477–491, ISSN 0006-3088

Buide, M.L.; Sánchez, J.M. & Guitián, J. (1998). Ecological characteristics of the flora of the Northwest Iberian Peninsula. *Plant Ecology*, Vol. 135, No.1, (March 1998), pp. 1–8, ISSN 1385-0237

Camps M.; Mourenza C.; Álvarez E. & Macías, F. (2004). Influence of parent material and soil type on the root chemistry of forest species grown on acid soils. *Forest Ecology and Management*, Vol. 193, No.3, (June 2004), pp. 307-320, ISSN 0378-1127

Carballeira, A.; Devesa, C.; Retuerto, R.; Santillan, E. & Ucieda F. (1983). *Galicia Bioclimatology*. Xunta de Galicia-Fundación Barrie de la Maza, ISBN 84-85728-27-0, Santiago de Compostela (A Coruña, Spain)

Castroviejo, M. (1988). *Phytoecology of the Buio Forests and Sierra del Xistral (Lugo)*. Xunta de Galicia-Fundacion Barrie de la Maza, ISBN 978-84-96713-90-1, Santiago de Compostela (A Coruña, Spain)

Direction General of Nature Conservation [DGCONA] (2001). *Tercer Inventario Forestal Nacional. Galicia.* Ministerio de Medio Ambiente, ISBN 978-84-80144-30-8, Madrid, Spain

Direction General of Nature Conservation [DGCONA] (2001). *Tercer Inventario Forestal Nacional. Principado de Asturias.* Ministerio de Medio Ambiente, ISBN 978-84-80144-30-8, Madrid, Spain

Díaz T.E. & Fernández J.A., (1994). La vegetación de Asturias. *Itinera Geobotanica,* Vol. 8, pp. 243–528, ISSN 15771814

Díaz-Maroto, I.J. & Vila-Lameiro, P. (2005). Seasonal evolution of the chemical properties and macro nutrients of the soil in natural strands of *Quercus robur* L. in Galicia, Spain. *Agrochimica,* Vol. 49, No.5-6, (Sep-Dec 2005), pp. 201–211, ISSN 0002-1857

Díaz-Maroto, I.J. & Vila-Lameiro, P. (2006). Litter production and composition in natural stands of *Quercus robur* L., Galicia, Spain. *Polish Journal of Ecology,* Vol. 54, No.3, (September 2006), pp. 429–439, ISSN 1505-2249

Díaz-Maroto, I.J.; Vila-Lameiro, P. & Silva-Pando, F.J. (2005). Autecology of oaks (*Quercus robur* L.) in Galicia (Spain). *Annals of Forest Sciences,* Vol.62, No.7, (November 2005), pp. 737–749, ISSN 1286-4560

Díaz-Maroto, I.J.; Fernández-Parajes, J. & Vila-Lameiro, P. (2006). Autecology of rebollo oak (*Quercus pyrenaica* Willd.) in Galicia (Spain). *Annals of Forest Sciences,* Vol. 63, No.2, (March 2006), pp. 157–167, ISSN 1286-4560

Food and Agriculture [FAO] 1977. *Soil Profiles Description Guide.* FAO, ISBN 92-5-100508-7, Rome, Italy

Guitián, L. (1995). Origen y evolución de la cubierta forestal de Galicia. Ph.D. Thesis, University of Santiago de Compostela, Spain.

Graae, B.J. & Sunde, P.B. (2000). The impact of forest continuity and management on forest floor vegetation evaluated by species traits. *Ecography,* Vol. 23, No.6, (December 2000), pp. 720–731, ISSN 0906-7590

Göransson, H.; Rosengren, U.; Wallander, H.; Fransson, A.M. & Thelin, G. (2006). Nutrient acquisition from different soil depths by pedunculate oak. *Trees–Structure and Function,* Vol. 20, No.3, (May 2003), pp. 292–298, ISSN 0931-1890

Hagen-Thorn, A.; Callesen, I.; Armolaitis, K. & Nihlgard, B. (2004). The impact of six European tree species on the chemistry of mineral topsoil in forest plantations on former agricultural land. *Forest Ecology and Management,* Vol. 195, No.3, (July 2004), pp. 373–384, ISSN 1286-4560

Hagen-Thorn, A.; Varnagiryte, I.; Nihlgard, B. & Armolaitis, K. (2006). Autumn nutrient resorption and losses in four deciduous forest tree species. *Forest Ecology and Management,* Vol. 228, No.1-3, (June 2006), pp. 33–39, ISSN 1286-4560

Hardtle, W.; Oheimb, G. von & Westphal, C. (2005). Relationships the vegetation and soil conditions in beech and beech-oak forests of northern Germany. *Plant Ecology,* Vol. 177, No.1, (March 2005), pp. 113–124, ISSN 1385-0237

Klute, A. (1986). *Methods of Soil Analyses* (2nd ed.), Madison Wisc.: American Society of Agronomy, ISBN 089118841X, Madison (Wisconsin, USA)

Landgraf, D.; Leinweber, P. & Makeschin, F. (2006). Cold and hot water-extractable organic matter as indicators of litter decomposition in forest soils. *Journal of Plant Nutrition and Soil Science*, Vol. 169, No.1, (February 2006), pp. 76–82, ISSN 1436-8730

Neirynck, J.; Mirtcheva, S.; Sioen, G. & Lust, N. (2000). Impact of *Tilia platyphyllos* Scop., *Fraxinus excelsior* L., *Acer pseudoplatanus* L., *Quercus robur* L., and *Fagus sylvatica* L. on earthworm biomass and physico-chemical properties of a loamy topsoil. *Forest Ecology and Management*, Vol. 133, No.3, (August 2000), pp. 275–286, ISSN 1286-4560

Mansson, K.F. & Falkengren-Grerup, U. (2003). The effect of N deposition on nitrification, carbon and N mineralization and litter C:N ratios in oak (*Q. robur*) forests. *Forest Ecology and Management*, Vol. 179, No.1-3, (July 2003), pp. 455–467, ISSN 1286-4560

Nornberg, P.; Sloth, L. & Nielsen, K.E. (1993). Rapid changes of sandy soils caused by vegetation changes. *Canadian Journal of Soil Science*, Vol. 73, No.4, (November 1993), pp. 459–468, ISSN 0008-4271

Peterken, G.F. & Game, M. (1984). Historical factors affecting the number and distribution of vascular plant-species in the Central Lincolnshire woodlands. *Journal of Ecology*, Vol. 72, No.1, (January 2004), pp. 155–182, ISSN 0022-0477

Rivas-Martínez, S. (1987). *Memoria y mapas de series de vegetación de España. 1:400.000* (ICONA), ISBN 84-85496-25-6, Madrid, Spain.

Rivas-Martínez, S.; Fernández, F.; Loidi, J.; Lousã, M. & Penas, A. (2001). Syntaxonomical checklist of vascular plants communities of Spain and Portugal to association level. *Itinera Geobotanica*, Vol. 14, pp. 5–341, ISSN 15771814

Rozados, M.J.; Silva-Pando, F.J.; Alonso, M. & Ignacio, M.F. (2000). Edaphic and foliar parameters in a *Quercus robur* L. forest in Galicia (Spain). *Investigación Agraria: Sistemas y Recursos Forestales*, Vol. 9, No.1, (March 2000), pp. 17–30, ISSN 1131-7965

Ruiz de la Torre, J. (1991). *Forest Map of Spain* (General Direction of Nature Conservation, Environmental Ministry), ISBN 978-84-80144-22-3, Madrid, Spain

Russell, J.S. & Moore, A.W. (1968). Comparison of different depth weightings in the numerical analysis of anisotropic soil profile data. Proceedings of the 9th International Congress on Soil Science. International Society of Soil Science and Angus and Robertson, Adelaide (Australia), August, 1968.

Sas Institute Inc. (2004). *SAS/STATw 9.1. User's guide* (SAS Institute Inc), ISBN 1590475135, Cary (North Carolina, USA)

Thomas, F.M. & Buttner, G. (1998). Nutrient relations in healthy and damaged stands of mature oaks on clayey soils: Two case studies in Northwestern Germany. *Forest Ecology and Management*, Vol. 108, No.3, (August 1998), pp. 301–319, ISSN 1286-4560

Van de Moortel, R.; Rampelberg, S. & Deckers, J. (1998). Condition of *Quercus robur* L. along a natural Luvisol microtoposequence on loess in Central Belgium. *Soil Use Management*, Vol. 14, No.3, (September 1998), pp. 184–186, ISSN 0266-0032

Walpole, R.E.; Myers, R.H. & Myers, S.L. (1999). *Probability and statistics for engineers* (Prentice Hall), ISBN 978-03-21694-98-0, México, D.F. (México)

Wilde, S.A. (1946). Forest soils and forest growth (Chronica Botanica Comp.), Waltham (Massachusetts, USA)

4

Impact of an Introduced Forage Legume and Grazing on Soil Fertility in Native Pastures of the Humid Tropics of Mexico

Braulio Valles-de la Mora, Epigmenio Castillo-Gallegos,
Jesús Jarillo-Rodríguez and Eliazar Ocaña-Zavaleta
Universidad Nacional Autónoma de México,
Facultad de Medicina Veterinaria y Zootecnia,
Centro de Enseñanza, Investigación y Extensión en
Ganadería Tropical (CEIEGT),
México

1. Introduction

About 45 %of the world's humid tropics are located in tropical Latin American. The majority (81 %) of soils in this region are acidic, classified as Oxisols, Ultisols and Dystropepts that are deficient in phosphorus (90 %), exhibit aliminium toxicity (73 %), often under drought stress (53 %) and have low nutrient reserves (50 %) (Sánchez, 1987). These constraints in soil fertility are a barrier to high and sustainable productivity and a risk factor that could lead to irreversible soil degradation. Recent research has identified several management options to overcome fertility limitations or prevent the rapid process of soil degradation in such tropical regions (Crowder and Chheda, 1982).

Agricultural practices adopted to improve livestock production, to meet the demand for food of the growing population, have led to an inadequate management of agroecosystems such as native pastures (rangelands). As a result, changes in physical, chemical and biological soil properties have ocurred, to the detriment of soil nutrients and plant growth.

In this Chapter we address the problem of soil fertility in tropical rangelands, through the use of an introduced forage legume (*Arachis pintoi*) and grazing management strategies. This legume was chosen due to its benefits for soil fertility (e.g., N_2 fixation) and good yield potential, mentioned by Kerridge (1994). Several experiments were carried out, in vegetated pots grown under glasshouse conditions and at the field-scale, including grazing experiments on pastures with native plants alone (*Paspalum* spp, *Axonopus* spp, mainly) or a mixture of native plants and the introduced forage legume *A pintoi*. The field experiments were conducted under a rotational grazing regime, in trials that lasted from 12 to 36 months.

Introduced forages contribute strongly to forage quality and productivity in pastures when they are grown in association with grasses, because these take advantage of the N_2 fixed by the legume (Rao and Kerridge, 1994). We believe that sustainable improvement of native rangelands in the Mexican humid tropics could be attained through the use of persistent legumes that increase animal production by raising the nutritional quality of the domestic ruminant diet. These legumes should also have beneficial effects on soil properties and overall soil fertility.

According to this reasoning, we tested the benefits of legumes related to N_2 fixation, considering that N soil deficiency is a major limitation to pasture productivity in tropical grass-based rangelands. In recent years several new tropical forage legumes have been tested to improve pasture productivity. Some of these genuses are (in alphabetical order) *Arachis, Centrosema, Desmodium* and *Stylosanthes*. As has been mentioned in a previous sentence, about the potential of *Arachis pintoi* (Kerridge, 1994), other authors (Argel, 1994; Cárdenas *et al.*, 1999) have suggested this species as promising for the humid tropics of Latin America.

The effect of soil acidity on legumes growing in tropical agricultural systems, including rangelands has scarcely been evaluated (Davidson and Davidson, 1993). The development of soil acidification in pastures containing legumes has been attributed to causes such as: an accumulation of organic matter and an increase in soil cation exchange capacity, to an imbalance in carbon and nitrogen cycles, particularly removal of alkalinity in farm products, net losses of nitrogen through nitrate leaching, and to the release of H^+ by legume roots due to excess uptake of cations over anions during N_2 fixation (Tang, 1998). As legumes (by nitrification process) releases H^+, according to the following soil-plant reaction: $NH_4^+ + 3/2$ $O_2^- + H_2O^+ + 2H^+$, we tested the hypothesis that the introduction of the forage legume *A. pintoi* CIAT 17434 in a tropical native rangeland will increase soil acidity.

There are few studies that evaluated soil fertility as impacted by grazing and/or livestock stocking rates under tropical conditions (Cadisch *et al.*, 1994; Thomas, 1995; Veldkamp *et al.*, 1999). In an innovative experiment, we tested the effect of increased stocking rates on selected soil biological, chemical and physical properties by measuring soil N mineralization, organic matter, C/N ratios, plant-available N, P and K concentrations, bulk density in native pastures, with and without the introduced forage legume *A. pintoi* CIAT 17434.

In all cases, the data were statistically analyzed, using appropriate methods, considering the particularities of each experiment. Results showed that is possible to make a positive impact on fertility of soils, applying technologies that are available and affordable to farmers, on lands that are devoted to livestock grazing.

2. Biological nitrogen fixation by the forage legume *Arachis pintoi* in a native pasture system in Veracruz, Mexico

Introduction of persistent legumes leads to improved soil fertility, which enhances dry matter yield and nutritive value of the rangelands. Legume intake improves the diet of the animal, leading to higher milk yields and better reproductive performance. *Arachis pintoi*, a pasture legume native to South America, has been shown to be persistent under heavy grazing, has a high nutritive value, is palatable to cattle, and can improve soil fertility. Agronomic and grazing trials conducted in Veracruz State, Mexico indicated that *A. pintoi* CIAT 17434 was a promising legume for the hot-humid climate and acid, infertile soils of the research station (Argel, 1994; Valles *et al.*, 1992; Vos, 1998).

2.1 Experimental site characteristics

The experiment was conducted at the Center for Teaching, Research and Extension in Tropical Animal Husbandry (Centro de Enseñanza, Investigación y Extensión en Ganadería Tropical –CEIEGT) of the Faculty of Veterinary Medicine and Animal Science of the National University of Mexico. The Center is located near to the town of Martinez de la Torre in the North-Central region of Veracruz State, at 20° 03' N, 97° 04' W, at at 112 m.a.s.l. The climate is warm and humid, with an annual rainfall of 1991 ± 392 mm. The average daily mean temperature is 24 ± 0.47 °C. Three seasons can be distinguished. "Dry" from March to June (30.6 ± 1.7 °C and 19.9 ± 1.7 °C of maximum and minimum temperatures, respectively, and 477 ± 343 mm of rainfall). "Rainy" from July to October (31.2 ± 1.3 °C and 20.5 ± 1.8 °C of maximum and minimum temperatures, respectively, and 1032 ± 612 mm of rainfall) and, "Winter" from November to February (25.0 ± 2.1 °C and 15.1 ± 1.3 °C of maximum and minimum temperatures, respectively, and 482 ± 268 mm of rainfall)

The soil has been classified as Ultisol and has moderate drainage. The A horizon has a maximum depth of 30 cm. There is no B horizon. Horizon C is an acid hardpan formed with material that was transported deposited and compacted. Due to the microrelief, soils are prone to waterlogging during the wet and northern wind periods and thus horizon A exhibits gleyization (Hernández, 1988). A chemical description of the soil in the experimental site is shown in Table 1.

Depth (cm)	Total N[1] (%)	δ 15N[2] (⁰/₀₀)	Total C[3] (%)	Delta 13C values[4]	P µg/g[5] soil	pH[6]
0-5	0.18	7.5	1.8	-18.3	1.6	5.31
5-15	0.11	5.8	1.1	-20.0	0.15	5.63
15-30	0.09	6.2	0.9	-19.4	0.14	5.87
15-60	0.05	7.8	0.5	-20.2	0.0	nd
60-100	0.03	6.8	0.3	-22.1	0.0	nd

[1] N(%): Nitrogen in soil >1.0=very high, 0.5-1.0=high, 0.2-0.5=medium, 0.1-0.2=low, <0.1=very low (Landon, 1984).
[2] Natural enrichment of the 15N isotope expressed as δ 15N (⁰/₀₀). Values could vary from -6 to + 16 δ 15N.
[3] C (%): Total soil carbon.
[4] δ PDB carbonate standard: Isotopic composition of 13C. The range in soils is -9 to -32 (⁰/₀₀). The values reflect the relative contribution of plant species (C_3 plants: -32 to –22 ⁰/₀₀, or C_4 plants: -9 to -17 ⁰/₀₀.) (Boutton et al., 1998).
[5] P, Bray I (µg/g soil): A range of 2.5 to 11.4 µg/g soil is considered as critical level for tropical legumes (CIAT, 1984).
[6] Soil pH (1:2.5 H_2O): 4.5-5.0 very strongly acidic, 5.0-5.5 strongly acidic, 5.5-6.0 moderately acidic (Landon, 1984).

Table 1. Chemical description of the native pasture soil.

2.2 Rangelands used

Paddocks were composed of native pasture-*Arachis pintoi* from an ongoing grazing experiment at the research station in Veracruz, Mexico. Paddocks with native pasture alone (mixed grasses, mostly *Paspalum* spp, and *Axonopus* spp,) were included as a control. The paddocks described in this chapter, are nested within one of seven permanent divisions of the mentioned native pasture-legume association and were originally planted to study the establishment and persistence of the accessions CIAT 17434, 18744 and 18748 under grazing. Three replicates of each ecotype were located within the pasture, and native grasses grew in mixtures with the legume ecotypes. These plots have been under rotational grazing (1-day grazing and 20-day rest periods), by sectioning the divisions into three temporary subdivisions. There have been 5 "fixed" cows per treatment and in all treatments, three additional cows per treatment were introduced in order to consume the extra forage produced in the rangelands during the rainy season. Thus, stocking rates were 2 cows/ha when only fixed cows grazed and 3.2 cows/ha when both fixed and additional cows grazed the experiment. These rangelands were not fertilised with mineral or organic fertilizer for at least five years prior to the planting of the legume, nor after legume planting. These grasses grew mixed in the ecotypes plots. Introduced legumes and adjacent native grasses were sampled at the beginning and at the end of each one of the three seasonal periods (dry, rainy and winter) during 1999.

2.3 N₂ fixation assessment and statistical data management

Dry matter yield of shoots was measured in legumes, as well as in grasses, and analysed for N (%) and $\delta^{15}N$ ($^0/_{00}$) using a Europa 20-20 isotope ratio mass spectrometer (Europa Scientific, Crewe, UK), coupled to an automated Roboprep C/N analyser (Europe Scientific, Crewe, UK).

The percentage of atmospheric N_2 fixed by the legumes (P) using the natural ^{15}N abundance technique was calculated using the following equation (Bergersen & Turner, 1983; Ledgard *et al.*, 1985):

$$P = \frac{\delta^{15}N\ ref - \delta^{15}N\ leg}{\delta^{15}N\ ref - B} \times 100 \tag{1}$$

where B is the $\delta^{15}N$ of the legume grown with atmospheric N_2 as the only N source, $\delta^{15}N$ ref and $\delta^{15}N$ leg are the $\delta^{15}N$ units ($^0/_{00}$) for reference plant and legume, respectively. B-values of shoots corresponding to each ecotype from the "Isotope fractionation" were used. These values were ($^0/_{00}$) 0.08, -0.18 and –0.29 for *A. pintoi* CIAT 17434, 18744 and 18748, respectively.

The total N_2 fixed for every *A. pintoi* ecotype was calculated using the following equation:

$$Total\ N_2 fixed\ \left(kg\ /\ ha\right) = Dry\ matter\ yield\ \left(kg\ /\ ha\right)\ x\left(\%N\ /\ 100\right)\ x$$
$$\left(\%N_2 fixed\ /\ 100\right)\ x\ \left(\%\ legume\ in\ pasture\ /\ 100\right) \tag{2}$$

In each pasture treatment, botanical composition of each component (%) was estimated using the dry weight rank (DWR) method of Mannetje & Haydock (1963) four times a year: March-April, June-July, September-October and December-January.

The experiment was analysed as a completely randomized design with two rangelands (native pasture (NP) and the association (NP+Ap) each with three replicates. The comparison among the *A. pintoi* ecotypes was done also as a completely randomized design with three treatments (the ecotypes) and two replications.

2.4 Results

2.4.1 N₂ fixation by *Arachis pintoi* ecotypes

Figure 1 shows the percentage of N_2 fixed by the three *Arachis pintoi* ecotypes using either *Axonopous* spp (Figure 1a) or *Paspalum* spp (Figure 1b) as reference plant, averaged for each climatic period. The analysis of variance did not find statistical differences in N_2 fixation for ecotype or season within by each reference plant, and *Ap* CIAT 18744 averaged 86 % of fixed N_2, compared to reference plants. When data were analysed considering both reference plants, statistical differences among ecotypes were found. *Ap* CIAT 18744 and 18748 were statistically similar (P ≤ 0.05) but different to the ecotype *Ap* CIAT 17434, e.g. %N fixed was 86 ± 2.3, 77 ± 3.1 and 69 ± 4.5, respectively.

2.4.2 Botanical composition

The highest proportion of *A. pintoi* (on average, 29%) was measured in the native pasture during the dry season, with less prominence (16-19%) on other sampling dates. The average legume content for the whole year was 20 %, while native grasses were 64 % of the stand. Weeds and a low proportion of native legumes (3%), and bare soil (13%) occupied the remaining area.

2.5 Discussion

The best N_2 fixation estimates (%) were obtained with *Arachis pintoi* CIAT 18744, regardless of the reference plant used (Figure 1); although the ecotype CIAT 18748 fixed a similar proportion of N_2 as ecotype CIAT 18744. The proportions of N_2 fixed were higher than the values reported by Peoples *et al.* (1992) for a variety of the grain legume *A. hypogaea*, in Australia. They estimated a N_2 fixation in two years of 22 to 31 % and 44 to 48 %, respectively using a non-nodulating peanut genotype as a reference plant. Likewise, Cadisch *et al.* (2000) reported percentages of N_2 fixed in *A. hypogaea* of 45 to 54 % in the first year and 21 to 16 % in the second year, also with a non-nodulating groundnut and maize as reference plants.

In Brazil, Behling-Miranda *et al.* (1999) reported that four non-inoculated *Stylosanthes* species fixed (on average) more N_2 using *P. plicatulum* as reference plant (on average, 80 %) than with *B. decumbens* (71 %). In our experiment, no major difference in N_2 fixation was calculated due to the choice of reference plant, e.g. *A. pintoi* ecotypes gave similar estimates of N_2 fixed whether using *Axonopus* spp or *Paspalum* spp were used as the non-N_2 fixing reference plant. In Colombia, Suarez *et al.* (1992) evaluated the N_2 fixation of non–inoculated *Arachis pintoi*, using the isotope dilution method and found that the legume derived 63 % of its total N from the air, which is lower than the results shown in this Chapter. In our case, although the legumes were not inoculated, the high levels estimated of N_2 fixation, mainly with the ecotypes *Arachis pintoi* CIAT 18744 and 18748 showed that the native *Rhizobium* population was effective in N_2 fixation.

A rough estimation of kg/ha/day of N_2 fixed by *A. pintoi* ecotypes could be obtained if we consider a DMY of 25 kg/ha/day for this legume under field conditions (average from two years of data collected by Valles *et al.* (1992) in Veracruz, Mexico). Using this assumption and considering the measured percentages of N_2 fixed by the *A. pintoi* ecotypes as well as the proportion of the legume in the associated pasture, the estimated amounts of N_2 fixed are shown in the Table 2.

The estimated N_2 fixation during the rainy season, considering 71, 93 and 77 % of N_2 fixed by *A. pintoi* ecotypes and using data from Table 2 for N % and presence of the legume (%) resulted in 140, 230 and 148 g N/ha/day for Ap 17434, 18744 and 18748, respectively. This is within the same range as Thomas *et al.* (1997), who reported for *A. pintoi* CIAT 17434 (23 % of the pasture stand and deriving 79 % of its N from N_2 fixation) a figure of 11.4 kg N/ha/8 weeks, e.g. approximately 200 g N/ha/day. Our values are also similar to Cadisch *et al.* (1989) for eight forage legumes in Colombia, e.g. on average 463 and 202 g N_2/ha/day for legumes that received or no P+K fertiliser, respectively.

2.6 Conclusions

The ecotypes of *Arachis pintoi* CIAT 18744 and 18748 had the highest % N derived from N_2 fixation. This represents a significant input of N to the pasture systems in the tropics. Although the ecotype CIAT 17434 showed a reasonabe percentage of N_2 fixation, its estimated N contribution to the pasture was lower, mainly due to its low presence in the sampled plots. On the other hand, all other ecotypes seemed to tolerate intensive grazing and competed successfully with aggressive grasses and weeds, e.g. maintained a population of about 20% of the stand in the native pasture.

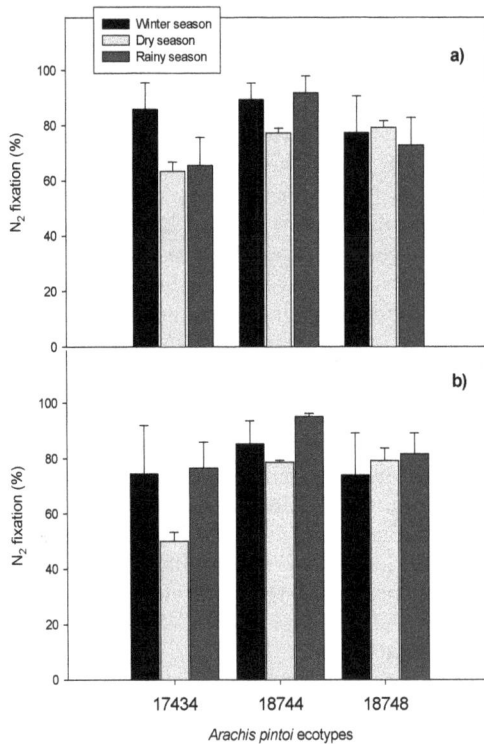

Fig. 1. Percentage of nitrogen fixed by three *Arachis pintoi* ecotypes considering a) *Axonopus* spp or b) *Paspalum* spp as reference plant, during three climatic periods in Veracruz, Mexico. Lines upon bars are standard error of means.

Ap Ecotype	DMY (kg ha/yr)	Shoots N (%)	N_2 fixed[2] (%)	Ap ecotype in the pasture (%)	N_2 fixed (kg/ha/yr)
17434	9125	2.84	69	20	36
18744	9125	2.80	86	35 1	78
18748	9125	2.72	77	36 1	69

[1] Estimations in the paddocks of *Ap* ecotypes during February 1999.
[2] Average values from Figure 1.

Table 2. Estimation of N_2 fixed by three *Arachis pintoi* ecotypes under field conditions in Veracruz, Mexico.

3. Changes in soil properties after the introduction of *Arachis pintoi* in native rangelands in the humid tropic of Mexico

In the Mexican humid tropics, native rangelands compose from 25 to 75 % of the pastureland, and are the main feed source for cow-calf and dual-purpose production systems. In practice, these lands are not fertilized or sown with productive legumes to add nitrogen (N) to the system. Since N is the element driving pasture growth, decades of use without adequate N fertilization (either from nutrient inputs or N_2 fixation) have led to considerable degradation of soil and grassland fertility. Legumes are well known as soil building plants. They enhance soil quality by adding organic matter, and improve soil structure and water infiltration. Earthworm populations are usually greater in fields planted to perennial forages than in fields planted to annual row crops, which may further enhance nutrient recycling in these pastures.

Arachis pintoi CIAT 17434 is a legume that has induced increases in milk and meat production when planted in association with introduced grasses like African stargrass (*Cynodon plectostachyus, C. nlemfuensis*) and *Brachiaria* spp (González *et al.*, 1996; Hernández *et al.*, 1995), and is known to raise soil N and C inventories (Ibrahim, 1995). The present study documented the effect of the perennial forage legume *Arachis pintoi* CIAT 17434 when introduced into native grasslands on some soil variables, considering soil physico-chemical properties important for pasture sustainability.

3.1 Experimental details

3.1.1 Climate and soil in the experimental site

The experiment was carried out at the Center for Teaching, Research and Extension Center in Tropical Animal Husbandry (Centro de Enseñanza, Investigación y Extensión en Ganadería Tropical - CEIEGT) of the Faculty of Veterinary Medicine and Animal Science (FMVZ) of the Universidad Nacional Autónoma de México (UNAM). The CEIEGT is located on the Gulf of Mexico coastal plain, 40 km west of the coast (20° 02' N, 97° 06' W) at 112 m above sea level. Climate (1980-2000) is hot and humid with rains year round, and an average annual rainfall of 1931 ± 334 mm. Mean monthly rainfall is highly variable (161 ± 132 mm), but is generally sufficient for pasture growth as dry periods are occasional and short. Average daily temperature during the study was 23.9 ± 6.4 °C, with average monthly maximums (29.2 ± 3.3 °C) and minimums (18.6 ± 3.9 °C) that are reasonably uniform from year to year. There are three seasons: rainy fromJuly to October, with high rainfall and temperature; winter or "northwind" from November to February, with rain and lower temperatures; and dry from March to June. Low winter temperatures and high evapotranspiration during the dry season do not favor forage production, because the shallow soils of the experimental site do not store much moisture.

Soils in the experimental fields are acidic, heavy clay, Ultisols (Durustults) with low P (3.5 ppm by Bray, 2.0 ppm by Olsen), S, Ca and K concentrations, low cation exchange capacity (10.5 meq/100 g) and aluminium saturation levels below the level toxic to plants (Arscott, 1978; Toledo, 1986). There is an impermeable soil layer from 0 to 25 cm depth that causes inadequate drainage during the rainy season and winter. Soil samples taken in the experimental area in 1998 showed that C content decreased from 1.02 ± 0.44 % at 0 to 15 cm, to 0.78 ± 0.15 % at 15 to 30 cm, and that N content also dropped, from 0.13 ± 0.06 at 0 to 15 cm, to 0.10 ± 0.04 % at 15 to 30 cm.

3.1.2 Treatments and soil variables assessed

Two treatments were applied, a native grass pasture (NG) and a NG associated to *A. pintoi* CIAT 17434 (NG+Ap); each one nested in a larger field (F1 and F2, respectively). Each field was divided into seven permanent sections measuring 22 m wide by 165 m long. These were then temporarily subdivided into three paddocks of 22 m by 55 m (21 paddocks per field), to allow a rotation grazing of 1 d of grazing and 20 d of recovery. Stocking rate (SR) was 2 cows/ha during the low forage production period and 3.2 cows/ha during the rest of the year. The SR was reduced when standing dry matter before grazing was less than 2,500 kg/ha in any paddock.

Soil samples were taken in April 1999 and September 2000 in transects located in the center and passing through the entire length of each of the seven divisions. Six soil samples were taken along each transect at even intervals, for a total of 336 soil samples. Samples were analyzed for the variables described below.

To estimate apparent density (AD, g/cm^3), a metal cylinder (50 mm long, 50 mm internal diameter) was inserted into the bare soil to collect a soil sample, which was dried at 65 °C until constant weight was reached (Anderson & Ingram, 1993). Samples for other variables were dried at room temperature, cleaned of all roots and other organic matter and then ground until they passed through a 2 mm mesh. A potentiometer was used to measure pH in a soil: distilled water (2:1) suspension. Organic carbon (C) was analyzed according to Walkley & Black (1934), and organic matter (OM, %) was calculated assuming a 58% C content in the OM (Anderson & Ingram, 1993). Total C for the 0-20 cm soil layer was calculated using $Ct = Vs \times DA \times C/100$, where: Vs is volume in 1 ha at 20 cm depth, AD is apparent density (kg/m^3) and C is as defined above. It was assumed that AD in the 0-5 cm layer was representative of the entire 0-20 cm layer. Nitrogen (N, %) was estimated with the Kjeldahl technique calculating total N (Nt, kg/ha) similar to Ct by substituting C for N.

Field 1 was evaluated every three months to determine the contribution of *A. pintoi* to the botanical composition. Samples for this analysis were taken from two paddocks, each within the three divisions chosen visually as the most representative of the experimental area. Sections b1, d1 and f1 had higher *A. pintoi* content compared to the other three (i.e. b3, d3 and f3), and were consequently classified as having high and low legume content, respectively.

3.1.3 Data statistical management

The division was used as the experimental unit and thus variation between sampling sites within a transect was not considered. Variance analysis was done separately for F1 and F2. The resulting linear model was:

$$Yijk = M + Tj + Di(Tj) + Ak + (T \times A)jk + Eijk, \tag{3}$$

where: Yijk is the response variable, recorded in the transect corresponding to division i, in treatment j, in year k; M is the general mean, common to all observations; Tj is the effect of treatment j (j: NG and NG+Ap); Di(Tj) is the variation between divisions within treatment j, used as an error to test the treatment effect; Ak is the effect of year k (k: 1999 y 2000); (T x A)jk is the combined effect, or interaction, of treatment by year; and, Eijk is residual variation, used as an error to test the effect of the year and interaction. The annual means were not included because the samples were taken in different seasons each year, and to concentrate on the effects of treatments.

3.2 Results and discussion

3.2.1 Effect of rangelands on soil variables

The T x A interaction did not affect (P>0.05) any response variable, meaning the effects of T and A were independent. The effect of T on pH and AD in F1 was not significant, was close to significant for the C:N ratio (P=0.0563) and was significant for all other variables (P<0.05). In contrast, T had no effect on any variable in F2 (Table 3). The different effect of treatments between can be explained by the length of time the legume had been growing in each field, meaning this legume needs a medium to long term to start manifesting its beneficial effects on the soil. Fields 1 and 2 were established in November 1996 and 1999, respectively.

Legumes are reported to reduce soil compaction at higher stocking rates (Alegre & Lara, 1991), though *A. pintoi* had no effect on compaction during the three-and-half year period of this study. In Costa Rica, Ibrahim (1994) found that an increase in stocking rate from 2 to 3 animal units/ha in a *Brachiaria* spp/*A. pintoi* association only slightly increased AD from 0.76 to 0.85 g/cm3 in the 0-5 cm layer.

Previous research (Thomas *et al.*, 1997; Valles, 2001) indicates that *A. pintoi* CIAT 17434 derives from 65 to 85% of its N_2 from fixation, suggesting N accumulation in the soil, probable NO_3 leaching losses and soil acidification (Haynes, 1983). In the present study, however, *A. pintoi* had not been planted long enough to influence pH. Valles (2001) found no signs of difference in pH in samples of the soil profile to 1 m depth in both treatments in F1, one year after *A. pintoi* had been established. However, in native grasslands associated with this legume for five years, the pH was significantly (P<0.05) lower at all depths than in grasslands without it. The pH values for the NG+Ap were 5.05 (0-5 cm), 5.07 (5-15 cm) and 5.52 (15-30 cm), and for the NG they were 5.66 (0-5 cm), 5.44 (5-15 cm) and 5.72 (15-30 cm). This suggests the possibility of a decrease in pH over time in the soils with the *A. pintoi*/native grasslands association.

Variable	Pasture	
	GN	GN+Ap
- - - - - Field 1 - - - - -		
pH	5.12±0.05[a]	5.17±0.04[a]
Bulk density, g cm^3	1.22±0.02[a]	1.24±0.02[a]
Organic matter, %	2.27±0.10[a]	2.53±0.09[b]
Organic Carbon, g kg^{-1}	1.32±0.06[a]	1.47±0.05[b]
Nitrogen, % x 10^{-2}	14.15±1.39[a]	22.43±2.02[b]
C:N relationship	12.40±1.38[a]	9.46±1.11[a]
Soil Carbon, kg ha^{-1}	31968±1285[a]	36197±1144[b]
Soil Nitrogen, kg ha^{-1}	3452±341[a]	5610±526[b]
- - - - - Field 2 - - - - -		
pH	5.03±0.04[a]	5.12±0.05[a]
Bulk density, g cm^3	1.17±0.03[a]	1.21±0.02[a]
Organic matter, %	2.61±0.12[a]	2.67±0.15[a]
Organic carbon, g kg^{-1}	1.52±0.07[a]	1.55±0.09[a]
Nitrogen, % x 10^{-2}	10.45±1.35[a]	11.99±1.27[a]
C:N relationship	22.19±2.41[a]	18.96±1.83[a]
Soil Carbon, kg ha^{-1}	35783±2179[a]	37847±2493[a]
Soil Nitrogen, kg ha^{-1}	2420±298[a]	2861±285[a]

[a, b] Means with different letters within a row are significantly different (P<0.05).

Table 3. Effect of pasture on soil variables in the two, grazed paddocks with natural grass (GN) or natural grass and the introduced forage *A. pintoi* (GN+Ap) in Veracruz, Mexico. Soil sampling en Field 1 was done 16 months after the establishment of *A. pintoi*; and in Field 2, sampling was 20 months before the establishment of the legume. Values are the mean ± standard error.

Accumulation of OM in the soil of pasture agroecosystems is due to continuous addition of roots and plant litter, coupled with a chronic N deficiency (Huntjes and Albers, 1978). Although in Field 1, N deficiency probably was not evidente, because the supply of this element by the N_2 fixation from the legume, changes in organic C and total N was probably due to the build-up of organic matter process, done by soil microorganisms on grassland vegetation.

The calculated initial (1998) Ct value for both treatments was 24.7 t/ha. Average annual Ct increase to the year 2000, by linear regression, was 3.7 t/ha/yr for NG and 4.8 t/ha/yr for NG+Ap, an acceptable differential increase. This coincides with observations reported by Fisher *et al.* (1994) for the Colombian savannah and by Ibrahim (1994) for the Costa Rican humid tropics. The C:N ratio is a soil quality indicator with an ideal value of 10:1 or less. In F1, the NG was above the ideal value, and the NG+Ap was slightly below it (Table 3). In F2, both treatments were higher than the ideal ratio. This difference is likely due to the different amounts of time the legume had been established in the two fields. Data from Costa Rica (Ibrahim, 1994) show a C:N ratio in a *Brachiaria* spp/*A. pintoi* association of 12.5:1, which is comparable to the present F1 control treatment but lower than the F2 results. This suggests that *A. pintoi* remains productive in this association and needs time to alter the soil C:N ratio.

3.2.2 Contribution of *Arachis pintoi* to soil C and N stocks

The contribution of *A. pintoi* to the botanical composition increased linearly in sections of Field 1 with high and low initial content of the legume. In the sections with high legume composition, its abundance continued to increase at a rate of 13.3% per year, whereas the sections with low initial legume content showed an increase of 7.8% per year (Figure 2). The respective contribution of the legume to the soil C and N stocks tended to be greater in sections with high legume content than those with lower legume presence, although not significant due to high variation (Table 4). An important aspect of productivity in tropical rangelands is to maintain or increase the stock of soil C and N in response to increases in biomass induced by the biological N_2 fixation, which implies long-term sequestration of CO_2 and improved soil fertility. There are also negative effects to biological fixation of N_2 , such as soil acidification, that are only apparent over the long term (Haynes, 1983).

3.3 Conclusions

Introduction of *Arachis pintoi* CIAT 17434 to native rangelands increased soil organic C and total N concentrations, but had a marginal effect on the apparent density and soil pH. This indicates that this legume can improve soil fertility, particularly from the perspective of building soil organic matter, thus maintaining a sustainable dual-purpose cattle system in the tropics.

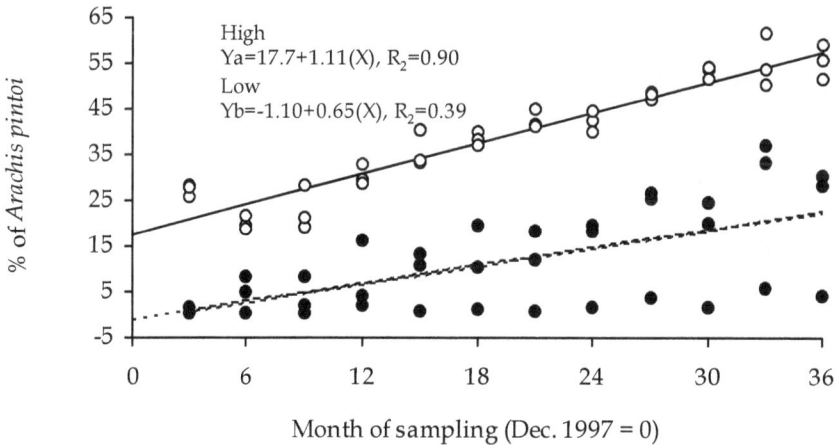

High
$Ya=17.7+1.11(X)$, $R_2=0.90$
Low
$Yb=-1.10+0.65(X)$, $R_2=0.39$

% of *Arachis pintoi*

Month of sampling (Dec. 1997 = 0)

Fig. 2. Increase in the contribution of *Arachis pintoi* to botanical composition in Field 1 over a three year period (1997-2000). Field 1 was a grazed paddock with a mixture of natural grasses and introduced *A. pintoi*, located in Veracruz, Mexico.

Inicial *Arachis pintoi* percentage	*A. pintoi* at soil sampling (%)	C at soil sampling (kg/ha)	N at soil sampling (kg/ha)
Low	17 ± 15	32524 ± 10871	5667 ± 2474
High	46 ± 11	43420 ± 5,424	4929 ± 2416

Table 4. Contribution of *Arachis pintoi* to botanical composition, and amounts (kg/ha) of soil organic carbon and total nitrogen at 0.2 m depth, 9 months after the introduction of this forage legume to a grazed paddock in Veracruz, Mexico. Values are the mean ± standard error-($n = 6$).

4. Evidences of topsoil acidification by the forage legume *Arachis pintoi*

Soils in tropical Latin American include 45 % of the world humid tropics, with a high proportion of acid soils. Oxisols, Ultisols and Dystropepts cover 81 percent of these areas. They are characterised as deficient in phosphorus (90 %), aluminium toxicity (73 %), drought stress (53 %) and low nutrient reserves (50 %) (Sanchez, 1987).

These constraints in soil fertility are a barrier to high and sustainable productivity and a risk factor that could lead to irreversible soil degradation. Recent research highlights several management options in order to overcome fertility limitations or prevent the very fast process of soil degradation. The introduction of legume-based rangelands is considered as one of the most desirable option to reduce the fertiliser requirements in animal production systems.

4.1 Implications of soil acidity

Basically there are two acidity sources: the atmosphere and the soil. In the soil, the decomposition of organic matter by mineralization, nitrification and leaching releases H^+ ions. In agricultural systems, the use and abuse of N fertilisers, particularly those containing ammonium ions, make large contributions to soil acidity. Fortunately for plants, there are buffering materials in soils that consume H^+ ions and maintain the soil pH at a more or less constant level.

4.2 Acidification by legumes

The soil acidity induced by an intensive use of legumes in animal production systems and other agricultural systems has been scarcely evaluated under tropical and sub-tropical conditions. In Australia, Davidson & Davidson (1993) cited reports of this problem dated since 1939 and Haynes (1983) makes mention of references since 1954. Considering the previous reasoning, we set up this experiment to determine if there is a relationship between pasture age and soil acidification under legume covers.

4.3 Soil sampling and rangelands used

The experimental fields were located at the Center for Teaching, Research and Extension in Tropical Animal Husbandry (CEIEGT) of the Faculty of Veterinary Medicine and Animal Science of the National University of Mexico. The soils were classified as Ultisols. Additional details about this site were described previously in this Chapter.

During the period June-August 1997, soil samples were taken at four sites containing native rangelands associated with *A. pintoi* CIAT 17434 of different ages (1, 5, 8 or 11 years) and corresponding controls (Table 5). The control areas, all adjacent to *A. pintoi* (*Ap*) sites, consisted of soils with native vegetation, which were under rotational grazing regime at the time of sampling, except the control site for the pasture 8 years *Ap* that remained without grazing The 11 years control site remained under grazing until 2 years before sampling.

The total size of each site was 7.5, 10.0, 0.5 and 0.25 ha for 1, 5, 8 and 11 year old *Ap* sites, respectively. In the first site, 7.5 ha of native rangelands were assigned to carry out a grazing experiment of which 5 ha remained unchanged and 2.5 ha were used to introduce the legume (1 year *Ap*). In the 5 years old *Ap* site, half of the site were used to introduce the legume mixed with the native pasture or associated with African star grass -*Cynodon nlemfuensis*. On the other sites (8 and 11 year old *Ap*) *Arachis pintoi* was planted to establish seed banks for further spread of the legume. No fertilisers were added during the establishing period, nor during the last five years.

Soil depth, cm	N %	$\delta^{15}N$ $^0/_{00}$	C %	Delta PDB	P µg/g soil	pH	Sand g/kg	Clay g/kg	Silt g/kg	Al sat. g/kg
Control 1 year										
0-5	0.18	7.5	1.8	-18.3	1.6	5.31	nd [1]	nd	nd	nd
5-15	0.11	5.8	1.1	-20.0	0.15	5.63				
15-30	0.09	6.2	0.9	-19.4	0.14	5.87				
15-60	0.05	7.8	0.5	-20.2	0.0	nd				
60-100	0.03	6.8	0.3	-22.1	0.0	nd				
Control 5 years										
0-5	0.18	7.4	1.9	-20.0	nd	5.50	nd	nd	nd	Nd
5-15	0.12	7.9	1.3	-20.4	nd	5.65				
15-30	0.09	8.2	1.0	-19.8	nd	5.70				
Control 8 years										
0-5	0.25	7.2	2.8	-20.2	nd	5.55	22.2	47.0	30.8	2.8
5-15	0.16	7.7	1.6	-20.7	nd	5.70	8.6	70.9	20.5	1.5
15-30	0.11	7.6	1.3	-20.6	nd	5.65	18.2	57.5	24.4	1.8
Control 11 years										
0-5	0.12	6.6	1.1	-17.3	nd	5.45	15.0	45.0	39.4	9.8
5-15	0.08	8.0	1.1	-16.3	nd	5.66	10.9	60.5	28.6	23.3
15-30	0.04	7.4	1.0	-15.3	nd	5.74	11.2	63.8	25.0	25.4

[1].- nd= not determined.

Table 5. Chemical description of four control soils in rangelands located in Veracruz, Mexico, in 1997.

At the time of soil sampling, these areas were divided each into four portions to obtain a more uniform sampling. In each quadrat, samples were taken using an auger at several depths . At the time of soil sampling, these areas were divided each into four portions to obtain a more uniform sampling. In each quadrat, samples were taken using an auger at several depths . All sites were sampled at 0 - 5, 5 - 15 and 15 - 30 cm depth, and the 1 year old site was additionally sampled at 30 - 60 and 60 - 100 cm depth.

All sites were used to estimate changes in soil fertility and acidity. The samples were immediately identified and air dried for 3 - 4 days before stored in black plastic bags until further analysis. In all cases, soils were passed through a 2 mm sieve.

4.4 Soil pH and statistical analysis

Measurements of pH were made by weighing 10 ± 0.1 g air-dried soil (< 2 mm) in a test tube and adding 25 ml of deionised water. Tubes were stirred for one minute, left to settle for one hour and stirred again for one minute. A pH meter with a glass electrode and previously calibrated for acid soils (pH=4) was used. Between each pH reading value, the electrode was washed with water and gently dried with a tissue. Every ten measurements, the pH meter was checked to ensure correct pH readings. Data were statistically analysed performing analysis of variance and differences among means were established by least significant differences, using the Genstat statistics program (Lawes Agricultural Trust, 1996). Regressions using "age" as a dependent variable were also considered but very low R^2 (determination coefficients) were obtained.

4.5 Results

4.5.1 Changes in soil pH

Changes in soil pH for every one of the paired sites (with and without *A. pintoi*) was analysed statisticall (Figure 3). There were no significant differences in pH between the 1 year old *Ap* based pasture and the control pasture in the topsoil (5.2), although the control soil was less acid at 15-30 cm. The 5 years old *Ap* pasture showed lower soil pH (P $\leq$ 0.001) than   the control site at all sampling depths. Larger differences in pH were observed between the two rangelands at 5 cm depth. Likewise, the soil pH increased significantly (P≤0.001) with increasing depth in the soil profile, average values being 5.3, 5.3 and 5.6 for the depths 0-5, 5-15 and 15-30 cm, respectively. The pH of the 8 year old *Ap* pasture (pH=5.3) was statistically different (P $\leq$ 0.001) from the control site (pH=5.6) (P $\leq$ 0.001).

The acidity level (soil pH) in the *Ap* pasture decreased from 5.4 at 15 - 30 cm to 5.2 at 0 - 5 cm depth while in the control pasture the values were similar with depth. In the case of 11 years old *Ap* pasture differences in soil pH between the two sites were not found, except for depth and the interaction site/depth (P $\leq$ 0.001). The pH at the lowest soil depth was higher in the legume based pasture.

Acidity was less pronounced at 15-30 cm of soil depth compared to the other depths; but the sites *5* and *8 years* under *A. pintoi* rangelands showed lower pH values with respect to the rest of the sites (5.4). In the case of the *control* site (native pasture) differences among pH values with soil depth were less evident than in sites with *Arachis pintoi*. The averages for each depth were 5.52, 5.67 and 5.67, respectively.

4.6 Discussion

The soil pH results suggested that apparently an acidification process occurred in most of the sites covered by the forage legume *A. pintoi* compared with nearby control sites. This effect was most evident in the first 5 cm of the soil profile and in pastures with 5 year old and 8 year old *A. pintoi* forage. However, the site covered by the legume for 11 years did not show an even greater increase in soil acidity.

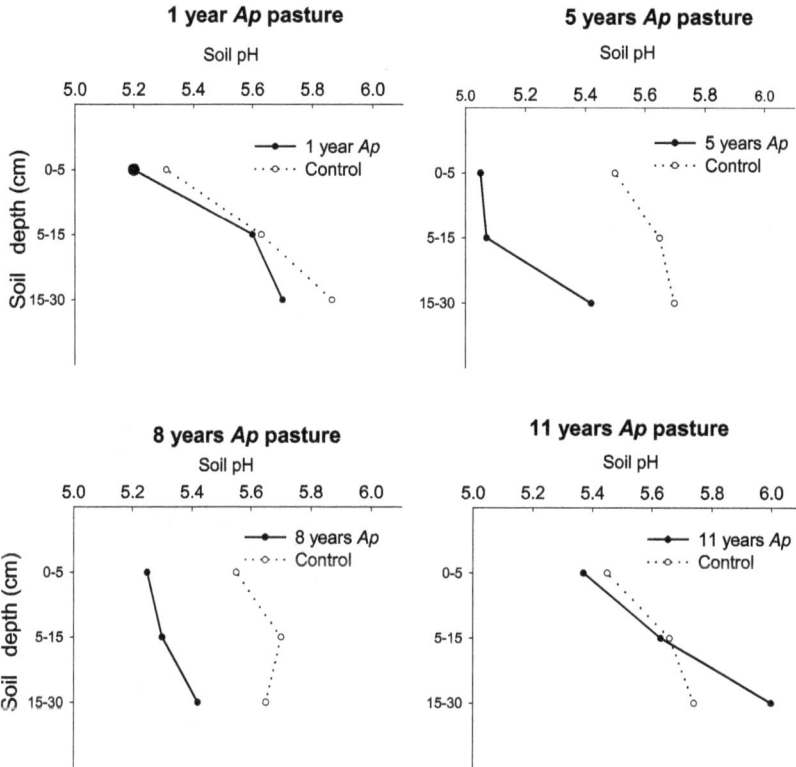

Fig. 3. Changes in soil pH with depth in four native rangelands mixed with *Arachis pintoi* at different ages, compared to nearbycontrol sites, located in Veracruz, Mexico

In all cases, the lowest pH values were found in the first five centimetres depth. This fact could be related with the largest presence of legume roots in this layer (CIAT, 1991), and because the addition of litter and excreta from grazing animals. Consequently, the exchange of cations/anions is more intense at this depth. Soil pH values at 15-30 cm did not show a consistent trend in acidity. The highest pH value (6.0) was found under "11 years *Ap* site" site. Marschner *et al.* (1986) mentioned that rhizosphere pH may be different (higher or lower) than the pH of the bulk soil. The results suggested that the magnitude and direction of pH change depended mainly on soil depth. Marschner *et al.* (1986) mentioned that shifts in the cation/anion uptake ratio with the age of root zones, and different rates of depletion and replenishment of various cations and anions could be presumably the main factors responsible for these pH differences. Comparing the control sites data with the *Ap* rangelands in each site (Figure 3) soil acidification was evident mainly in the cases of 5 and 8 years old *Ap*. Soil pH values of 5.6 for native rangelands at 0-5 cm depth were also reported by Bosman *et al.* (1990) for this region, including a site in the same research station, which are in accordance with our observations.

Several researchers have reported an acidifying effect of legumes. Noble *et al.* (1997) assessed acidification in Australian soils under *Stylosanthes* spp. They found the highest rate of acidification in an irrigated *Stylosanthes* seed production system. In a pot experiment, Yan *et al.* (1996) studied soil pH changes produced by field beans (*Vicia faba*). In a period of 45 days they observed that soil pH decreased significantly from 6.00 to 5.64.

Nyatsanga & Pierre (1973) mentioned that acidity developing in the soil from N_2 fixation will depend on whether all or only a part of the crop is removed in the harvest. If only the grain of any leguminous crop is harvested, the N_2 fixed would have a relatively small effect on the soil acidity because most of the N_2 fixed is in the seed. But in the case of a legume used as green manure or a grazed legume-based pasture, N_2 fixation would result in larger increases of acidity because the N_2 fixed will be acid-forming when completely nitrified in the soil. This could partially explain the low acidification under the 11 year old *A. pintoi* pasture since this was extensively harvested for planting material.

Considering the importance of tropical forage legumes to animal production systems, the acidifying effect is an important subject. Several options have been proposed to overcome this problem. Application of lime is probably the most popular recommendation, but in economically depressed areas, this strategy may not be a viable option, even if transportation difficulties were overcome. In Veracruz, Mexico, farmers known that lime application has only short-term effects on soil acidity because of high rainfall and high temperatures. More investigation into the subject is necessary.

4.7 Conclusions

The presence of *Arachis pintoi* in grazed rangelands resulted in a decrease of soil pH in several rangelands. This acidification was strongest in the topsoil (0-5 cm). Also, soil acidification was most pronounced in 5 and 8 year old *Ap* rangelands. Because of the limitations in the experimental design, these are preliminary results and further investigation will be necessary to establish the acidification problem and devise possible solutions for legume-based tropical rangelands.

5. Effect of stocking rate on soil properties in a native pasture of the humid tropic of Veracruz, Mexico

5.1 Alternatives to intensification of pasture management

In the search of alternatives for improving animal production, intensification of the system is proposed, in order to meet the demand for food of animal origin for the population. Unfortunately, this intensification has led to inadequate management of agroecosystems, resulting in deterioration of physical, chemical and biological soil properties that affect the productivity of the system to the detriment of the animals reliant on soil nutrients for plant growth (Lal, 2000).

The stocking rate as grazing strategy has been developed to control the use of pasture, in order to obtain an optimal response of vegetation to grazing (Mousel *et al.*, 2005). Thus, a more efficient management of the stocking rate could significantly improve the nutrient use efficiency in rangelands, resulting in an improvement in system productivity (Dubeux, 2005).

In the humid tropic of Mexico, studies concerning the effect of stocking rate on physical and chemical characteristics of soil are scarce.

5.2 Details of the site and experimental area

The work began in September 2005 and ended in August 2007, in the Center for Teaching, Research and Extension in Tropical Animal Husbandry (CEIEGT) of the Faculty of Veterinary Medicine and Animal Science of the National University of Mexico. Details about this site had been described previously in this Chapter.

We used a native pasture (*Paspalum* spp and *Axonopus* spp) area; since February 2002, it has been under rotational grazing with a pattern of use-rest of 3-27 days and a stocking rate of 2, 3 and 4 cows/ha. The treatments were stocking rates of 2 (low), 3 (average) and 4 (high) cows/ha, each housed in 5.0, 3.3 and 2.5 ha, where each area is divided into 10 paddocks.

5.3 Variables measured

In 100 quadrats of 0.25 m^2, we estimated, by direct observation, the percentage of bare soil. Two samples per paddock (replicates) in two cycles of grazing per season, within two years of evaluation were considered, which produced a total of 144 samples for the three treatments.

The length of roots (mm/cm^2) and root density (mg/cm^2) was measured in the last two cycles of grazing periods of rainy, dry and north season of the second year (Sept-2006/Jun-2007). We used a metal cylindrical tube of 10 cm of diameter and 20 cm long (Rowell, 1997). Eighteen samples were obtained by repetition, randomly at depths of 0-10 cm and 10-20 cm, respectively. After sampling, the soil was placed in trays to wash with water and retrieve the root in a 2 mm mesh sieve. Subsequently, the fresh root sample was placed in an oven for forced air drying at 60 ° C for 72 h, and then weighed to obtain the dry weight of the sample. Values are expressed in terms of organic matter of roots. The root length was determined by the linear intercept method (Tennant, 1975). The data were expressed in terms of mg of root root/cm^3 (density) and mm/cm^3 (length). The number of samples per repetition was 12, at two soil depths (0 - 10 and 10 to 20 cm), in the three seasons of the year, resulting in a total of 432 samples. Bulk density (BD, g/cm^3) was calculated at a depth of 0 - 5 cm, using a cylindrical metal auger of 5 cm in diameter. Ten samples were taken for treatment in each of the two repetitions, in the three seasons and for two years, making a total of 360 samples. The samples were weighed fresh and then dried at 100 ° C for 72 h, to obtain the dry weight of soil, and with the volume of the cylinder to calculate the BD (weight/volume). To determine soil compaction (kg/cm^2), 400 measurements were made randomly per repetition by stocking rate, in a single sampling in the rainy season (October) 2007 before grazing, using a penetrometer (Dickey-John Soil Compaction Tester), which produced a total of 2400 observations.

The rate of soil nitrogen mineralization was determined by anaerobic incubations of soil samples, collected at the season of the "nortes" (January 2007) at depths of 0-5 and 5-15 cm, using the technique of Waring & Bremner (1964) with slight modification, where we use 10 g of soil instead 5 g; as well as, a solution 3 M KCl to rinse the test tubes, replacing 2 M solution.

5.4 Experimental design and statistical data management

The experimental design was completely randomised. The analysis of variance was done considering the effect of treatment, repetition and the season. We used the PROC MIXED procedure of SAS (1999) for repeated measures and considered the covariance structure of symmetrical components, as the best fit to the data (Little *et al.*, 1998). The comparison of treatment means was performed using LSMEANS (SAS, 1999). In all cases, the variation between replicates was used to generate the experimental error. In the case of the N mineralization rate, we considered the effect of treatment, repetition and soil depth; and likewise, the data of soil compaction (kg/cm^2) were analyzed per treatment and repetition, with the PROC GLM (SAS, 1999).

5.5 Results and discussion

5.5.1 Root density and root length

The density and root length were not affected by the stocking rate (P> 0.05) at any level of soil depth (Table 6). However, the season affected (P <0.05) the density and root length in both soil depths (P <0.05). Thus, the dry season was detrimental to both soil depths, compared with the other seasons. Results of Chen *et al.* (2006) showed that root density decreased with increasing stocking rate, but the presence of moisture encouraged root growth. This coincides with the results obtained here, where the seasons wetter roots generated higher densities.

The root length was not affected by stocking rate for any of the two soil depths, but in the first 10 cm, there was a trend of greater length in the highest stocking rate, compared with the low stocking rate. On the coast of Texas in a pasture with common Bermuda grass and Coastal Bermuda grass, Rouquette, and Florence (1986) reported that the increase in stocking rate (high, medium and low) in terms of forage allowance affected the root mass in these grasses at 350, 460 and 477 kg / ha of roots. Also, no difference in root length, by effect of stocking rate could be related to the presence of hardpan at shallow depths (<30 cm), which could have affected the expansion (length and density) of roots in the soil.

Variable	Soil depth			
	0 – 10 cm		10 – 20 cm	
Level	Density (mg OM/cm³)	Length (mm/cm³)	Density (mg OM/cm³)	Length (mm/cm³)
Season				
Rainy	18.8 ± 0.2[b]	40.5 ± 0.8[a]	15.7 ± 0.1[b]	19.4 ± 0.3[a]
Winter or "Nortes"	21.5 ± 0.2[a]	43.1 ± 2.2[a]	17.7 ± 0.2[a]	14.9 ± 0.9[a]
Dry	16.47 ± 0.2[c]	30.2 ± 0.9[b]	14.4 ± 0.1[c]	16.8 ± 0.9[a]
Stocking rate				
2 cows/ha	19.1 ± 0.3[a]	39.2 ± 2.4[a]	15.4 ± 0.2[a]	14.8 ± 1.2[a]
3 cows/ha	18.6 ± 0.3[a]	34.3 ± 1.7[a]	16.0 ± 0.2[a]	17.5 ± 1.1[a]
4 cows/ha	18.9 ± 0.3[a]	40.2 ± 2.0[a]	16.4 ± 0.2[a]	18.8 ± 0.8[a]

Different letters for season or stocking rate within column are statistically different (P≤0.05).

Table 6. Effect of the stocking rate and year season over root density and root length (mean ± standard error) in grazed native rangelands, in the humid tropics of Mexico.

5.5.2 Soil physico-chemical changes

The experiment did not detect any significant change in the soil physical and chemical properties between the beginning and end (Table 7), except the organic matter. The soil organic matter increased from the low to medium stocking rate, and decreased in the high stocking rate. Beare *et al.* (2005) mentioned that the SOM does not necessarily increase with increasing dry matter production, which is consistent with the results observed here.

Variable	Stocking rate		
	2 cows/ha	3 cows/ha	4 cows/ha
sand (g kg^{-1})	31.9±1.6^a	37.4±1.2^a	35.2±1.5^a
Silt (g kg^{-1})	43.6±2.4^a	38.2±0.8^a	41.0±1.2^a
Clay (g kg^{-1})	24.4±1.6^a	25.2±1.7^a	23.7±1.9^a
pH, 1:2	5.6±0.1^a	5.4±0.2^a	5.5±0.2^a
OM (g kg^{-1})	2.3±0.2^a	3.2±0.04^b	2.3±0.1^a
N (g kg^{-1})	0.04±0.007^a	0.03±0.004^a	0.04±0.009^a
P (mg kg^{-1})	2.8±0.5^a	1.5±0.4^a	2.5±0.1^a
K$^+$ (mg kg^{-1})	4.2±1.1^a	6.5±1.3^a	7.5±2.3^a
Ca^{3+} (mg kg^{-1})	3.2±0.7^a	6.7±2.8^a	5.3±1.9^a
Mg^{2+} (mg kg^{-1})	2.9±0.4^a	5.6±1.7^a	4.4±0.9^a

Means with different letters within rows are significantly different (P≤0.05).

Table 7. Effect of the stocking rate over chemicals and physical variables, in a soil covered by native rangelands, in the humid tropic of Mexico.

5.5.3 Bulk density and soil compactation

In both periods, we observed that increasing the stocking rate tended to increase the bulk density, averaging 1.27 ± 0.01, 1.3 and 1.3 ± 0.02 ± 0.01 for 2, 3 and 4 cows/ha, although we found no effect on it throughout the experiment. The pooled bulk density data were fitted to the linear equation y = 0.019x + 1.23 (R^2 = 0.83), which indicates that every unit increase in stocking rate (x) increased the bulk density (y) to 0.019 g/cm^3. Castillo *et al.* (2003), who measured the effect of native grass, alone and associated with *A. pintoi*, three and a half years of its establishment, under similar soil and climatic conditions, reported an average value of bulk density of 1.21 g/cm^3, slightly lower than reported in this trial for any stocking rate or season.

5.5.4 Bare soil

The evaluation of bare soil indicated significant differences due to the stocking rate (P <0.0001) and season (P <0.0001), as well as their interaction (P <0.0001, Table 8). The proportion of bare soil increased as the stocking rate increased. With increased stocking rate, the return of the animal at a given point of the pasture is more frequent (Stewart, 2003). Another aspect that could explain this difference is related to the increased presence in the rainy season of introduced grasses, which may favor the bare soil due to its erect growth habit, compared with stoloniferous species (Barrios *et al.* , 2004).

5.5.5 Soil nitrogen mineralization rate

A trend of increased mineralization as the stocking rate increased was observed (Table 9), but the effect was not significant (P> 0.05) for this variable.

The equation y = 68.09 + 19.59 ln (x) (R^2 = 0.51), described the change in the mineralization of N (y) in the 0-5 cm soil depth, which increased with the stocking rate (x), while in the 5-15 cm soil depth, the equation was y = 21.29 + 9.55 ln (x) (R^2 = 0.77). The highest rate of soil N mineralization was observed from 0-5 cm, and decreased with soil depth. These results match those of Valles *et al.* (2008) who found the highest rate of soil N mineralization within the first 5 cm of soil depth, although their values were lower (5.8 - 18.7 mg NH_4-N g/soil) than the values measured in this study.

In this experiment, the increase in the rate of N mineralization with increasing stocking rate was consistent with other reports. Dubeux *et al.* (2006), who evaluated the effect of three levels of N from litter of *Paspalum notatum* (40, 120 and 360 kg N/ha) with three stocking rates (1.3, 2.7 and 4.0 animal units/ha), also found that the N mineralization rate increased with increasing stocking rate.

Season of the year	Stocking rate (cows/ha)		
	2	3	4
Rainy	9.5 ± 0.5 c	16.4 ± 0.7 a	13.0 ± 0.6 b
Winter or "Nortes"	4.0 ± 0.2 c	7.6 ± 0.5 a	6.6 ± 0.4 b
Dry	8.3 ± 0.2 b	5.9 ± 0.4 c	16.0 ± 0.4 a

Different letters for the season x stocking rate interaction indicate significant differences (P≤0.05).

Table 8. Bare soil (%, mean ± standard error) in a native rangeland during three climatic seasons and three stocking rates, in the humid tropic of Mexico.

Soil depth, 0-5 cm			Soil depth, 5-15 cm		
Stocking rate (Cows/ha)					
2	3	4	2	3	4
63.6±70.4	93.8±105.2	82.0±95.8	20.1±24.4	31.2±40.6	29.7±37.1

Table 9. N Mineralization rate (µg NH_4-N g soil/day, means ± standard error) in a native pasture soil, at three stocking rates and two depths in the humid tropic of Mexico.

5.6 Conclusions

There was no effect of stocking rate on most variables measured. Soil compaction did not change due to stocking rate. The length and root density did not change, however, the trend of increased root length coincided with the increasing trend in the rate of soil N mineralization with increasing stocking rate, suggesting a more dynamic nutrient cycling in pastures with high stocking rate. It is possible that the absence of differences indicate variability in the response of native rangelands to cattle grazing and trampling activities. There may be sufficient, inherent adaptation in native rangelands to offset the negative effect of increased stocking rate at the levels evaluated in this study.

6. References

Agricultural Trust Lawes (1996). *Genstat 5*. (Second ed.), release 3.2. Lawes Agricultural Trust, IACR, ISBN 1-85206-183-9, Rothamsted, UK.

Alegre, J.C. & Lara, P.D. (1991). Efecto de los animales en pastoreo sobre las propiedades físicas de suelos de la región tropical húmeda del Perú. *Pasturas Tropicales* Vol.13, No.1, (Enero 1991), pp. 18-23. ISSN 1012-7410

Anderson, J.M. & Ingram, J.S.I. (1993). *Tropical Soil Biology and Fertility: Handbook of Methods*, (2nd ed.), CAB International, ISBN-10 0851988210, Wallingford, U.K.

Argel, P. J. (1994). Regional experience with forage *Arachis* in Central America and Mexico. In: *Biology and Agronomy of Forage Arachis pintoi*. P. C. Kerridge & B. Hardy (Eds.), pp. 134-143. Centro Internacional de Agricultura Tropical, ISBN 958-9183-96-4, Cali, Colombia.

Arscott, T.G. (1978). *Los suelos del CIEEGT. Informe de Consultoría en Suelos al Proyecto: Investigación, Enseñanza y Extensión en Ganadería Tropical*, FAO/CIEEGT, FMVZ, UNAM, Martínez de la Torre, Veracruz, México (Circulación interna).

Barrios, R., Fariñas, J., Díaz, A. & Barreto, F. (2004). Evaluación de 11 accesiones de leguminosas utilizadas como cobertura viva en palma aceitera en el estado Monagas, Venezuela. *Bioagro*. Vol.16, No.2, (Mayo 2004), pp. 113-119. ISSN 1316-3361

Beare, M.H., Curtin, D., Thomas, S., Fraser, P.M. & Francis, G.S. (2005). Chemical components and effects on soil quality in temperature grazed pastures systems. In: *Optimizations on Nutrient Cicling and Soil Quality for Sustainable Grasslands*. S.C. Jarvis, P.J. Murray & J.A. Roker (Eds.), pp. 25-36, Wageningen Academic publishers.ISBN 978-90-76998-72-5, Wageningen, The Netherlands.

Behling-Miranda, C. H., Fernandes, C. D., & Cadisch, G. (1999). Quantifying the nitrogen fixed by *Stylosanthes*. *Pasturas Tropicales* Vol.21, No.1, (Enero 1999), pp. 64-69. ISSN 1012-7410

Bergersen, F. J., & Turner, G. L. (1983). An evaluation of ^{15}N methods for estimating nitrogen fixation in a subterranean clover perennial ryegrass sward. *Australian Journal of Agricultural Research* Vol.34, No.4, (July 1983), pp. 391-401. ISSN 1836-0947

Bosman, H. G., Castillo, E., Valles, B., & De Lucía, G. R. (1990). Composición botánica y nodulación de leguminosas en las pasturas nativas de la planicie costera del Golfo de México. *Pasturas Tropicales* Vol.12, No.1, (Enero 1990), pp. 1-8. ISSN 1012-7410

Boutton, T.W., Archer, S.R., Midwood, A.J., Zitzer, S.F., & Bol, R. (1998). δ ^{13}C values of soil organic carbon and their use in documenting vegetation change in a subtropical savanna ecosystem. *Geoderma* Vol.82, No.1, (February 1998), pp. 5-41. ISSN 0016-7061

Cadisch, G., Bradley, R. S., & Nosberger, J. (1989). ^{15}N-Based estimation of nitrogen fixation by eight tropical forage legumes at two levels of P:K supply. *Field Crops Research* Vol.22, No.3, (November 1989), pp. 181-194. ISSN 0378-4290

Cadisch, G., Schunke, R. M., & Giller, K. E. (1994). Nitrogen cycling in a pure grass pasture and a grass-legume mixture on a red latosol in Brazil. *Tropical Grasslands* Vol. 28, No.1 (March 1994), pp. 43-52. ISSN: 1051-0761

Cadisch, G., Hairiah, K., & Giller, K. E. (2000). Applicability of the natural ^{15}N abundance technique to measure N_2 fixation in *Arachis hypogaea* grown on an Ultisol.

Netherlands Journal of Agricultural Science Vol.48, No.1, (January 2000), pp. 31-45.ISSN 1573-5214

Cárdenas, E. A., Maass, B. L., & Franco, L. H. (1999). Evaluación de germoplasma nuevo de *Arachis pintoi* en Colombia. 2. Bosque muy húmedo-Premontano (zona cafetera), Caldas. *Pasturas Tropicales* Vol. 21, No. 2, (Mayo 1999), pp. 42-59. ISSN 1012-7410

Castillo, G.E. (2003). Improving a native pasture with the legume *Arachis pintoi* in the humid tropics of México, PhD Thesis Wageningen, University, The Netherlands. 165 p.

CIAT (1984). Selección y evaluación de pastos tropicales en condiciones de alta concentración de aluminio y bajo contenido de fósforo disponible. Guía de estudio, 47 p., Centro Internacional de Agricultura Tropical, Cali, Colombia.

CIAT (1991). *CIAT (Centro Internacional de Agricultura Tropical) Report/Informe CIAT 1991.* Centro Internacional de Agricultura Tropical, Cali, Colombia.

Crowder, L.V., & Chheda, H. R. (1982). *Tropical Grassland Husbandry.* Longman, ISBN: 0582466776, London, UK.

Davidson, B. R., & Davidson, H. F. (1993). *Legumes: The Australian Experience.* Taunton: Research Studies Press, ISBN-10: 0471942316, Somerset, England.

Dubeaux, J.C.B., Sollengerger, L.E., Comerford, N.B., Ruggieri, A.C. & Portier, K.M. (2005). Characterization of soil organic matter from Pensacola bahiagrass pastures grazed for four years at different management intensities. In: *Optimizations on Nutrient Cicling and Soil Quality for Sustainable Grasslands.* S.C. Jarvis, P.J. Murray & J.A. Roker (Eds.), p. 87 Wageningen Academic publishers.ISBN 978-90-76998-72-5, Wageningen, The Netherlands.

Dubeaux, Jr J.C.B., Sollenberger, L.E., Interrante, S.M., Vendramini, J.M.B. & Stewart, Jr R.L. (2006). Litter decomposition and mineralization in Bahiagrass pastures managed at different intensities. *Crop Science.* Vol.46, No.3, (May 2006), pp. 1305–1310. ISSN 0931-2250

Fisher, M.J., Rao, I.M., Ayarza, M.A., Lascano, C.E., Sanz, J.I., Thomas, R.J. & Vera, R.R. (1994). Carbon storage by introduced deep-rooted grasses in the South American savannas. *Nature* Vol.371, No.6494, (15 September 1994), pp. 236-238. ISSN : 0028-0836

González, M.S., Van Heurck, L.M., Romero, F., Pezo, D.A. & Argel, P.J. (1996). Producción de leche en pasturas de estrella africana (*Cynodon nlemfuensis*) solo y asociado con *Arachis pintoi* o *Desmodium ovalifolium*. *Pasturas Tropicales* Vol. 18, No.1, (Enero 1996), pp. 2-12. ISSN 1012-7410

Haynes, R.J. (1983). Soil acidification induced by leguminous crops. *Grass Forage Science* Vol. 38, No.1, (January 1983), pp. 1-11. ISSN 0142-5242

Hernández, G. S. (1988). Estudio de los suelos del CIEEGT y de sus módulos de influencia. FAO (Project: Mex/87/020), CIEEGT, FMVZ-UNAM, Tlapacoyan, Ver. México.

Hernández, M., Argel, P.J., Ibrahim, M.A., & 't Mannetje, L. (1995). Pasture production, diet selection and liveweight gains of cattle grazing *Brachiaria brizantha* with or without *Arachis pintoi* at two stocking rates in the Atlantic Zone of Costa Rica. *Tropical Grasslands* Vol. 29, No.3, (September 1995), pp. 134-141. ISSN: 1051-0761

Huntjes, J.L.M. & Albers, R.A.J.M. (1978). A model experiment to study the influence of living plants on the accumulation of soil organic matter in pastures. *Plant and Soil* Vol.50, No.2, (December 1978), pp. 411-418. ISSN 0032-079X

Ibrahim, M.A. (1994). Compatibility, persistence and productivity of grass-legume mixtures for sustainable animal production in the Atlantic Zone of Costa Rica. PhD Thesis, Wageningen Agricultural University, Wageningen, The Netherlands. 140 P.

Kerridge, P. C. (1994). Future prospects for utilization and research in forage *Arachis*. In: *Biology and Agronomy of Forage Arachis pintoi*. P. C. Kerridge & B. Hardy (Eds.), pp. 199-206. Centro Internacional de Agricultura Tropical, ISBN 958-9183-96-4, Cali, Colombia.

Lal, R. (2000). Soil management in the developing countries. *Soil Science*. Vol.165, No.1, (January 2000), pp. 57-72. ISSN 0038-075X

Landon, J. R. (1984). *Booker Tropical Soil Manual*, Longman, Inc., ISBN 0-582-46049-2, New York, USA.

Ledgard, S. F., Simpson, J. R., Freney, J. R., & Bergersen, F. J. (1985). Field evaluation of ^{15}N techniques for estimating nitrogen fixation in legume grass associations. *Australian Journal of Agricultural Research* Vol.36, No.2, (March 1985), pp. 247-258. ISSN 1836-0947

Littell, R.C., Henry, P.R. & Ammerman, C.B. (1998). Statistical analysis of repeated measures data using SAS Procedures. *Journal of Animal Science*. Vol.76, No.4, (April 1998), pp. 1216–1231. ISSN 0021-8812

Mannetje, L.'t, & Haydock, K. P. (1963). The dry-weight-rank method for the botanical analysis of pasture. *Journal of the British Grassland Society* Vol.18, No.4, (December 1963), pp. 268-275. ISSN 0142-5242

Marschner, H., Romheld, V., Horst, W. J., & Martin, P. (1986). Root-induced changes in the rhizosphere - Importance for the mineral nutrition of plants. *Zeitschrift Fur Pflanzenernahrung Und Bodenkunde* Vol.149, No.4, (July-August 1986), pp. 441-456. ISSN: 1436-8730

Mousel, E.M., Schacht, W.H., Moser, L.E. & Zanner C.W. (2005). Root and Vigor Response of Big Bluestem to Summer Grazing Strategies. XX International Grassland Congress 2005. June 26-July 1, 2005. ISBN 978-90-76998-81-7, Dublin, Ireland, 975 p.

Noble, A. D., Cannon, M., & Muller, D. (1997). Evidence of accelerated soil acidification under *Stylosanthes*-dominated pastures. *Australian Journal of Soil Research* Vol.35, No.6, (November-December 1997), pp. 1309-1322. ISSN 0004-9573

Nyatsanga, T., & Pierre, W. H. (1973). Effect of nitrogen fixation by legumes on soil acidity. *Agronomy Journal* Vol.65, No.6, (November-December 1973), pp. 936-940. ISSN 0002-1962

Peoples, M. B., Bell, M. J., & Bushby, H. V. A. (1992). Effect of rotation and inoculation With *Bradyrhizobium* on nitrogen fixation and yield of peanut (*Arachis hypogaea* L, Cv Virginia Bunch). *Australian Journal of Agricultural Research* Vol.43, No.3, (March 1992), pp. 595-607. ISSN 1836-0947

Rao, I. M., & Kerridge, P. C. (1994). Mineral nutrition of forage *Arachis*. In: *Biology and Agronomy of Forage Arachis pintoi*. P. C. Kerridge & B. Hardy (Eds.), pp. 71-83. Centro Internacional de Agricultura Tropical, ISBN 958-9183-96-4, Cali, Colombia.

Rowell, L.D. (1997). *Soil Science Methods and Applications*. Longman Group UK Limited. ISBN 0-582-087848, Essex, England

Sanchez, P. A. (1987). Management of the acid soil in the humid tropics of Latin America, *Proceedings of an IBSRAM Inaugural Workshop*, pp. 63-107, ISBN 9747614391 IBSRAM, Bangkok, Thailand.

SAS. (1999). SAS OnlineDoc®, Version 8. SAS Institute Inc., Cary, NC., U.S.A.

Suarez, V. S., Wood, M., & Nortcliff, S. (1992). Crecimiento y fijacion de nitrógeno por *Arachis pintoi* establecido con *Brachiaria decumbens*. *Crenicafé(Colombia)* Vol.43, No.1, (Enero 1992), pp. 14-21. ISSN 0120-0275

Tang, C. (1998). Factors affecting soil acidification under legumes I. Effect of potassium supply. *Plant and Soil* Vol.199, No.2, (February 1998), Pp. 275–282, ISSN 0032-079X

Tennant, D. (1975). A test of a modified line intersect method of estimating root length. *Journal of Ecology*. Vol.63, No.3, (November 1975), pp. 995-1001. ISSN 1365-2745

Thomas, R. J. (1995). Role of legumes in providing N for sustainable tropical pasture systems. *Plant and Soil* Vol.174, No. 1-2 (July 1995), pp. 103-118. ISSN 0032-079X

Thomas, R. J., Asakawa, N. M., Rondon, M. A., & Alarcon, H. F. (1997). Nitrogen fixation by three tropical forage legumes in an acid-soil savanna of Colombia. *Soil Biology and Biochemistry* Vol.29, No.5-6, (May-June 1997), pp. 801-808. ISSN 0038-0717

Toledo, JM. (1986). *Plan de Investigación en Leguminosas Tropicales para el CIEEGT*, Martínez de la Torre, Veracruz, México. Informe de Consultoría en Pastos Tropicales al Proyecto: Enseñanza y Extensión para la Producción de Leche y Carne en el Trópico. FAO/CIEEGT, FMVZ, UNAM, Martínez de la Torre, Veracruz, México (Circulación interna).

Valles, B., Castillo, E., & Hernández, T. (1992). Producción estacional de leguminosas forrajeras en Veracruz, México. *Pasturas Tropicales* Vol.14, No.2, (Mayo 1992), pp. 32-36. ISSN 1012-7410

Valles, M.B. (2001). Contribution of the forage legume *Arachis pintoi* to soil fertility in a tropical pasture system in Veracruz, México, PhD Thesis. University of London, London, U.K. 261 p.

Veldkamp, E., Davidson, E., Erickson, H., Keller, M., and Weitz, A. 1999. Soil nitrogen cycling and nitrogen oxide emissions along a pasture chronosequence in the humid tropics of Costa Rica. *Soil Biology and Biochemistry*, Vol.31, No.3, (March 1999), pp. 387-394. ISSN 0038-0717

Vos, C. (1998). Quality and quantity of forage from a native pasture alone or associated with *Arachis pintoi* CIAT 17434. Thesis, Agricultural University of Wageningen, The Netherlands.

Walkley, A. & Black, T.A. (1934). An examination of the Degtjareff method for determining soil organic matter and a proposed modification of the chromic acid titration method. *Soil Science*, Vol.37, No.1, (January 1934), pp. 29-38. ISSN 0038-075X

Waring, S.A. & Bremner, J.M. (1964). Ammonium production in soil under waterlogged conditions as an index of nitrogen availability. *Nature*. Vol.201, No.4922, (29 February 1964), pp. 951-952. ISSN 0028-0836

Yan, F., Schubert, S., & Mengel, K. (1996). Soil pH changes during legume growth and application of plant material. *Biology and Fertility of Soils* Vol.23, No.3, (October 1996), 236-242. ISSN 1432-0789

5

Effects on Soil Fertility and Microbial Populations of Broadcast-Transplanting Rice Seedlings in High Standing-Stubble Under No-Tillage in Paddy Fields

Ren Wan-Jun, Huang Yun and Yang Wen-Yu
Sichuan Agricultural University, Wenjiang, Sichuan,
China

1. Introduction

No-tillage broadcast-transplanting of rice is a new cultivation technology developed in recent years (Xia, 2003; Liu et al., 2002). Our system involves no-tillage of rice paddies, which leaves high standing stubble, and raising seedlings on dryland beds. Then, we developed the technology of broadcast-transplanting seedlings in the field with high standing-stubble under no-tillage condition (Fig.1~4), which has several advantages such as lower fuel costs, savings in labour and time (yang et al., 2000; Liu, 2006; Ren et al., 2008). The first experiments of broadcast-transplanting of rice began in 1950s, reported by Peiris (1956) in Sri Lanka. In the 1980s, broadcast-transplanting of rice was successfully used in China (Zhang et al., 1993). From the middle and late 1990s, Guangdong province (Liu et al., 2002) and Guangxi province (Jiang et al., 2005) experimented with no-tillage broadcast-transplanting of rice under two rice crops and Sichuan province (Liu and Li, 2002) under single indica rice crop. Compared with conventional tillage broadcast-transplanting, the no-tillage broadcast-transplanting rice grew slower and had fewer tillers at the early stage (Liu et al., 2002) , as well, root biomass and length were reduced during the seedling standing period (Jiang et al., 2005), but it brought to a higher spike-bearing rate and more grains.

Residue management is an important component of this new rice production system (broadcast-transplanting seedlings into high standing-stubble under no-tillage condition). In addition to the 20 to 50 cm of high stubble left upon harvest of the previous crops (wheat or rape) harvested, the farmer is encouraged to return the harvested stubble from previous crops to the field (yang et al., 2000; Ren et al., 2003). This practice increases the soil fertility: soil organic matter, total N and K, and available N and K concentrations were higher than those of conventional tillage paddy fields, either with conventional rice transplanting or broadcast-transplanting rice in one study (Ren et al., 2007). This promising result needs to be verified in other paddy fields, to determine the effects of broadcast-transplanting seedlings into no-tillage fields with high standing-stubble on soil fertility and microbial populations in paddy field. The objectives of this study were to clarify the influence of the new technology on soil fertility and microbial populations of paddy field, and to provide a scientific basis and give some evidence in the introduction – what type of yields can be achieved with the broadcast-transplanting method in a no-tillage field?

Fig. 1. Field covered with high standing-stubble from the previous rape crop

Fig. 2. Broadcasting rice seedlings into high standing-stubble

Fig. 3. Early growth stage of rice seedlings

Fig. 4. Straw decay and soil surface conditions after rice harvest

2. Materials and methods

2.1 Plant material and condition

Ya' an experiment: The hybrid rice combinations Gangyou22, Iyou162 and Kyou047 were the cultivars used in this study.The experiments were conducted in the paddy at Daxing town of Ya' an city, Sichuan province, China (lat. 29°59 N, long. 102°59 E) in 2003. The experiment was on a sandy-loam soil with the following chemical properties: organic matter 23.66 g kg-1; total N 1.422 g kg-1, alkali hydrolysable N 109.36 mg kg-1; total P 0.458 g kg-1, available P 12.01 mg kg-1; total K 11.66 g kg-1, available K 30.22 mg kg-1. Previous crop was rape.

Pixian experiment: The hybrid rice combination Iyou21 were used as plant materials in this study. The experiment was conducted in the paddy at Gucheng town of Chengdu city, Sichuan province, China (lat. 30°93 N, long. 103°91 E) in 2005. The experiment was on a fluviatile loamsandy soil, and its chemical properties were as follows: organic matter, 18.73 g kg-1; total N, 1.18 g kg-1, alkali hydrolysable N, 78.26 mg kg-1; total P 0.21 g kg-1, available P, 43.53 mg kg-1; total K, 22.73 g kg-1, available K, 42.21 mg kg-1. Previous crop was wheat.

2.2 Experimental treatment and cultivation management

Ya' an experiment: The experiment was laid out in a split plot design with tillage and cultivation methods in the main plots and hybrid combination in the sub-plots with three replicates. Plot area was 15.84 m2. In the tillage and cultivation methods, there were three levels, including as conventional tillage and transplanting (CTT), conventional tillage and broadcast-transplanting of seedlings (CTB) and broadcast-transplanting of seedlings in the field with high standing-stubble under no-tillage condition (BSNT). The hybrid combination tested were Gangyou22, Iyou162 and Kyou047. The seedlings of CTT were raised on wet land and seeded at a rate of 30 g m-2; those of CTB and BSNT were raised on plastic trays with 428 holes and each hole was seeded about 2 seeds. Seeds were sown on March 30, then transplanted on May 9. When previous rape was harvested, 50 cm of stubble was left in the field. The soil was plowed and plugged after removing the rape stubble in tillage treatments; for no-tillage treatments, a herbicide was used 5 days after harvesting rape. Then paddies were irrigated to reach soil water saturation prior to broadcast-transplanting. Fertilizer inputs as a basal dressing were at the following rates: 90 kg P_2O_5 ha^{-1} as superphosphate, 45 kg K_2O ha^{-1} as KCl and 105 kg N ha^{-1} as urea. Additional 45 kg N ha^{-1} was broadcasted at the tillering stage and 45 kg K_2O ha^{-1} at the booting stage. The transplanting density was 30×10^4 hills ha^{-1}. Irrigation water was kept at 1-5 cm depth for 1 week following transplanting to aid in seedling establishment. Pests were controlled according to the standard recommendation and other rice management was similar as that in the paddy field (Ren et al., 2003).

Pixian experiment: The experiment was carried out in a randomized plot design with four treatments: No-tillage+Straw (NT+S), No-tillage (NT), Tillage+Straw (T+S) and Tillage (T). There were three replicates of each treatment, in plots of 30 m^2 . Ridges were built between plots to avoid fertilizer and water movement to adjacent plots. The +S treatment involved returning straw to soil that was removed when the previous wheat crop was harvested (leaving 25 cm of stubble in the field). After rice seedlings was survived, the straw was returned and scattered evenly in the plots. Returned straw provided an organic residue input of about 5-6 t ha^{-1}. In the NT+S treatment, straw was left on the soil surface, but the T+S treatment involved cultivating and mixing straw with soil. For tillage treatments, we plowed and plugged the soil after removing the wheat stubble. In the no-tillage treatments, a herbicide was used 5 days after harvesting wheat. Then we irrigated the paddy until it reached soil water saturation prior to broadcast-transplanting. Seeds were sown on March 20, then transplanted on May 23. Fertilizer inputs as a basal dressing were at a rate of 75 kg P_2O_5 ha^{-1} as superphosphate, 75 kg K_2O ha^{-1} as KCl and 105 kg N ha^{-1} as urea. Additional 30 kg N ha^{-1} was broadcasted at the tillering stage and 75 kg K_2O ha^{-1} and 15 kg N ha^{-1}at the booting stage. The transplanting density was 25.5×10^4 hills ha^{-1}. Irrigation water was kept at 1-5 cm depth for 1 week following transplanting to aid in seedling establishment. Pests were controlled according to the standard recommendation and other rice management was similar as that in the paddy field (Ren et al., 2003).

2.3 Soil and plant analysis

2.3.1 Soil fertility

Samples for nutrient measurement were sampled randomly with soil drill (25 mm in diameter).Five cores were taken after rice harvest from the 0-10 cm (upper layer) and 10-20 cm (deep layer) from each plot and pooled together. Large pieces of plant and animal resideues, gravel, *etc.*, were removed by sieving soil through a 2-mm mesh, mixing and a subsample was taken for analysis. Concentrations of soil organic matter, total N, P and K, alkali hydrolysable N, available P and K were tested by the methods described by Lu (2000).

2.3.2 Soil microorganism enumeration

Soil samples for microbiological assessment and cellulose decomposition intensity were collected from the Pixian experiment only at the tillering stage, elongation stage, booting stage and maturity stage of rice from the 0-10 cm (upper layer) and 10-20 cm (deep layer) from each plot and pooled together. Samples were cooled with ice packs in the field immediately after collection, and were stored at 4℃ for later analysis. Microbial communities were enumerated using the dilution plate method. Bacteria was isolated by BF medium, fungi was cultured by Martin agar medium and actinomyces were determined by modified Goss I medium (Department of Microbiology, 1985). Cellulose decomposition intensity was tested by the method of burying fabric into situ in the laboratory (Department of Microbiology, 1985).

2.3.3 Rice development and yield

Sixty (60) hill plants from each plot were investigated to study panicle development. Five (5) hills from each plot were harvested at maturity to estimate the yield components. The grain filling percentage was determined according to Zhu et al (1995). All plants from each plot were harvested at maturity for the determination of grain yield.

2.4 Statistical method

Statistical analyses were made by Office Excel 2003 and the SAS-Stat package (SAS Institute Inc. 1996). The Duncan test and least significant difference test at the 90% and 95% confidence levels were used to compare treatment means.

3. Results

3.1 Effects on soil fertility of tillage and residue management in a system with broadcast-transplanting of rice seedlings

Soil fertility was increased when rice seedlings were broadcast-transplanted among high standing-stubble under no-tillage condition (BSNT), as the organic matter, total N and K, and available N and K concentrations were higher than those of conventional tillage systems with conventional transplanting or broadcast-transplanting (Table1).

Treatment	Organic matter (g·kg⁻¹)	Total N (g·kg⁻¹)	Total P (g·kg⁻¹)	Total K (g·kg⁻¹)	Alkali hydrolysable N (mg·kg⁻¹)	Available P (mg·kg⁻¹)	Available K (mg·kg⁻¹)
CTT	23.42	1.444	0.604	14.24	109.89	15.88	33.45
CTB	23.31	1.506	0.583	12.55	105.29	13.66	36.59
BSNT	30.04	2.375	0.524	15.18	131.72	14.25	61.93
Initial soil fertility	23.66	1.422	0.458	11.66	109.36	12.01	30.22

CTT: conventional tillage and conventional transplanting; CTB: conventional tillage and broadcast-transplanting of seedlings; BSNT : broadcast-transplanting of seedlings in the field with high standing-stubble under no-tillage condition.

Table 1. Effects of different tillage and transplanting methods on soil fertility (0-20 cm depth) in paddy field. The experiment was conducted near Ya' an city, Sichuan province, China in 2003.

As shown in Table 2, in the upper 0-10 cm soil layer, the organic matter content for 'no-tillage + returning straw' treatment was 5.33, 2.79 and 5.37 g·kg⁻¹ higher than that for 'no-tillage', 'tillage + returning straw' and 'tillage' treatment, respectively. However, in the deep 10-20 cm layer, content of organic matter in 'tillage + returning straw' and 'tillage' treatments were higher than others. Since residues are mostly left on the surface in the no-tillage treatment, there organic matter accumulation was found in the surface soil, whereas a tilled soil had organic matter incorporated into the deeper soil layer. The maximum difference of two soil layers was noted for the 'no-tillage + returning straw' treatment, and there was little difference in soil organic matter in the 'no-tillage' treatment, and 'tillage' treatment without the extra straw residue input.

In upper soil layer, total and available N, P and K concentrations were greatest in the 'no-tillage + returning straw' treatment, those in the 'no-tillage' treatment and 'tillage + returning straw' treatment followed, and those in the 'tillage' treatment were the lowest. In the deep layer, the soil fertility indicators for 'tillage + returning straw' treatment were greater than those for other treatments. At the same time, the difference of nutrient status between two soil layers was the maximum for 'no-tillage + returning straw' treatment, therefore, the 'no-tillage + returning straw' treatment enriched soil fertility in the surface soil layer.

Soil layer	Treatment	Organic matter (g·kg⁻¹)	Total N (g·kg⁻¹)	Total P (g·kg⁻¹)	Total K (g·kg⁻¹)	Alkali hydrolysable N (mg·kg⁻¹)	Available P (mg·kg⁻¹)	Available K (mg·kg⁻¹)
Upper layer	NT+S	30.02a	1.67a	0.260a	29.114a	120.250a	82.593a	81.063a
	NT	24.69b	1.62ab	0.255a	26.556a	113.436b	70.739b	71.095ab
	T+S	27.23ab	1.59b	0.254a	22.900b	110.966c	69.013bc	64.718ab
	T	24.65b	1.56b	0.244b	21.920c	101.778d	64.001c	60.497b
Deep layer	NT+S	20.69a	1.35b	0.223a	11.255b	85.162a	40.305a	48.983a
	NT	18.12b	1.34b	0.221a	10.241b	84.572a	40.622a	46.781a
	T+S	22.20a	1.42a	0.224a	18.784a	89.686a	42.890a	54.674a
	T	22.20a	1.40ab	0.226a	16.345a	87.918a	40.755a	51.640a

NT+S: No-tillage+Straw; NT: No-tillage; T+S: Tillage+Straw; T: Tillage. Values followed by different small letters meant significant difference at 0.05 level, respectively (LSD test).

Table 2. Effect of different tillage methods on soil fertility in paddy field with broadcast-transplanting of rice. The upper soil layer was 0-10 cm depth and the lower soil layer was 10-20 cm depth. The experiment was conducted at Pixian near Chengdu city, Sichuan province, China in 2005.

3.2 Effects on soil microorganisms of tillage and residue management in a system with broadcast-transplanting of rice seedlings

3.2.1 Bacterial numbers

The average numbers of soil bacteria under different treatments are given in Table 3. The results showed that soil bacterial numbers for 'returning straw to soil' treatments were higher than that for 'no returning straw' treatments, and those in upper soil layer were higher than in deep layers too. In upper soil layer, the highest soil bacteria' numbers appeared in the 'no-tillage + returning straw' treatment at five growth stages, and the lowest bacteria' numbers were in the soil under 'tillage'. Bacterial numbers for 'no-tillage + returning straw' treatment were 15.86%, 14.0%, 2.53% and 40.44% higher than that for 'no-tillage', and 22.62%, 13.25%, 6.32% and 29.05% higher than that for 'tillage + returning straw' treatment at the tillering stage, elongation stage, booting stage and maturity stage of rice, respectively. In the deep soil layer, bacterial numbers appeared lower at tillering stage, and increased during elongation to booting stage, reaching a maximum value at rice booting and declined thereafter. In this soil layer, the highest bacterial numbers appeared in the 'tillage + returning straw' treatment, and the lowest bacteria' numbers were in the soil treated with 'tillage', too.

3.2.2 Fungal numbers

The data in Table 3 showed that fungal populations assessed by direct counting were lower than the other microbial groups. Fungal numbers increased during the rice growing season. Soil fungi were more abundant in the upper soil layer than in the deep layer, and greater in 'returning straw to soil' treatments than that for 'no returning straw' treatments. In the upper soil layer, fungal numbers were the highest in the 'no-tillage + returning straw' treatment, followed by the 'no-tillage' treatment, and the lowest fungal numbers were in the soil treated with 'tillage'. In deep soil layer, the order of fungal numbers in all treatments was 'tillage + returning straw'>'no-tillage + returning straw' >'no-tillage'> 'tillage'.

Soil layer	Treatment	Bacteria ($\times 10^3$ CFU $\cdot g^{-1}$)				Fungi ($\times 10^3$ CFU $\cdot g^{-1}$)				Actinomyce ($\times 10^3$ CFU $\cdot g^{-1}$)			
		Tille-ring	Elonga-tion	Booti-ng	Matu-rity	Tille-ring	Elonga-tion	Booti-ng	Matu-rity	Tilleri-ng	Elonga-tion	Booti-ng	Matu-rity
Upper layer	NT+S	168	171	202	191	2.6	2.89	5.16	6.23	130	108	155	189
	NT	145	150	197	136	1.6	2.64	4.35	5.29	135	90	120	134
	T+S	137	151	190	148	1.2	1.7	3.38	4.82	134	72	104	108
	T	85	103	157	94	0.8	1.44	2.56	4.45	104	44	96	97
Deep layer	NT+S	72	89	132	94	1	1.1	2.7	3.1	50	30	67	57
	NT	75	78	138	98	0.5	0.7	2.3	2.2	46	27	59	48
	T+S	83	99	143	100	1.1	1.5	2.9	3.9	64	41	77	66
	T	64	70	109	63	0.3	0.5	1.5	2.7	32	12	48	39

NT+S: No-tillage+Straw; NT: No-tillage; T+S: Tillage+Straw; T: Tillage.

Table 3. Effect of different tillage methods on soil microbial population in paddy field with broadcast-transplanting of rice. The upper soil layer was 0-10 cm depth and the lower soil layer was 10-20 cm depth. The experiment was conducted at Pixian near Chengdu city, Sichuan province, China in 2005.

3.2.3 Actinomycete numbers

Actinomycete numbers are affected by gas permeability of soil because that this microbial group are facultative aerobes. The data in Table 3 showed that actinomycete numbers were the lowest at elongation stage, increasing thereafter and reaching a peak at booting stage (deep soil layer) or maturity stage (upper soil layer) of rice. In upper soil layer, 'no-tillage + returning straw' treatment appeared to possessaerobic microsites, therefore, it had the highest actinomycete numbers except at tillering stage. The lowest actinomycete numbers were in the 'tillage' treatment, presumably because conditions were not favorable for aerobic microorganisms, particularly in the deep soil layer, which had fewer actinomycetes than in the upper soil layer. In deep soil layer, actinomycetes abundance in four treatments followed the order 'tillage + returning straw'>'no-tillage + returning straw' >'no-tillage'> 'tillage'.

3.3 Effects on cellulose decomposition intensity of tillage and residue management in a system with broadcast-transplanting of rice seedlings

The cellulose decomposition intensity under as affected by residue management and tillage at the Pixian experiment was shown in Fig. 5. The highest cellulose decomposition intensity appeared at the tillering stage and booting stage during different growth periods of rice. At the same time, cellulose decomposition intensity in 'returning straw to soil' treatments were higher than 'no returning straw' treatments. In upper soil layer, cellulose decomposition intensity in 'no-tillage + returning straw' treatment was always higher than that in other treatments. At the tillering stage of rice, cellulose decomposition intensity in 'no-tillage + returning straw' treatment was 28.20%, 39.76% and 52.93% higher than that in 'tillage + returning straw', 'no-tillage' and 'tillage' treatment; furthermore, 25.51%, 46.27% and 91.62% higher at elongation stage of rice; 38.76%, 68.60% and 79.71% higher at booting stage of rice; and 26.44%, 79.01% and 98.15% higher at maturity stage of rice, respectively. There was little difference between 'no-tillage' and 'tillage' treatment. In the deeper soil layer, the highest cellulose decomposition intensity was the treatment of 'tillage + returning straw', probably due to incorporation of straw into the deep soil layer, which enhanced the activity of cellulose decomposing bacteria.

3.4 Correlation between soil microorganisms and soil fertility in paddy field with broadcast-transplanting of rice seedlings in no-tillage and conventional tillage plots

The relationship between soil microorganisms and soil fertility at the Pixian experiment was analyzed (Table 4). There were significant positive correlations between soil bacterial numbers and the soil organic matter, total N, total P, total K, alkali hydrolysable N and available P concentrations. In contrast, soil fungi numbers were not correlated with soil fertility indicators. There were significant positive correlations between soil actinomycete numbers and the soil organic matter, total N, total P, total K, alkali hydrolysable N, available P and K concentrations. The cellulose decomposition intensity was significantly and positively correlated with the soil fertility indicators, which indicated that high cellulose decomposition intensity was related to improved soil nutrient status.

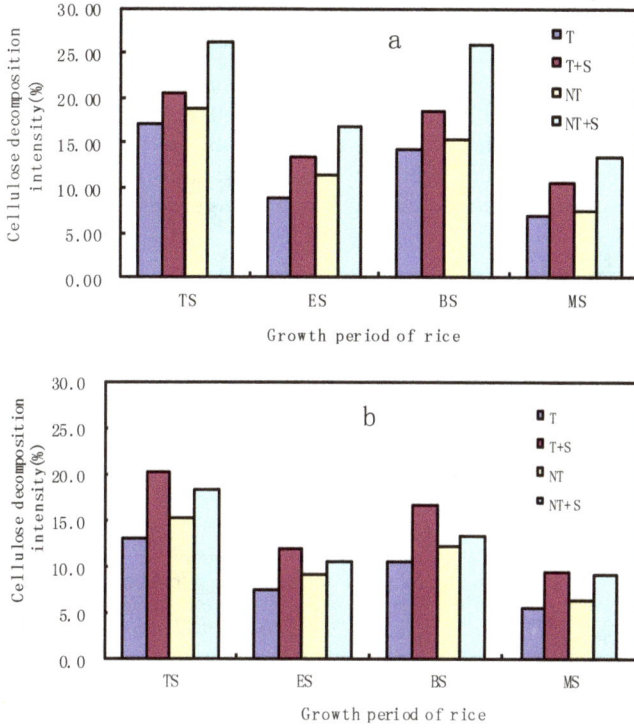

TS: Tillering stage; ES: Elongation stage; BS: Booting stage; MS: Maturity stage. NT+S: No-tillage+Straw;
NT: No-tillage; T+S: Tillage+Straw; T: Tillage.

Fig. 5. Effect of different tillage and residue management methods on cellulose
decomposition intensity in paddy field with broadcast-transplanting of rice. Results are for
(a) the upper soil layer, 0-10 cm depth and (b) the lower soil layer, 10-20 cm depth. The
experiment was conducted at Pixian near Chengdu city, Sichuan province, China in 2005.

Factor	Total K	Total N	Total P	Available K	Alkali hydrolysable N	Available P	Organic matter
Bacteria	0.5858*	0.6447*	0.7107**	0.5253	0.6610*	0.5625*	0.7139**
Fungi	0.3956	-0.0925	0.5613	-0.6073	-0.4090	-0.5324	0.0360
Actinomycete	0.8217**	0.8310**	0.8393**	0.7307**	0.8404**	0.7552**	0.6110*
Cellulose decomposition intensity	0.6781*	0.7854**	0.8848**	0.8535**	0.8339**	0.8623**	0.8818**

All the data was used for correlation analysis.*$P<0.05$; **$P<0.01$.

Table 4. Correlation (r values) between soil microorganisms and soil fertility in paddy field
with broadcast-transplanting of rice seedlings in no-tillage and conventional tillage plots. The
experiment was conducted at Pixian near Chengdu city, Sichuan province, China in 2005.

3.5 Effects on rice yield and its components in paddy field with contrasting tillage and transplanting methods, including broadcast-transplanting of rice seedlings

Analysis of variance showed that the main effect of yield by tillage and transplanting methods was significant at $P<0.1$ ($F_A=3.17$, $F_{0.05}=3.63$, $P=0.069$) and yield followed the order: CTT>BSNT>CTB (Table 1). There was no difference due to the hybrid grown ($F_B=0.36$, $F_{0.05}=3.63$) and interaction of the hybrid by tillage and transplanting method ($F_{A\times B}=1.99$, $F_{0.05}=3.01$) was not significant. Among three tillage and transplanting methods,

Treatment		Effective panicle (No. m⁻²)	Spikelet (No. ear⁻¹)	Seed-setting rate (%)	Filled grain percentage (%)	1 000-grain weight (g)	Yields (kg.m⁻²)
CTT	Gangyou22	175.2 b	167.6 a	87.69 a	77.67 a	26.31 a	0.683 a
	Ilyou162	189.0 b	179.4 a	77.23 b	73.71 a	26.65 a	0.689 a
	Kyou047	214.7 a	140.1 b	89.20 a	77.93 a	26.46 a	0.697 a
	Mean	193.0 ab	162.4 b	84.71 a	76.44 ab	26.47 a	0.690 a
CTB	Gangyou22	177.4 b	155.5 a	82.16 a	73.56 b	25.57 a	0.650 ab
	Ilyou162	184.6 b	172.4 a	72.90 b	72.20 b	25.49 a	0.596 b
	Kyou047	227.0 a	131.8 b	80.90 a	78.47 a	25.68 a	0.676 a
	Mean	196.3 a	153.2 b	78.65 b	74.74 b	25.58 b	0.641 b
BSNT	Gangyou22	172.7 b	187.6 a	82.35 b	77.58 ab	26.77 a	0.676 a
	Ilyou162	172.0 b	201.0 a	79.48 b	75.29 b	26.98 a	0.677 a
	Kyou047	205.5 a	141.1 b	87.12 a	80.57 a	26.60 a	0.628 a
	Mean	183.4 b	176.6 Aa	82.99 a	77.81 a	26.78 a	0.660 ab

CTT: conventional tillage and transplanting; CTB: conventional tillage and broadcasting seedlings; BSNT:broadcasting seedlings in the field with high standing-stubbles under no-tillage condition.

Table 5. Effect on rice yield and yield components of rice of different tillage and transplantingmethods, and hybrid types, in a paddy field. The experiment was conducted near Ya' an city, Sichuan province, China in 2003.

effective panicles were affected as follows: CTB>CTT>BSNT, and there were significantly more effective panicles in CTB than in BSNT ($F_A=4.06$, $F_{0.05}=3.63$, $F_{0.01}=6.23$). Under three tillage and transplanting methods, effective panicles of Kyou047 were the highest among three combinations. There were significant differences in the number of spikelets among different tillage and transplanting methods and between hybrids ($F_A=12.30$, $F_B=50.68$, $F_{0.01}=6.23$), and BSNT had 8.74% and 15.27% more spikelets than CTT and CTB, respectively.

Among three combinations, Ilyou162 had the greatest number of spikelets, and Kyou047 had the fewest. CTB had the lowest seed-setting rate, at 6.06% and 4.34% less seed-setting rate than CTT and BSNT, and this was significant ($P<0.05$). Among three hybrids, Ilyou162 had the lowest seed-setting rate. BSNT had the highest percentage of filled grain, which was significantly ($P<0.05$) higher than that of CTB; filled grain percentage of Kyou047 was also significantly ($P<0.05$) higher than Gangyou22 and Ilyou162. The 1000-grain-weight values were ranked BSNT>CTT>CTB , and those of BSNT and CTT were 1.20 g and 0.89 g higher than CTB, but no significant difference due to or hybrid combinations was detected for 1000-grain-weight.

4. Discussion

Our results showed that the method of broadcast-transplanting seedlings in the field with high standing-stubble and extra straw residues under no-tillage condition ('no-tillage + returning straw' treatment) promoted soil microbial populations at the surface soil layer because it returned about 5-6 t ha[-1] of straw to soil. Larger soil microbial populations are attributed to a more favorable soil ecological environment, with good conditions of water, nutrients, gas exchange and heat, that were beneficial for soil microorganisms. Therefore, the populations of bacteria, fungi, actinomycetes and cellulose decomposition intensity in the upper soil layer (0-10 cm depth) of the 'no-tillage + returning straw' treatment were the highest among four tillage and residue management systems. However, the cultivation method of 'tillage + returning straw' had a highest soil microbial communities in deep soil layer because it incorporated the straw input into the deeper soil layer (10-20 cm depth) and promoted soil microbial growth and activity.

Soil quality is a concept that generally refers to the soil's ability to sustain productivity, environmental quality, human and animal health (Doran and Parkin, 1994). Therefore, analysis of soil quality should consist of a minimum data set that includes measures of soil physical, chemical, and biological properties (Papendick et al., 1994). Soil microorganisms are important for maintaining soil quality due to their role in decomposition of organic matter and nutrient cycling and storage, and potentially represent a very sensitive biological marker (Turco et al., 1994). Morris and Boerner (1999) suggested that the spatial distributions of soil microorganisms and the factors affecting them should be further investigated since both soil chemistry and vegetation are affected by soil microbial communities. Different tillage systems have significantly effect on soil microorganism (Gao et al., 2004; Balesdent et al., 1990; Wei et al., 1993). Ridge no-tillage was advantageous to improve soil ecological environment and soil fertility in paddy field (Fan and Liu, 2002).

The straw return significantly improved soil structure and increased soil nutrient concentration (Fan and Liu, 2002), and increased soil organic matter and available N, P, and K (Luo and Zhang, 1999), furthermore, the straw's cellulose, hemicelluloses, lignin and other components were decomposed slowly to release nutrients in the field with no-tillage condition, so there was a higher soil fertility later in the rice growth period. In the upper soil layer, the organic matter content for 'no-tillage + returning straw' treatment was 5.33, 2.79 and 5.37 g·kg[-1] higher than that for 'no-tillage', 'tillage + returning straw' and 'tillage' treatment, respectively. Also, the concentration of total and available N, P and K were greatest in the 'no-tillage + returning straw' treatment, followed by the 'no-tillage' treatment and 'tillage + returning straw' treatment, and the lowest concentrations were in the 'tillage' treatment. In the deep layer, the fertility indicators for 'tillage + returning straw' treatment were higher than those for other treatments. The populations of bacteria, actinomycetes and cellulose decomposition intensity were significantly and positively correlated with the soil fertility indicators, which indicated that soil bacteria and actinomycetes were especially important for improving soil nutrient availability. The method of broadcast-transplanting seedlings in the field with high standing-stubble under no-tillage condition increased soil nutrient concentrations and soil microbial communities, and delayed senescence of root and shoots in the later growth stage (Xiao et al., 2009), and prolonged the grain-filling time, therefore, the spikelets of per ear, seed-setting rate, and 1 000-grains weight were increased, in comparison with conventional tillage and broadcasting seedlings. The findings provide insight into the physiological and ecological mechanisms whereby stable and high yield of broadcast-transplanted rice can be achieved in paddy field with high standing-stubble under no-tillage condition.

5. Acknowledgements

This study was supports by Chinese Ministry of Science and Technology (2004BA520A05; 2006BAD02A05; 2011BAD16B05).

6. References

[1] Xia J Y. Development and countermeasure about no-tillage cast-transplanted technology of high quality rice. *China Agricultura Technology Extension*, 2003, (6): 9-11 (in Chinese)

[2] Liu J, Huang H, Fu H, *et al*. Physiological mechanism of high and stable yield of no-tillage cast-transplanted rice. *Sci Agric Sin*, 2002, 35: 152-156 (in Chinese)

[3] Yang W Y, Ren W J. The new technology of broadcasting in the field with high standing-stubbles under no-tillage condition. *Sichuan Agricultural Sciences and Technology*, 2000, (4): 13-14 (in Chinese)

[4] Liu D Y. The results and experience about extension of no-tillage cast-transplanted rice in Sichuan province. *China Rice*, 2006, (1): 54-55 (in Chinese)

[5] Ren W J, Liu D Y, WU J X, *et al*. Effect of broadcasting rice seedlings in the field with high standing-stubbles under no-tillage condition on yield and some physiological characteristics. *Acta Agronomica Sinica*, 2008, 34(11): 1994-2002 (in Chinese)

[6] Peiris M E. Broadseedling- a promising new technique in paddy cultivation. *Trop Agriculrist*, 1956, 112: 105-108

[7] Zhang H C, Dai Q G, Zhong M X, *et al*. Studies on the yield formation and ecological characters of scattered-planting rice. *Sci Agric Sin*, 1993, 26(3): 39-49 (in Chinese)

[8] Jiang L G, Li R P, Wei S Q, et al. Root growth and stanging characteristics of Jinyou253 seedlings under no-tillage with cast transplantation. J Guangxi Agric Biol Sci, 2005, 24(1): 30-34 (in Chinese)

[9] Liu J Z, Li Y K. Studies on the no-tillage and seedling-throwing culture in hybrid rice. Hybrid rice, 1999, 14(3): 33-34 (in Chinese)

[10] Ren W J, Yang W Y, Liu D Y, et al. The technology of broadcasting in the field with high standing-stubbles under no-tillage condition. China Rice, 2003, (2): 22-23 (in Chinese)

[11] Ren W J, Yang W Y, Fan G Q, et al. Effect of different tillage and transplanting methods on soil fertility and root growth of rice. J Soil Water Conserv, 2007, 21: 108-110,162 (in Chinese)

[12] Lu R-K. Analytical Methods of Soil and Agro-Chemistry. Beijing: China Agricultural Science and Technology Press, 2000 (in Chinese)

[13] Department of Microbiology, Institute of Soil Science, Chinese Academy of Sciences. Analytical Methods of Soil Microbe. Beijing: Science Press, 1985 (in Chinese)

[14] Zhu Q S, Wang Z Q, Zhang Z J, et al. Study on indicators of grain-filling of rice. J Jiangsu Agric coll, 1995, 16(2): 1-4 (in Chinese)

[15] Doran J W, Parkin T B. Defining and assessing soil quality. In: Doran J W, Coleman D C, Bezdicek D F and Stewart B A ed. Defining Soil Quality for a Substainable Environment. Madison, W I, USA: SSSA Special Publication Number 35, 1994, 3-21

[16] Papendick R.I., Parr J.F., and J. van Schilfgaarde. Soil quality: New perspectives for a sustainable agriculture. In Proceedings for International SoilConservation Organization. New Delhi, India, December 4-8, 1994

[17] Turco R.F., Kennedy A.C., Jawson M.D. Microbial indicators of soil quality. In J.W. Doranet al. (ed.) Defining soil qualityfor a sustainable environment. SSSA Spec. Publ. 35. SSSA, Madison, WI. 1994, 73-90

[18] Morris S.J., and Boerner R.E.J. Spatial distribution of fungal and bacterial biomass in southern Ohio hardwood forest soils: Scale dependency and landscape patterns. SoilBiol. Biochem. 1999, 31:887-902

[19] Gao M, Zhou B-T, Wei C-F, et al. Effect of tillage system on soil animal, microorganism and enzyme activity in paddy field. Chinese Journal of Applied Ecology, 2004, 15(7): 1177-1181 (in Chinese)

[20] Balesdent J, Mariotti A, Boisgontier D. Effect of tillage on soil organic carbon mineralization estimated from abundance in maize fields. Soil Science, 1990, 41: 587-896

[21] Wei C-F, Gao M, Che F-C, et al. A study on infiltrated-ridged paddy soil ecosystem. Chinese Journal of Ecology, 1993, 12(3): 26-30 (in Chinese)

[22] Fan J-H, Liu M. Effect of different utilization methods on microorganism and its activation. Journal of tarim universivty of abricultural reclamation, 2002, 4(1): 15-17 (in Chinese)

[23] Luo A-C, Zhang Y-S. Effect of organic manure on the numbers of microbes and enzyme activity in rice rhizosphere. *Plant nutrition and fertilizer science*, 1999, 5(4): 321-327 (in Chinese)
[24] Xiao Q-Y, Ren W-J, Yang W-Y, *et al.* Effect of cultivation method of broadcasting rice seedlings in the field with standing-stubbles under no-tillage condition on senescence characteristics of leaves during late stages of rice development. *Acta Agronomica Sinica*, 2009, 35(8): 1562-1567 (in Chinese)

Part 2

Organic and Inorganic Amendments for Soil Fertility Improvement

6

The Potential of Organic and Inorganic Nutrient Sources in Sub-Saharan African Crop Farming Systems

Felix K. Ngetich, Chris A. Shisanya, Jayne Mugwe,
Monicah Mucheru-Muna and Daniel Mugendi
Kenyatta University,
Kenya

1. Introduction

Soil nutrient balance studies in Africa show evidence of widespread nutrient mining (Speirs and Olsen, 1992). Amount of nutrients annually taken away in the form of harvested crops, crop residues transferred out of fields or lost through leaching, erosion and volatilization are higher than the amount of nutrient inputs through fertilizers, deposition and Biological N_2 fixation (Smaling and Braun, 1996). Soil nutrient mining has been estimated to average 660 kg of nitrogen (N), 75 kg of phosphorus (P) and 450 kg of potassium (K) per hectare per year during the last 30 years from about 200 million hectares of cultivated land in 37 countries in Africa (Smaling et al., 1997). Continuous nutrient depletion and low soil fertility has not only led to the development of integrated soil fertility management technologies that offer potential for improving soil fertility in Africa (Scoones and Toulmin, 1998), but almost simultaneously triggered extensive studies on nutrient balance in various African farming systems. Thus, this chapter reviews the potential of organic and inorganic nutrient sources to counteract nutrient mining in Sub-Saharan African (SSA) crop farming systems.

Lack of plant nutrients is one of the principal causes for low agricultural productivity and food insecurity in Africa (Sanchez, 2002). In SSA, smallholder farmers have been experiencing declining agricultural productivity, mostly due to soil fertility depletion, leading to food insecurity. The impacts of smallholder-induced nutrient depletion express themselves in form of continued declines in crop yields, which can be abrupt or gradual depending on soil type (Bekunda et al., 2010). Low and declining soil fertility arises from continuous cultivation where levels of soil replenishment, by whatever means, are too low to mitigate the process of soil nutrient mining, whereby the soil fertility is not restored by new inputs (Shisanya et al., 2009). Intensively cultivated highlands in East Africa lose an estimated 36 kg N ha^{-1} yr^{-1}, 5 kgP ha^{-1} yr^{-1}, and 25 kg K ha^{-1} yr^{-1}, while croplands in the Sahel decline by 10, 2, and 8 kg ha^{-1}, respectively (Bekunda et al., 2010). Hence, decline of soil fertility is seen as the most important constraint to crop production in SSA, where most agro-ecosystems remove more nutrients than are provided by external inputs making it a fundamental biophysical root cause for declining food security in the smallholder farms (Sanchez and Jama, 2002). As a result, it is widely appreciated that protecting and improving the soil makes economic and social sense (Bekunda et al., 2010).

Fertilizer use in the SSA region is very low compared with the world average, despite many African farmers being aware of the beneficial contribution of mineral fertilizers to crop production. Average rate of inorganic fertilizer use in SSA, excluding South Africa, is 10 kg ha^{-1} as compared to 87 kg ha^{-1} for the developed countries (Bationo et al., 2006). With 9% of the world's population, SSA accounts for less than 1.8% of the global fertilizer use and less than 0.1% of global fertilizer production (FAO, 2008). This laxity is attributed to farmers' lack of confidence in the economic returns of fertilizing food crops, and lack of knowledge as to which kinds and rates of fertilizers are recommended for their specific crops, soils, and agro-climatic conditions (Vlek, 1990). Further, due to high price of imported fertilizers at farm gate and delays in delivery due to poor infrastructure, smallholder farmers often apply very low rates of inorganic fertilizer late in the growing season, leading to poor crop-yield responses (Heisey and Mwangi, 1996).

Sustaining the nutrient base calls for measures that combat the loss of nutrients. These include: application of modest amounts of the appropriate fertilizer, complying with recommendations that are specific to both crop and agro-ecological zones; efficient use of animal manure and household waste; adopting more nitrogen-fixing species in cropping systems, leaving residues as mulch or ploughing them into the soil, properly timed or split application of mineral fertilizers to combat leaching; and appropriate tillage and soil conservation measures to combat erosion (Smaling et al., 1992).

2. Nutrient flows and balances in SSA crop farming systems

The use of nutrient audits and nutrient balances to assess the changes in soil nutrient status and the prospects for future food production is becoming increasingly important (Sheldrick et al., 2003). Nutrient flows and balances are currently and increasingly being used as powerful tools for estimating nutrient depletion or accumulation. The relevance of soil nutrient balances to agricultural potential of land has been emphasized by many scientists (Stoorvogel et al., 1993; Smaling, 1993; Van den Bosch et al., 1998). Soil nutrient flow is the amount of plants nutrients that flow in and out of a system (Nkonya et al., 2004). The difference between nutrient inflow (sum of nutrient inputs) and outflow (sum of nutrient outputs) is the nutrient balance. Nutrient output flows comprise removal of economic crop products and crop residues, leaching, gaseous losses, runoff and erosion. A situation where inputs exceed outputs is termed as surplus nutrient accumulation; when outputs exceed inputs, this is nutrient depletion. Negative nutrient balances indicate that a system is losing nutrients; on the contrary, excess nutrient accumulation may lead to extended losses as a result of toxicities. The net difference between inputs and outputs of nutrients expressed in kg nutrients integrated over a certain area and time gives the net soil nutrient budget (Stoorvogel and Smaling, 1990). Balanced or equilibrium nutrient levels occur when inputs equals outputs. Hence, a summary of nutrient inputs and outputs from a defined system over a defined period of time is the nutrient budget for that spatio-temporal unit (Oenema et al., 2003).

The concept of nutrient depletion is derived from quantifying nutrient flows resulting in nutrient balances and/or stocks (Vlaming et al., 2001). In assessing nutrient depletion through use of nutrient balances at any given time, a number of inputs and outputs are considered. Nutrient input processes are: application of mineral fertilizer and organic manure, atmospheric deposition, Biological N_2 fixation and sedimentation by irrigation and flooding. Even though a number of nutrient flows can easily be quantified and valued in monetary terms based on the input and output processes (partial balance), other flows are hard to quantify hence often estimated on the basis of transfer functions, for instance those developed for sub-Saharan Africa by Stoorvogel et al., (1993), adopted and critically reviewed by others (Smaling, 1998; Scoones and Toulmin, 1998; FAO, 2003). A partial nutrient balance considers direct nutrient inflows from mineral fertilizers and organic materials and outflows resulting from harvested products and crop residues (Harris, 1998).

Nutrient budgets are important indicators of potential land degradation, for optimizing nutrient use, and designing policy to support improved soil fertility management by smallholder farmers. They have been used extensively for improving natural resource management and/or for policy recommendations over the last decades (Grote et al., 2005). At a relatively small scale, nutrient budgeting can be used to assess the level of nutrient sources and flows, opportunities for improved use efficiency and scope for possible interventions.

Most studies for SSA point at declining nutrient stocks whereby outputs exceed inputs hence raising question on the impact of declining nutrient stocks on agricultural production in the region (Smaling, 1998). Regional and national estimates of nutrient balances are negative in most of sub-Saharan Africa. Consistently negative nutrient balances, be it full or partial, have been reported. The negative balances can be attributed to the several nutrient outflow channels especially through harvest and soil erosion. For instance, Stoorvogel et al. (1993) estimated annual N losses from arable land to be 31 kg ha^{-1} in Zimbabwe, 68 kg ha^{-1} in Malawi, 112 kg ha^{-1} in Kisii, Kenya and 27 kg ha $^{-1}$ in Tanzania. Similar results have been found in Mali, where aggregate N losses are reported for all districts, except the cotton growing zone where imports of inorganic fertilisers compensate for harvest exports and other losses (Van der Pol and Traore, 1993). For Africa as a whole, low level of inputs relative to outputs results in a consistently negative balance (Stoorvogel and Smaling, 1990; Stoorvogel et al., 1993).

3. Nutrient inputs in soils of SSA crop farming systems

From the previous discussion, it is apparent that crop yields cannot increase nor can yields be sustained over time without adequate levels of soil fertility and nutrient inputs are needed on most soils in SSA, due to their low inherent fertility. Sustained soil fertility replenishment in Africa requires increased use of both inorganic fertilizers and organic manures (Palm et al., 1997). Successful design and implementation of soil fertility replenishment initiatives in Africa requires an understanding of basic rationale of small-scale farming on the continent (Omamo et al., 2002). To achieve sufficient levels of nutrients in most soils, organic and inorganic fertilizers must be applied, and in any locality, the optimal mix between the two will depend on their availability and water supplies (Larson 1996). The most common nutrient inputs in SSA crop farming systems consists of inorganic fertilizers, Biological N_2 fixation, farmyard and animal manure and agroforestry trees.

3.1 Inorganic fertilizers

Whereas in developed world, excess applications of fertilizer and manure have damaged the environment, low use of inorganic fertilizer is one of the main causes of environmental degradation in Africa (Bationo et al., 2006). Except for countries where governments subsidize fertilizers use in cereal production, inorganic fertilizers account for about one-third of inputs in Africa (Smaling, 1993). However, they are used largely in mechanized agriculture and on export crops. By the turn of the century, fertilizer use in Africa was only 8 kg/ha, compared with 96 kg/ha in East and Southeast Asia and 101 kg/ha in South Asia (Morris et al., 2007). At the present time, Africa accounts for less than 1% of global fertilizer consumption (Denning et al., 2009).

Millions of smallholder farmers throughout Africa use fertilizers, most of which are imported. Limited use of fertilizers is determined by a variety of reasons including high costs, especially after market reforms removed subsidies, inefficient marketing systems, and restricted markets for outputs that constrain investment opportunities (Bekunda et al. 2010). Because of the high price of imported fertilizers at farm gate and delays in delivery due to poor infrastructure, smallholder farmers often apply very low rates of fertilizer and very late in the growing season, leading to poor crop-yield responses (Heisey and Mwangi, 1996).

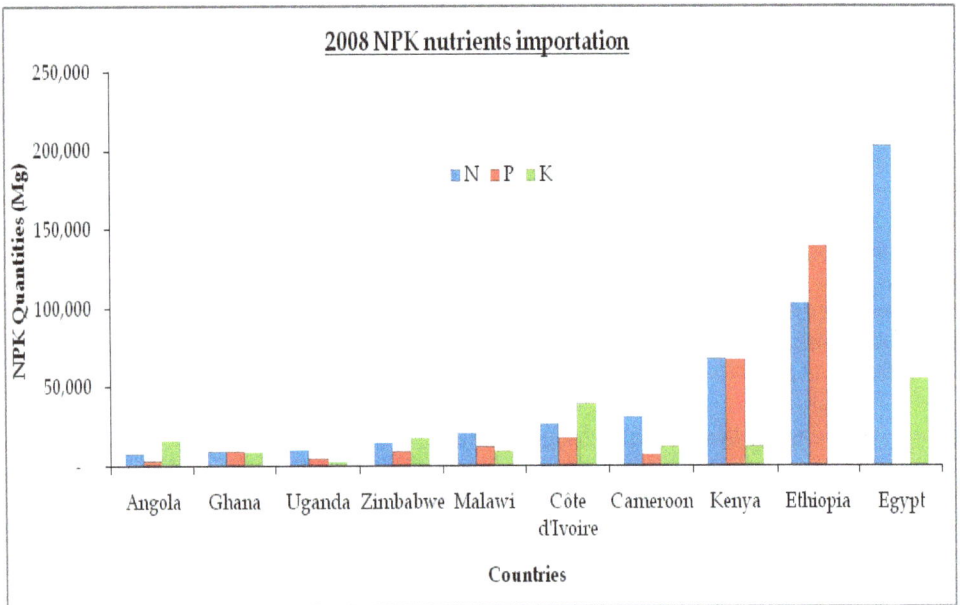

Fig. 1. Quantities of NPK nutrients imported by various African countries in 2008 (Source: FAO 2011)

Overall, inorganic fertilizer use in Africa is very low compared to other parts of the world. Based on the amount of N, P and K imported in the year 2008, Egypt leads in the amount of N imported followed by Ethiopia, and Kenya was in the third position (Fig 1). Ethiopia was the highest importer of phosphatic fertilizers, followed by Kenya while Côte d'Ivoire was third. Egypt imports negligible amounts, or in some years nil amounts, of P fertilizers but at the same time they are leading importers of K. The trend observed in the 2008 nutrient importation is on average identical to that observed as early as 2003. Given the high population in Ethiopia and Egypt, the large amounts of nutrient importations might not necessarily imply higher fertilizer consumption compared to other countries. Also, in Egypt, the crop production is mainly irrigated farming system implying high mechanization and agricultural intensification that facilitates better utilization of external inorganic inputs compared to other SSA countries where they depend on rainfed subsistence farming systems. Studies in Eastern and Central Kenya, for example, have shown that inorganic fertilizer application rates are sometimes as low as 5 kg ha^{-1} of N, P and K (Gakuo et al., 1996). With donor driven liberalization policies of 1990s in many African states, subsidies on fertilizer purchase and distribution were abandoned and the private sector was encouraged to take over supply systems, leading to high costs and sometimes shortages in the market (Winpenny et al., 1995), which dramatically affected nutrient management practices for some time. These economic disincentives to the use of inorganic fertilizers resulted to reduction in agricultural productivity. The trend has however been reversed with the introduction of smart subsidies by most African governments like Malawi, Kenya among others and the prospects of increase in the use of mineral fertilizers is promising. If average application rates of inorganic fertilizer in sub-Saharan Africa rose to 50 kg/ha, there can be substantial impact on agricultural yields (Larson 1996). There is therefore need to increase average fertilizer application rates in sub-Saharan Africa from the current10 kg/ha to 50 kg/ha, which can lead to substantial increase in agricultural productivity in the region. Without substantial additions of nutrients in the form of inorganic fertilizers, the conditions created through improved soil conservation (Larson 1996), improved crop varieties and field management practices cannot be exploited for sustained yield gains. Organic inputs, improved crop rotations, and soil conservation measures can all contribute significantly to many, but not all of Africa's plant nutrient requirements. Increased productivity will require that organic manures be supplemented with nutrients from inorganic sources for crops to receive a proper balance of nutrients. Therefore, soil fertility management must combine biological nutrient sources with inorganic fertilizer to maintain sustainable production

3.2 Biological N$_2$ fixation

Substantial nitrogen input into global agriculture comes from N$_2$ fixation. It has been estimated that global rate of nitrogen fixation has doubled during the last few decades, mostly through agricultural activities such as fertilizer manufacturing and increased use of dinitrogen (N$_2$) fixing crops (Galloway et al., 1995). About 60% of the total N demand by a nodulating leguminous crop can be supplied through symbiotic N$_2$ fixation (Stoorvogel & Smaling, 1990). Biological N$_2$ fixation becomes an N input after atmospheric N$_2$ gas is converted into plant N by symbiotic plants and the whole plant or its residues are incorporated into the soil. Potentially, this is an important N input into tropical African agro-ecosystems where legumes are a major component of the cropping systems (Giller and Cadisch, 1995). With judicious management of biological N$_2$ fixation in traditional cropping systems, most of the N required to maintain agricultural productivity in SSA can be achieved. The incorporation of nitrogen fixing legumes into cereal cropping systems either as intercrop or in rotation is common practice throughout SSA. It offers considerable benefits because of the legumes' ability to fix atmospheric nitrogen biologically in symbiosis with the *Rhizobium* bacteria. The contribution of N from legumes to the soil N supply can offset or reduce the amount of additional N fertilizers required for high cereal yields. It has been observed that Biological N$_2$ fixation from legumes can sustain tropical agriculture at moderate levels of output (Giller et al., 1997; Giller, 2001). According to Giller (2001), biological N$_2$-fixation can contribute as much as 300 kg N ha^{-1} in a season in grain legumes while Harris (1998) estimated nitrogen fixation by legumes to contribute up to 48 kg N ha[-1] within individual fields, depending on the density of legumes in the field.

About 60% of the total N demand by a nodulating leguminous crop can be supplied through symbiotic N_2 fixation (Stoorvogel & Smaling, 1990). Several factors such as high cost have collectively limited N fertilizer use by subsistence and small-scale farmers throughout Africa. Consequently, most of the N required for agricultural productivity in the continent comes largely from a judicious management of biological N fixation mostly through cereal-legume intercropping systems. Due to this importance, biological N_2-fixation by legumes has been extensively researched as a potential source of N for many smallholder farming systems (Giller, 2001). It has substantially contributed to promotion and adoption of cereal-grain legume cropping systems either as intercrops or in rotation as a means of soil nitrogen enhancement strategy.

Even when legumes grow well, the contribution to soil fertility depends on the amount of N_2 fixed in relation to the amount removed from the system in the crop harvest, reflected in the N harvest index (Giller & Cadisch, 1995). Multi-purpose varieties of grain legumes such as cowpea and soybean produce a lot of biomass while still giving reasonable grain yields (Mpepereki et al., 2000). For an equivalent grain yield of 2 ton ha[-1], promiscuous soybean varieties can produce up to 150 kg N ha[-1] in stover compared with new varieties that have about 40 kg N ha[-1] in their stover (Sanginga et al., 2001). Sanginga et al., (2004) observed doubling in yields of maize grown after the promiscuous varieties compared to those of maize grown after the new varieties. Despite the benefits, the system is faced with various challenges such as inadequate soil moisture, soil fertility status, rhizobia inoculants production, quality, storage and distribution related issues that affect N_2 fixation of field legumes.

Biological N_2 fixation process is probably the cheapest and most effective tool for maintaining sustainable yields in African agriculture (Dakora & Keya, 1997). Therefore, there is need for comprehensive estimation of N balances of various leguminous crops to help guide their adoption and integration into legume-cereal rotation or intercrop systems.

3.3 Animal and farmyard manure

In SSA, animal and farm yard manure are important and frequently used resources to improve soil fertility. For instance, farmyard manure is common in central highlands of Kenya and has been estimated that more than 95% of smallholder farmers growing maize use it (Harris, 1998). Baijukya et al., (2005) reported that manure production in some farming systems was a major reason indicated by smallholder farmers for keeping cattle, whereas in other systems, such as arid areas of Zimbabwe, manure is a potential resource for nutrient recycling that is hardly used (Mapfumo and Giller, 2001). Studies have shown that use of manure can improve crop yields considerably (Lekasi et al., 1998). Crop yield responses to manure can be seen in crops for several years after application when the manure is supplied in sufficiently large amounts (Mugwira and Murwira, 1997).

On most farms, animal manure used is mainly from cattle (65%) with the rest coming from diverse sources such as sheep and goats (6%) and poultry (4%) (Lekasi et al., 2001). The amount and quality of manure available to a farmer depends on many factors: source, herd size and management system and seasonal climatic changes, which determine availability of feeds to livestock (Mugwira and Murwira, 1997). Given that external/free range grazing is the predominant livestock feeding system, the quality of manure with regard to nutrient release and crop uptake is mostly poor, posing a challenge to smallholder farmers. For instance, Dejager et al. (1998) reported an average N balance per head of different livestock activities where cattle in zero grazing system had 11 kg head^{-1}, cattle in semi-grazing had 7 kg head^{-1}, cattle in external grazing system had 3.7 kg head^{-1} while poultry had 0.2 kg head^{-1}. Also, Baijukya (2005) reported N as an inflow from cattle manure ranging from 0 to as high as 63 kg ha^{-1} yr^{-1} in a banana-based farming in Bukoba, Tanzania.

The variability in quality can also be attributed to management of manure produced (Mugwira and Murwira, 1997). For example, Lekasi et al. (1998) reported that when manure was stored in heaps or in pits until application, the buried manure had substantially greater contents of N. Similarly, Mugwira and Murwira (1997) reported that manures stored in pits had significantly higher N content compared to the manure stored in heaps. This could be due to large ammonia losses that occur throughout decomposition period, which was associated with alkaline conditions possibly due to mineralization of N and microbial NH_4^+ consumption under aerobic conditions. Quality of manure can be improved through provision of high quality feed such as calliandra and leucaena to animals and through proper management of manure (Delve, 2000). Such manure can provide substantial amounts of nutrients into the agro-ecosystem. For instance, in intensively managed smallholder farms in Kisii District of Kenya, application of manures to fields from cattle enclosures average 23 kg N ha^{-1}, or about one-third of total N inputs (Smaling, 1993).

Besides the soil fertility amelioration perspective, the contribution of manure in building up soil organic carbon (SOC), and total soil carbon stocks, is an important extra benefit of using manure as a soil amendment (Rufino et al., 2006). This is especially critical given that most SSA soils are inherently low in organic carbon (<20 to 30 mg/kg) due to: low root growth of crops and natural vegetation, continuous cultivation of crops and rapid turnover rates of organic materials with high soil temperature and microfauna (Bationo et al., 2006; Bationo et al., 1995). The loss of organic matter consequently results in soil acidity, nutrient imbalance and low crop yields (Ayoola et al. 2007). Based on field experiments, there are indications that additions of 5-10 t ha $^{-1}$ year $^{-1}$ of manure are sufficient to maintain SOC close to contents of soil under undisturbed savanna vegetation in West Africa (de Rouw and Rajot, 2004). In an experiment in Saria, Burkina Faso (Mando et al., 2005), additions of 2 tons of C ha^{-1} year $^{-1}$ to plots cropped with sorghum (Sorghum bicolor L.) and hand-hoed over a 10 year period resulted in a net increase of approximately 3.5 t ha^{-1} to the SOC pool.

Although manure is important for resource poor farmers in improving crop and soil productivity, drawbacks exists in its use as a sole source of nutrients for plants given that the quantities produced at farm level are more often inadequate to meet nutritional demands for various crop enterprises (Makokha et al., 2001). Integrated soil fertility management (ISFM) approach, which advocates for the combined use of organic residues and mineral fertilizers, has potential to resolve this practical limitation of input availability, and may also benefit crop-nutrient synchrony and nutrient loss reduction through interactive effects between both types of inputs (Gentile et al., 2009). High crop yield can be obtained with judicious and balanced inorganic fertilization combined with organic matter amendments (Palm et al., 1997).

There has been little research on manure handling and manure storage in Africa where most studies of soil N mineralization from manures comprise mainly laboratory incubations, with very few field experiments (Rufino et al., 2006). Limited research has been carried on the effects of manure on soil water retention and agricultural productivity. Potential benefits of animal and farmyard manure not only in improving soil fertility status but also boosting soil moisture retention, especially during dry spells, holds promise to address and correct water stress effects in smallholder farming systems.

3.4 Green manure

Green manures play a key role in providing subsequent crops with nutrients, maintaining soil quality, and helping to control weeds and pests (Krauss et al., 2010). Green manuring involves spreading plant material with high nitrogen content on fields and sometimes also working it into the soil (van der Werff et al., 1995). The most important features of a green manure are large dry matter production and high ability to fix nitrogen (Wivstad, 1997). Green manure can be used directly or after composting as a nutrient input that will, after decomposition, be taken up by crops to produce biomass and grain. The effectiveness of green manuring as a soil fertility management technology depends on the stage of crop during incorporation, placement (incorporated or surface placement), nutrient content and carbon to nitrogen ratio of material which influences decomposability and mineralization. Young and succulent plant materials tend to decompose at faster rate compared to materials from mature plants, while incorporation enhances mineralization relative to surface applied plant material. Using green manures from deep rooting perennial agro-forestry trees allows organic matter to be tapped and nutrients to be drawn from deeper levels of soil more than is possible from animal or annual plant manures (Larson 1996). For instance, use of Crotalaria as green manure has been found to improve productivity of maize-bean cropping systems in eastern Africa (Fischler et al., 1999).

As highlighted by Byerlee and Heisey (1992), green manures may not provide sufficient and balanced nutrients (such as phosphorus, potassium, etc) required by subsequent crop for improved productivity although they are a good source of nitrogen. The key to achieving the maximum benefit from green manure is synchronization of nutrient release from decomposing green manure with demands of subsequent crop. In addition to moisture, temperature and aeration of soil, rate of nutrient release from green manure debris is influenced by quantity, quality, placement and degree of incorporation into the soil (Fischler et al., 1999).

Herbaceous green manure legumes like mucuna grown specifically for soil fertility restoration have not been widely adopted by farmers in SSA. Lack of a direct usable food product is the principal disincentive in farmers readily adopting green manuring in most SSA farming systems. Given that green manures compete for land resources with other food crops, and do not contribute directly to income or food security (Snapp et al., 2002) and given that land is very limited, it pauses a challenge as this as this is a soil fertility management option that might not always or readily fit in farmers cropping cycles. A potentially practical solution to this challenge is utilization of alternative plants such as using water hyacinths. Gunnarsson and Petersen (2007) in a review entitled "water hyacinths as a resource in agriculture and energy production" concluded that dried water hyacinths was a feasible alternative as green manure in many developing countries since hyacinths can be rich in nitrogen, have up to 3.2% of dry matter and have a carbon to nitrogen (C/N) ratio of about 15. Chemical analyses have indicated a high nutrient content of water hyacinth, 20% crude protein and very high dry matter production (Abdelhamid and Gabr, 1991), making water hyacinth a potentially suitable alternative to traditional green manure crops.

3.5 Agroforestry trees

Agroforestry is a soil fertility enhancement system that combines agricultural and tree crops of varying longevity, arranged either temporally or spatially, to maximize and sustain agricultural yield and minimize degradation of soil and water resources (Lal, 1990). According to Nair (1989), agroforestry can be classified into various systems and practices such as: (i) agrisilvicultural system characterized by improved fallows, biomass transfer, the taungya/shamba system, hedgerow intercropping, tree gardens, trees/shrubs on farmlands, shelterbelts, soil conservation hedges, (ii) silvopastoral systems involving cut and carry fodder banks, live fences of fodder trees and hedges, trees and shrubs on pasture land, and (iii) agrosilvopastoral systems comprising of woody hedges for browse mulch, green manure and soil conservation among other systems such as apiculture, aqua-forestry etc. Ecological interactions between trees and crops in agroforestry system are beneficial because: leguminous trees have a beneficial effect on soil fertility through nitrogen fixation, greater organic matter production, and recycling of nutrients, a combination of annual crops and trees increases biomass production because differences in rooting depth enable uptake of more water and nutrients (Young, 1986). Finally, agroforestry trees act as a protective barrier against soil erosion and as windbreaks.

The most common agroforestry systems in most smallholder farms in SSA were designed for soil improvement, particularly for nutrient recycling and as a nutrient source for cropland. Among those that have received the most attention are alley cropping, improved fallows and biomass transfer.

3.5.1 Alley cropping

Alley cropping is an agro-forestry practice in which perennial crops are grown simultaneously with an arable crop (Alary et al., 2007). The perennial crops provide ecosystem functions (food production, nutrient cycling, erosion control, water conservation, habitats for other biota, etc.) The harvested aboveground biomass is either incorporated into the soil or applied as mulch on the soil surface. The potential of alley cropping to improve yields depends on the nutrients added or recycled through hedgerows if the local constraint is nutrient supply, and on the ability to retain soil water by increased infiltration and reducing runoff if the crop production is constrained by water availability (Mathuva et al., 1998). Alley farming has been successfully demonstrated in Benin and produced best yield, compared to annual legume cover planting and maize groundnut rotational farming (Versteeg et al., 1998). Shisanya et al (2009) in their study on the "effect of organic and inorganic nutrient sources on soil mineral nitrogen and maize yields in central highlands of Kenya" recommended that farmers in the area should be encouraged to incorporate non-competitive fast growing trees in their farming systems that would assist in 'capturing' the leached nutrients and recycling them back to the system. The net benefit of the system to crop production is however determined by the extent of competition between hedges and crops and by potential improvements in soil nutrient and water status (Mathuva et al., 1998).

3.5.2 Improved fallows

Traditionally, farmers have relied on long fallow periods to replenish the soil fertility depleted through cropping, but due to declining farm sizes, increase in human and animal population and land-use pressure, fallows have been reduced both in length and area or even abandoned in many farming systems (Kaya et al. 2000). In order to overcome the limitations and still enjoy benefits of natural fallows, improved fallows have been designed to achieve the same results as natural fallows in shorter time periods through purposeful selection and management of leguminous plants. Improved fallows are the deliberate planting of fast-growing trees, shrubs, and herbaceous legume species for rapid replenishment of soil fertility (Sanchez, 1999). Unlike the natural grass and bushy fallows, improved fallows consist of deliberately planted species, usually legumes, with the primary purpose of fixing nitrogen as part of a crop-fallow rotation (Sanchez, 1999). Short rotation improved fallows using various leguminous crops have been shown to replenish soil fertility (Kwesiga et al., 1999). Such crops include: sesbania (*Sesbania sesban*), pigeon pea (*Cajanus cajan*), tephrosia (*Tephrosia vogelii*), gliricidia (*Gliricidia sepium*) and Leucaena (*Leucaena leucocephala*) (Mafongoya & Dzowela, 1999) and increase agricultural productivity. Several studies in SSA countries have evaluated and ascertained the potentials of improved fallows for soil fertility replenishment (Palm and Sanchez, 1991; Mugendi and Nair, 1997). Legume fallow technologies improve nutrient cycling by enhancing availability of nutrients resulting from production and decomposition of substantial quantities of biomass, and deep roots of planted trees or shrubs acting as "safety nets" for nutrients, reducing leaching losses (van Noordwijk et al., 1996). Leguminous fallows accumulate nitrogen in the biomass, recycle other nutrients in the soil, and smother weeds (De Rouw, 1995). On nutrient-deficient sites, additional nitrogen, phosphorus and potassium supply from leguminous biomass may markedly improve crop vigour and resistance to pests and diseases (Schroth et al., 2000). For example, in southern Benin, improved tree fallow of *Acacia auriculiformis,* and a short season fallow of mucuna *(Mucuna puriens)* have been successfully applied to address soil fertility problems (Versteeg et al., 1998).

Merits of improved fallows include the diversity of farm sizes where improved fallows are used, the advantage of sequential versus simultaneous systems, the ability of fallow crops to grow during dry seasons that are unfavourable for crop production, the comparative advantages of woody versus herbaceous leguminous fallows, the magnitude of N accumulation, and the strategic use of N fertilizers. Other key services provided by fallows include fuel wood production, recycling of other nutrients besides N, provision of a C supply to soil microorganisms, weed suppression, Striga control, and improved soil water storage (Sanchez, 1999). Total farm production can be greater with improved fallow-crop rotations than with continuous cropping even though crop production is skipped for one or more seasons with improved fallows (Sanchez & Leakey, 1997). This is especially true for low input, family-based smallholder farming systems that are characterized by limited resources, technology and information, and rely on subsistence rather than producing surpluses for market. Although the overall productivity might be equal or slightly lower than that achievable with inorganic fertilizers, the improved fallow strategy is still a feasible option to resource poor smallholder farmers who operate in fragile environment where their integrated activities are strongly constrained by socioeconomic, biophysical and institutional factors.

Although large yield responses by subsequent crop have been reported after 9 to 18 months of improved monoculture-species fallows, with substantial residual benefits for the following crops (Jama et al., 1998), efficiency of improved fallows can be further enhanced through mixing species of leguminous plants. For example, mixed-species of improved legume fallows have potential to provide maize yields comparable or even above those obtained with 100 kg N ha⁻¹ of mineral N fertilizer for at least two cropping season (Ndufa et al., 2009). Sileshi and Mafongoya (2003) recommend the use of mixed-species fallows of sesbania, pigeon pea and tephrosia for improvement of maize production in areas with nutrient-deficient soils in Zambia rather than crotalaria (*Crotalaria grahamiana*) (monocrop) fallows. Furthermore, mixing high quality fallow residues with polyphenol-rich legume material (e.g., from calliandra) may also result in the formation of complexes with proteins and carbohydrates that are resistant to microbial degradation (Handayanto et al., 1997) and hence increase soil organic matter and long-term benefits of such systems (Cadisch and Giller, 2001). The increased soil organic matter can improve soil physical conditions, root development and thereby the water and nutrient status of subsequent crops (Mafongoya and Dzowela, 1999; Chirwa et al., 2002). According to Sanchez (1999), the main limiting factor in the spread/adoption of improved fallows in Africa is the supply of germplasm of improved fallow species and accessions that must be overcome through seed orchards and nursery development before large scale impact on soil fertility is realized.

3.5.3 Biomass transfer

Biomass transfer is one of the nutrient source input in smallholder farms in SSA although it is an internal flow in the nutrient balance and flow. This technology involves ex-situ production of biomass away from the cropping land in designated areas, such as hedges around or within the farm. The biomass is then harvested, transferred to cropping land and applied as mulch or incorporated into the soil before and/or after the crop is planted. In this technology, interaction between the components is only through the nutrient release from applied mulch and uptake by crop. Research on biomass transfer is done mainly by scientists with interest in agroforestry tree species. Studies have been conducted in many countries in Africa (Mugendi et al., 2004). Some of the plant species that have been used for biomass transfer include *Tithonia diversifolia, Calliandra calothyrsus, Leucaena trichandra* and *Sesbania sesban.* For instance, tithonia applied at 5 t ha⁻¹ on a dry weight basis contains about 60 kg N ha⁻¹ (Gachengo, 1996). Jama et al. (1999) reported an average of 3.5% N in green biomass of tithonia on a dry matter basis. This coupled with its ease of decomposition and releases of nutrients make it an important source of N for crops. Leguminous trees like calliandra and leucaena have also been widely used as a source of nutrients, with positive results on crop yields (Mugwe et al., 2009). These trees produce high amounts of biomass (Delve et al., 2000). Biomass transfer offer opportunities on increasing production due to increased nutrients but, except for N_2 fixing trees, it is usually not an addition to the farm unit. It is a matter of transferring N from one part of the farm to the other.

3.6 Integrated Soil Fertility Management

Integrated Soil Fertility Management (ISFM) refers to making the best use of inherent soil nutrient stocks, locally available soil amendments (for instance, crop residues, compost, animal manure, green manure), and inorganic fertilisers to increase productivity while maintaining or enhancing the agricultural resource base (IFDC, 2003; TSBF, 2003). It is a holistic approach to soil fertility research that embraces the full range of driving factors and consequences; biological, physical, chemical, social, economic and political, of soil degradation (Barrios et al. 2006). Strategically targeted fertilizer use together with organic nutrient resources to ensure fertilizer use efficiency and crop productivity at farm scale are basic principles of ISFM (Vanlauwe and Giller, 2006). Although ISFM recognizes the absolute necessity of mineral fertilizer use (Vanlauwe et al., 2010), it advocates the best combination of available nutrient management technologies that are economically profitable and socially acceptable to different categories of farmers (Vanlauwe, 2004). It is rapidly becoming more accepted by development and extension programs in SSA, and, most importantly, by small-holder farmers (Place et al., 2003). Beneficial effects of ISFM on soil fertility have been shown to increase nutrient use efficiency associated with combined nutritional and non-nutritional effects of organic and inorganic inputs compared to inorganic fertilizer applied alone (Fofana et al., 2005; Wopereis et al., 2005). ISFM is a market-led technology build on the hypothesis that better market opportunities provide incentives for farmers to invest in the technologies (Sanginga et al., 2007). This market-led hypothesis is increasingly becoming a key pillar not only in ISFM research (TSBF, 2005) but also other areas of integrated agricultural research for development (Sanginga et al., 2007). It is about expanding the choice set of farmers by increasing their awareness of the variety of options available and how they may complement or substitute for one another (Place et al. 2003).

Effectiveness and beneficial effects of ISFM practices are likely to differ from place to place in Sub-Saharan Africa, due to the many heterogeneous agroecological zones in the region (Kato & Place, 2011). Available reviews on effects (Place et al. 2003) are mixed and inconclusive given that limited empirical evidence exists on the potential of ISFM for improving yields and profitability that can be used to support the arguments for use of ISFM as an alternative to high doses of fertilizers to maintain favourable nutrient balances and soil quality (World Bank, 2007).

In spite of development and availability of excellent methods, technologies and a wide range of tools, large scale adoption of ISFM remains a hurdle (Pound et al., 2003). Challenges in implementation of ISFM technologies include differing perceptions between farmers', needs and research which is limited due to time constraints and other resources (Misiko & Ramisch 2007). For instance, the farmers' immediate need for food might override perceived long-term soil fertility benefits associated with ISFM. There is also complexity of diversity in different farm situations relative to adaptive methods. Problems are inherent in the fact that broader use of ISFM concepts requires scaling up of knowledge itself, which is not the case with the spread of more simple technologies or goods (Misiko & Ramisch 2007). This calls for intensive training of farmer and extension service providers on the principles of ISFM, innovation and experimentation to tailor generic management technologies to diverse prevailing local conditions.

4. Challenges in utilization of organic and inorganic nutrient sources

Overall growth in agricultural research investments in Africa has effectively stagnated over the past two decades (Beintema & Stads, 2006). This among other major impediments to improved soil fertility management in the SSA include low levels of farmers' human, physical and financial capitals, low agricultural commodity prices relative to fertilizer and other input prices, lack of pro-agriculture policies, and the failure to view the maintenance of soil fertility as an important public good (Smaling and Dixon, 2006).

4.1 Low external nutrient strategies

Organic fertilizers will continue to play an important role in maintaining soil structure, moisture control, and nutrient levels, yet under farm conditions, they cannot supply all of the additional nutrients needed to sustain rapid yield growth (Larson and Frisvold, 1996) and are mostly labour intensive. Use of livestock manure to replenish nutrients for agricultural production is a land-extensive production method. About 23 ha of pasture per 1 ha of cropland would be needed in the Sahel to produce enough manure just to maintain, but not improve upon, current levels of soil fertility (Speirs and Olsen, 1992). This level of manure production may be unrealistic in SSA. Adverse effects of deforestation and increased demand for land limit the possibility of extending various agroforestry systems, such as improved fallows and alley cropping, in densely populated areas. These concerns, coupled with the continuous nutrient depletion and low soil fertility, have led to the development of ISFM technologies that offer potential for improving soil fertility in Africa (Scoones and Toulmin, 1998). The ISFM approach advocates for the combined use of organic residues and mineral fertilizers, which may resolve the practical limitation of input availability, and may also benefit crop N synchrony and N loss reduction through interactive effects between both types of inputs (Gentile et al., 2009). High crop yield can be obtained with judicious and balanced inorganic fertilization combined with organic amendments (Palm et. al., 1997). Donovan and Casey (1998) showed that technologies that combine mineral fertilizers with organic nutrient sources can be considered as better options in increasing fertilizer use efficiency, and providing a more balanced supply of nutrients. More specifically, Mugendi et al. (1999) reported higher crop yields in Leucaena biomass combined with mineral fertilizer treatment compared to sole use of mineral fertilizer or sole leucaena biomass.

4.2 Environmental considerations

Excessive fertilizer use, whether organic or inorganic, can create environmental problems associated with leaching of nitrogen into groundwater and deposition of phosphorous in surface waters through soil erosion (Larson, 1996), especially in regions with continuous high levels of fertilizer applications (often over 200 kg/ha) and large livestock operations. Over-intensification is not a widespread problem in SSA and will not become one; rather, future predictions indicate that the main environmental concerns in agriculture in sub-Saharan Africa will stem from the lack of intensification (Larson, 1996). Bationo et al. (2006) argues that, whereas in the developed world, excess applications of fertilizer and manure have damaged the environment, the low use of inorganic fertilizer is one of the main causes for environmental degradation in Africa. They advocate for increased inorganic fertilizer use which will, in addition to increased productivity, benefit the environment by reducing the pressure to convert forests and other fragile lands to agricultural uses and, by increasing biomass production, helps increase soil organic carbon content. Past studies document that simple physical availability of fertilizers to farmers, in the appropriate quantity, packages and at the appropriate time of year, remains a main constraint on increased fertilizer use in SSA (Larson and Frisvold, 1996). A strategic approach recognizes that a large number of technical, social, environmental, and market conditions must be met to achieve sustained agricultural growth at a national or regional level.

5. Policies

Population growth, the limited land for extensive agricultural production and the reduced external aid to agriculture now calls for governments to better utilize the continent's internal resources for intensification (Bekunda, 2010). This calls for enactment of enabling and conducive policies that will promote the use of both organic and inorganic inputs. For example, Sanginga and Woomer (2009) explain that policies supporting traditional rights to free grazing contradict the establishment of land tenure and increasing land use intensification with ISFM. Communal rather than freehold land tenure systems in most semi-arid regions and other areas with low agricultural potential in SSA countries also inhibit adoption of organic and inorganic inputs.

It is apparent that, for a successful "African Green Revolution" soil health needs to be restored, through agroforestry techniques and use of organic and mineral fertilizers among other solutions. As pointed out earlier, low levels of inorganic fertilizer use is one of the major cause of soil fertility depletion. This calls for deliberate efforts, policies and legislations that will boost the use inorganic inputs at farm level. Therefore, a policy objective for SSA should be to increase fertilizer application rates significantly. It is necessary to increase average application rates of inorganic fertilizer to around 50 kg/ha, which would provide the foundation for further sustained agricultural growth.

Because of high price of imported fertilizers at farm gate, most small holder farmers, although aware of benefits of nitrogenous fertilizers, cannot afford cost of standard 50 kg bag of fertilizer. While there is no immediate substitute for nutrient supplied by inorganic fertilizer, there is need to improve the access and efficiency of its use.

Inorganic fertilizer demand can significantly improve, if a lower cost effective means of getting many small bags of fertilizer to farmer's doorstep were found. With good technical training and access to credit to maintain stocks, local retailers can stimulate inorganic fertilizer demand (FAO, 2007). In a number of countries, including Malawi, Nigeria, and Zambia, retailers repackage fertilizer into smaller packs, for which they typically charge a premium of 14–15 percent (World Bank, 2007). In Zimbabwe, agro-dealers retailers are encouraged to repackage fertilizers in smaller amounts that are less financially daunting to farmers (World Agroforestry Centre, 2008). This encourages farmers to use small amounts of fertilizers to increase production rather than being turned off completely by high cost of bulky and expensive packaged fertilizers. Although adding value for the retailers and improving convenience for small-scale farmers, this practice adds to already high retail prices. There have been a variety of success stories in this area since market reforms were introduced, mostly associated with selling fertilizers in small packages. For these "small-pack" programmes to be successful, governments and manufacturers often need to change legislation or company rules that prohibit breaking down larger packages (World Bank, 2007). For instance, Kenya repealed a law in the early 1990s that prohibited sales of fertilizer in less than 50 kg bags; by 1996, 46% of fertilizer sales to smallholders were in 10 kg bags (Kelly et al., 2003). Authorized re-bagging raises the issue of quality control. Re-bagging increases the possibility for both intentional and unintentional adulteration at the retail level, yet most verification of fertilizer quality tends to be done much higher up the supply chain. Part of the wide fluctuations in the nutrient concentration (quality) in fertilizers can be accounted for by the absence of effective measurement, quality control and calibration facilities at the retail level.

Regulations relating to the chemical composition of fertilizer are important to deter product adulteration, which can easily happen when fertilizer is re-bagged than it is with sales of standard 50 kg sacks. This calls for appropriate introduction of regulations establishing clearly defined assay standards for a standardized and limited set of fertilizers, as well as penalties for distributors whose products do not conform to those standards. Because regulations are meaningless if they cannot be enforced, the introduction of product quality controls will often need to be accompanied by supporting investments in facilities for carrying out rapid, low-cost testing (World Bank, 2007). It also calls for: regular inspection of fertilizer at all stages of the marketing chain, compound fertilizers formulations to meet location-specific plant nitrogen requirements, fertilizer pack sizes adapted to local needs and information on packaging in local languages. In the long run, farmers in all countries need to acquire the skills to evaluate their own situation and make informed judgments about the most appropriate doses and combinations of inputs; this implies significant improvements in basic education as well as in extension.

6. Conclusions and recommendations

Maintaining soil quality and improving yields in SSA calls for conscious moves. First, external sources of inorganic fertilizers and seed-fertilizer technologies play a major role and could also lead to steady yield increases in rain-fed agriculture. Another approach is to make better use of organic fertilizer sources and adopting low-input strategies such as improved crop rotations to maintain internal recycling of nutrients and soil fertility. These approaches are not mutually exclusive as each has its merits and limitations, therefore the best strategy would be to combine them subject to local economic and environmental circumstances. Additionally, fertilizers remain a key element for sustained agricultural development in SSA. It is apparent that inorganic fertilizers are a prerequisite for sustained growth, but not just a simple solution. Other changes in the production and policy environment, education levels, such as seeds, water management, pest control, land tenure, taxes and regulations, are also required. However, without adequate fertilizers, returns from these other changes will be small and unsustainable.

Sustainability of agroforestry systems is challenge that needs to be addressed by evaluation of productivity from long-term experiments in relation to evolution of soil properties and environmental quality. Although combining various organic and inorganic inputs improves agricultural productivity and soil physical properties, their interactive effects with soil moisture especially during dry spells needs to be further researched, not only in the light of soil fertility replenishment but also as a potential means of mitigating low soil water availability to alleviate water stress effects in smallholder farming systems.

Replenishing soil fertility in SSA is paramount in the quest of food security. It is a necessary but not sufficient condition, and must be pursued in tandem with policies that may involve: new legislation and their enforcement, improving health, education, governance, infrastructure, improving market information systems, offering attractive credit schemes, or generally becoming more enabling, allowing land users to produce in a more profitable and environmentally benign way (UN Millennium Project, 2005), and trade. It may also require policy change at the global level, especially at international trade level, allowing SSA to be part of global sustainable development by developing policies that reduce the disparity between the world market and the prices paid by African farmers for mineral fertilizers.

7. References

Abdelhamid, A. M. & Gabr, A. A. (1991). Evaluation of water hyacinth as feed for ruminants. *Archives of Animal Nutrition,* 41, 745-756.

Alary, V., Nefzaoui, A. & Jemaa, M.B. (2007). Promoting the adoption of natural resource management technology in arid and semi-arid areas: Modelling the impact of spineless cactus in alley cropping in Central Tunisia. *Agricultural Systems,* 94,573-585.

Ayoola, O.T. & Makinde E.A. (2007). Complementary Organic and Inorganic Fertilizer Application : Influence on Growth and Yield of Cassava/maize/melon Intercrop with a Relayed Cowpea . *Systems Research,* 1, 187-192.

Baijukya, F.P., de Ridder, N., Masuki, K.F. & Giller, K.E. (2005). Dynamics of banana-based farming systems in Bukoba District, Tanzania: changes in land use, cropping and cattle keeping. Agric. *Ecosyst. Environ.,* 106, 395-406.

Barrios, E. Delve, R.J., Bekunda M., Mowo, J. Agunda , J. Ramisch, J. Trejo, M.T. & Thomas a R.J. (2006). Indicators of soil quality: A South – South development of a methodological guide for linking local and technical knowledge. *Geoderma,* 135,248-259.

Bationo, A., Buerkert, A., Sedogo, M.P., Christianson, B.C. & Mokwunye, A.U. (1995). A critical review of crop residue use as soil amendment in the west African semi-arid tropics. In: Powell, J.M., Ferna´ndez-Rivera, S.,Williams, T.O., Renard, C. (Eds.), *Livestock and Sustainable Nutrient Cycling in Mixed Farming Systems of Sub-Saharan Africa. Proceedings of ILCA,* 22–26 November 1993, Addis Ababa, Ethiopia, pp. 305–322.

Bationo, A., Hartemink, A., Lungu, O., Naimi, M., Okoth, P., Smaling, E. & Thiombiano, L. (2006). African soils: their productivity and profitability of fertilizer use. In: *Proceedings of the African Fertilizer Summit.* June 9–13, 2006, Abuja, Nigeria, pp. 29.

Beintema, N.M. & Stads, G.J. (2006). Agricultural R&D in Sub-Saharan Africa: An era of stagnation. Background report, Agricultural Science and Technology Indicators (ASTI) Initiative. International Food Policy Research Institute, Washington DC, USA, pp 44.

Bekunda, B., Sanginga, N. & Woomer, P. L. (2010). Restoring Soil Fertility in Sub-Sahara Africa. *Advances in Agronomy,* 108,184-236.

Byerlee, D. & Heisey, P. (1992). Strategies for Technical Change in Small-Farm Agriculture, with Particular Reference to Sub-Saharan Africa. In Policy Options for Agricultural Development in Sub-Saharan Africa, eds N.C. Russell and C.R. Dowswell, pp. 21-52. Centre for Applied Studies. In *International Negotiations (CASIN)/Global 2000, Conference Proceedings,* Airlie House, Virginia, 23-25 August.

Cadisch, G. & Giller, K.E. (2001). Soil organic matter management: T e roles of residue quality in C sequestration and N supply. p. 97–111. In R.M. Rees (ed.) *Sustainable management of soil organic matter.* CAB Int., Wallingford, UK.

Chirwa, T.S., Mafongoya, P.L., Chintu, R. & Matibini, J. (2002). Mixed fallows of coppicing and non-coppicing tree species for degraded Acrisols in eastern Zambia. In: *Proceedings of the Agroforestry Impacts on Livelihoods in Southern Africa: Putting Research into Practice,* Warmbath, South Africa, 20–24 May 2002.

Cobo, J.G., Dercon, G. & Cadisch G. (2010). Nutrient balances in African land use systems across different spatial scales: A review of approaches, challenges and progress. *Agriculture, Ecosystems and Environment,* 136,1-15.

Dakora, F. D. & Keya, S. O. (1997). Contribution of legume nitrogen fixation to sustainable agriculture in sub-Saharan Africa. *Soil Biol. Biochem.,* 29, 809–817.

De Jager, A. Kariuki, I., Matiri, F.M. Odendo, M. & Wanyama, J.M. (1998). Monitoring nutrient flows and economic performance in African farming systems (NUTMON) IV . Linking nutrient balances and economic performance in three districts in Kenya. *Agriculture, Ecosystem and Environment* 71, 81-92.

De Jager, A., Onduru, D., van Wijk, M.S. Vlaming, J. & Gachini G.N. (2001). Assessing sustainability of low-external-input farm management systems with the nutrient monitoring approach: a case study in Kenya. *Agricultural Systems,* 69,99-118.

De Rouw, A. & Rajot, J.L. (2004). Soil organic matter, surface crusting and erosion in Sahelian farming systems based on manuring or fallowing. *Agric. Ecosyst. Environ.,* 104, 263-276.

De Rouw, A. (1995). The fallow period as a weed-break in shifting cultivation (tropical wet forest). *Agriculture, Ecosystems and Environment,* 54, 31–43.

Delve, R., Gachengo, C., Adams, E., Palm, C.A., Cadisch, G. & Giller, K.E. (2000). The Organic Resource Database. In: *The Biology and Fertility of Tropical Soils*: TSBF Report 1997- 1998. Pp 20-22.

Denning, G., Kabambe, P., Sanchez, P., Malik, A. & Flor, R. (2009). Input subsidies to improve smallholder maize productivity in Malawi: Toward an African Green Revolution. *PLoS Biol 7,* 1-10).

Donovan, G. & Casey, H., 1998. Soil fertility management in sub-Saharan Africa. World Bank, Technical Paper No. 408, Washington, D.C.

FAO, (2011). FAOSTAT. 13.07.2011. Available from: http://faostat.fao.org/default.aspx.

FAO, (2003). Assessment of soil nutrient balance. Approaches and methodologies. FAO Fertilizer and Plant Nutrition Bulletin 14. FAO, Rome.

FAO, (2007). Policies and actions to stimulate private sector fertilizer marketing in sub-saharan africa. Agricultural management, marketing and finance occasional paper 15.

Fischler, M., Wortmann, C. & Feil, B. (1999). Crotalaria (*C. ochroleuca G. Don.*) as a green manure in maize-bean cropping systems in Uganda. *Field Crops Research,* 61,97-107.

Fofana, B., Tamélokpo A., Wopereis M.C.S., Breman H., Dzotsi K. & Carsky R.J. (2005). Nitrogen use efficiency by maize as affected by a mucuna short fallow and P application in the coastal savanna of West Africa. *Nutrient Cycling in Agro-ecosystems,* 71,227-237.

Gachengo, C. (1996). *Phosphorus release and availability on addition of organic materials to Phosphorus fixing soils.* MPhil thesis, Moi University, Eldoret, Kenya.

Gakuo, A., Kebe, D. & Traore, B. (1996). Soil management in Mali: Final report. ODA, London/world bank. Washington.

Galloway, J.N., Schlesinger, W.H., Levy II, H., Michaels, A. & Schnoor, J.L., (1995). Nitrogen fixation: anthropogenic enhancement– environmental response. *Global Biochem. Cycles,* 9,235-252.

Gentile, R., Vanlauwe, B. van Kessel, C. & Six, J (2009). Agriculture , Ecosystems and Environment Managing N availability and losses by combining fertilizer-N with different quality residues in Kenya. *Agriculture, Ecosystems and Environment,* 131,308-314.

Giller, K.E. & Cadisch, G. 1995. Future benefits from biological nitrogen-fixation–an ecological approach to agriculture. *Plant Soil*, 174,255-277.

Giller, K.E., 2001. *Nitrogen Fixation in Tropical Cropping Systems*, CAB International, Wallingford, UK.

Giller, K.E., Cadisch, G., Ehaliotis, C., Adams, E., Sakala, W.D. & Mafongoya, P.L. (1997). Building soil nitrogen capital in Africa. In: Buresh, R.J., Sanchez, P.A., Calhoun, F. (Eds.) (1997). *Replenishing Soil Fertility in Africa*. SSSA Special Publication, vol. 51. *Soil Science Society of America*, Madison, WI, pp. 151-192.

Grote, U., Craswell, E. & Vlek, P. (2005). Nutrient flows in international trade: ecology and policy issues, *Environmental Science and Policy*, 8,439-451.

Gunnarsson, C. & Petersen, M. (2007).Water hyacinths as a resource in agriculture and energy production: A literature review. *Waste Management*, 27,117–129.

Handayanto, E., Cadisch, G. & Giller, K.E. (1997). Regulating N mineralization from plant residues by manipulation of quality. In: Cadisch, G., Giller, G.E. (Eds.), *Driven by Nature: Plant Litter Quality and Decomposition*. CAB International, UK, pp. 175–186.

Harris, F.M.A. (1998). Farm-level assessment of nutrient balance in northern Nigeria. *Agriculture, Ecosystems and Environment*, 71:201-214.

Heisey, P.W. & Mwangi, W. (1996). Fertilizer Use and Maize Production in Sub- Saharan Africa. CIMMYT Economics Working Paper 96-01, CIMMYT, Mexico.

IFDC (2003). Integrated soil fertility management. 09.07.2011. Available at: www.ifdc.org/PDF_Files/ISFM%20Brochure.pdf.

Jama, B., Buresh, R.J., Ndufa, J.K. & Shepherd, K.D. (1998). Vertical distribution of roots and soil nitrate: tree species and phosphorus effects. *Soil Sci. Soc. Am. J.* 62,280–286.

Jama, B., Palm, C.A., Buresh, R.J., Niang, A., Gachengo, C., Nziguheba, G. & Amadalo, B. (1999). Tithonia diversifolia green manure for improvement of soil fertility: A review from western Kenya. ICRAF, Nairobi.

Kato, E. & Place, F.M. (2011). *Heterogeneous Treatment Effects of Integrated Soil Fertility Management on Crop Productivity: Evidence from Nigeria*. IFPRI Discussion Paper 01089. Pp 20.

Kaya, B., Hildebrand, P.E. & Nair, P.K. (2000). Modeling changes in farming systems with the adoption of improved fallows in southern Mali. *Agricultural Systems*, 66,51-68.

Kelly, V., Adesina, A. & Gordon. A. (2003). Expanding Access to Agricultural Inputs in Africa: A Review of Recent Market Development Experience. *Food Policy*, 28,379-404.

Krauss, M., Berner, A. Burger, D. Wiemken, A. Niggli U. & Mader. P. (2010). Reduced tillage in temperate organic farming: Implications for crop management and forage production. *Soil Use Management*, 26,12-20.

Kwesiga, F.R., Franzel, S., Place, F., Phiri, D. & Simwanza, C.P. (1999) Sesbania sesban Improved fallows in eastern Zambia: Their inception, development and farmer enthusiasm. *Agroforestry systems* 47,49-66.

Lal, R. (1990). *Soil Erosion in the Tropics: Principles and Management*. New York: McGraw-Hill, Inc.

Larson, B.A., (1996). Fertilizers to support agricultural development in sub-Saharan Africa: what is needed and why *. *Food Policy*, 21,509-525.

Lekasi, J. K., Tanner, J.C., Kimani, S.K. & Harris, P.J.C. (1998). *Manure management in the Kenya highlands: Practices and Potential*. Emmerson Press, Kenilworth, UK

Lekasi, J.K., Tanner, J.C., Kimani, S.K., & Harris, P.J.C., (2001). *Managing Manure to Sustain Smallholder Livelihoods in the East African High- lands*. HDRA, Coventry, UK.

Mafongoya, P.L., Dzowela, B.H., (1999). Biomass production of tree fallows and their residual effect on maize yield in Zimbabwe. *Agroforestry systems,* 47,139–151.

Makokha, S., Kimani S., Mwangi, W., Verkuijl, H. & Musembi, F. (2001). *Determinants of Fertilizer and Manure use in Maize Production in Kiambu District, Kenya.* Mexico, D.F: CIMMYT and KARI.

Mando, A., Ouattara, B., Somado, A.E.,Wopereis,M.C.S., Stroosnijder, L. & Breman, H., (2005). Long-term effects of fallow, tillage and manure application on soil organic matter and nitrogen fractions and on sorghum yield under Sudano-Sahelian conditions. *Soil Use Management,* 21,25–31.

Mapfumo, P. & Giller, K.E. (2001). Soil Fertility Management Strategies and Practices by Smallholder Farmers in Semi-arid Areas of Zimbabwe. ICRISAT/FAO, Patancheru, AP, India.

Mathuva, M.N. Rao, M.R., Smithson, P.C. & Coe R. (1998). Improving maize (*Zea mays*) yields in semiarid highlands of Kenya: agroforestry or inorganic fertilizers ? *Field Crops Research.* 55,57-72.

Misiko, M. & Ramisch, J., (2007). Integrated Soil Fertility Management Technologies : review for scaling up. In A. Bationo (eds.), *Advances in Integrated Soil Fertility Management in Sub-Saharan Africa: Challenges and Opportunities, Soil Biology,* 873-880.

Morris, M., Kelly, V.A., Kopicki, R.J. & Byerlee, D. (2007) Fertilizer use in African agriculture: Lessons learned and good practice guidelines. The World Bank. 12.07.2011 Available: http://www- wds.worldbank.org/servlet/main?menu.

Mpepereki, S., Javaheri, F., Davis, P. & Giller, K.E. (2000). Soyabeans and sustainable agriculture: 'promiscuous' soyabeans in southern Africa. *Field Crops Research,* 65,137-149.

Mugendi, D.N., Kanyi M.K., Kung'u J.B., Wamicha, M.N. & Mugwe, J.N. (2004). The role of agroforestry trees in intercepting leached nitrogen in the farming systems of the central highlands of Kenya. *East African Agricultural and Forestry Journal,* 69,69-79.

Mugendi, D.N.,Nair, P.K.R., Mugwe, J.N., O'Neill, M.K. & Woomer, P.L. (1999). Callian- dra and Leucaena alley cropped with maize. Part 1. Soil fertility changes and maize production in the sub-humid highlands of Kenya. *Agroforest. Systems,* 46,39–50.

Mugwe, J., Mugendi D. N., Mucheru-Muna, M., Odee, D. & Mairura, F. (2009). Effect of selected organic materials and inorganic fertilizer on the soil fertility of a Humic Nitisol in the central highlands of Kenya. *Soil Use and Management,* 25,434-440.

Mugwe, J., Mugendi, D., Mucheru-Muna, M., Merckx, R., Chianu, J. & Vanlauwe, B. (2009). Determinants of the decision to adopt integrated soil fertility management practices by smallholder farmers in the central highlands of Kenya. *Experimental Agriculture,* 45,61-75.

Mugwira, L.M. & Murwira, H.K., (1997). Use of cattle manure to improve soil fertility in Zimbabwe: past and current research and future research needs. Network Working Paper No. 2. Soil Fertility Network for Maize-Based Cropping Systems in Zimbabwe and Malawi, CIMMYT, Harare, Zimbabwe.

Nair P.K. (1989). Agroforestry systems, practices and technologies. In: Nair P.K (ed.), *Agroforestry Systems in the Tropics.* Kluwer Academic publishers in co-operation with ICRAF, Dordrecht, Netherlands and Nairobi Kenya, pp 39-52.

Ndufa, J.K. Gathumbi, S.M., Kamiri, H.W. Giller, K.E. & Cadisch, G. (2009). Do Mixed-Species Legume Fallows Provide Long-Term Maize Yield Benefit Compared with Monoculture Legume Fallows? *Agronomy Journal.* 101,1352-1362.

Nkonya, E., Pender, J., Jagger, P., Sserunkuuma, D., Kaizzi, C. & Ssali, H. (2004). Strategies for sustainable land management and Poverty reduction in Uganda. International Food Policy Research Institute (IFPRI) Research Report 133, Washington, DC.

Oenema, O., Kros, H. & de Vries, W. (2003). Approaches and uncertainties in nutrient budgets: implications for nutrient management and environmental policies. 20. *European Journal of Agronomy*, 20,3-16.

Omamo, S., Williams, J., Obare, G. & Ndiwa, N. (2008). Soil fertility management on small farms in Africa: evidence from Nakuru District, Kenya. *Food Policy*, 27 159-170

Palm, C.A. & Sanchez, P.A. (1991). Nitrogen release from the leaves of some tropical legumes as affected by their lignin and polyphenolic contents. *Soil Biol. Biochem.* 23,83-88.

Palm, C.A., Myers, R.J.K. & Nandwa, S.M. (1997). Combined use of organic and inorganic nutrient sources for soil fertility maintenance and replenishment. In: Buresh, R.J., Sanchez, P.A., Calhoun, F. (Eds.), *Replenishing Soil Fertility in Africa*. Pp 56-84. *Soil Science Society of America*, Publication 51. SSSA and ASA, Madison, WI, USA.

Place, F. Barrett, C.B., Freeman, H.A. Ramisch, J.J. & Vanlauwe, B. (2003). Prospects for integrated soil fertility management using organic and inorganic inputs: evidence from smallholder African agricultural systems. *Food Policy*, 28,365-378.

Pound P., Snapp S., McDougall C. and Braun A. (Eds.) 2003 Managing natural resources for sustainable livelihoods. London: Earthscan.

Rufino, M.C., Rowea,E.C., Delve, R.J. & Giller, K.E. (2006). Nitrogen cycling efficiencies through resource-poor African crop – livestock systems. *Agriculture, Ecosystems and Environment*, 112, 261-282.

Sanchez, P. A., Shepherd, K.D., Soule, M.J., Place, F.M., Buresh, R.J., Izac, A-MN., Mokwunye, A.U., Kwesiga, F.R., Ndiritu, C.G. & Woomer, P.L. (1997). Soils fertility replenishment in Africa: An investment in natural resource capital. In: Buresh RJ, Sanchez PA & Calhoun F (eds) *Replenishing Soil Fertility in Africa*, pp 1–46. *Soil Science Society of America* Special Publication 51. SSSA and ASA, Madison, WI, USA

Sanchez, P., & B. Jama. (2002). Soil fertility replenishment takes off in east and southern Africa. p. 23–46. In B. Vanlauwe, J. Diels, N. Sanginga, and R. Merckx (ed.) *Integrated plant nutrient management in sub-Saharan Africa: From concept to practice*, CABI, Wallingford, UK.

Sanchez, P.A. & Leakey, R.R.B. 1997. Land use transformation in Africa: three determinants for balancing food security with natural resource utilization, *European Journal of Agronomy* 7:15-23.

Sanchez, P.A. (2002). Soil fertility and hunger in Africa. *Science* 129, 2019– 2020

Sanchez, P.A., (1999). Improved fallow comes of age in the tropics. *Agroforestry Systems* 47,3–12.

Sanginga P. C., Kaaria, S., Muzira, R., Delve, R., Njuki, J., Vanlauwe, B. Chiarm, J. & Sanginga, N. 2007. The Resources-to-Consumption System: A Framework for Linking Soil Fertility Management Innovations to Market Opportunities. Challenges, 979-992. In A. Bationo (eds.),. *Advances in Integrated Soil Fertility Management in Sub-Saharan Africa: Challenges and Opportunities*, 979–992.

Sanginga, N & Woomer, P.L. (eds) 2009. *Intergrated Soil Fertility Management in Africa: Principles, Practice and Developmental Process*. Tropical Soil Biology and Fertility Institute of International Centre of Tropical Agriculture. Nairobi. 266 pp.

Sanginga, N., Dashiell, K., Diels, J., Vanlauwe, B., Lyasse, O., Carsky, R.J., Tarawali, S., Asafo-Adjei, B., Menkir, A., Schulz, S., Singh, B.B., Chikoye, D., Keatinge, D. & Rodomiro, O. (2004). Sustainable resource management coupled to resilient germplasm to provide new intensive cereal–grain legume–livestock systems in the dry Savanna. *Agriculture Ecosystem & Environment,* 100,305-314.

Sanginga, N., Okogun, J.A., Vanlauwe, B., Diels, J. & Dashiell, K. (2001). Contribution of nitrogen fixation to the maintenance of soil fertility with emphasis on promiscuous soybean maize-based cropping systems in the moist savanna of West Africa. *In:* Tian, G., Ishida, F., Keatinge, J.D.H. (Eds.), *Sustaining Soil Fertility in West Africa.* ASA, Wisconsin, pp. 157-178.

Schroth, G., Krauss, U., Gasparotto, L., Duarte Aguilar, J.A. & Vohland, K. (2000). Pests and diseases in agroforestry systems of the humid tropics. *Agroforestry Systems,* 50,199–241.

Scoones, I. & Toulmin, C. (1998). Soil nutrient balances: What use for policy? *Agriculture, Ecosystems & Environment,* 71,255-267.

Sheldrick , W., Syers, K. & Lingard, J. (2003). Contribution of livestock excreta to nutrient balances. *Nutrient Cycling in Agroecosystems* 66,119–131.

Sheldrick, W., Syers, J. K. & Lingard, J. 2003. Contribution of livestock excreta to nutrient balances. *Nutrient Cycling in Agroecosystems.* 66,119-131.

Shisanya, C.A., Mucheru, M.W., Mugendi, D.N. & Kung'u, J.B. (2009). Effect of organic and inorganic nutrient sources on soil mineral nitrogen and maize yields in central highlands of Kenya. *Soil & Tillage Research,* 103,239-246.

Sileshi, G. & Mafongoya, P.L. (2003). Effect of rotational fallows on abundance of soil insects and weeds in maize crops in eastern Zambia. *Applied Soil Ecology,* 23,211-222.

Smaling E.M. & Dixon, J. (2006). Adding a soil fertility dimension to the global farming systems approach, with cases from Africa. *Agriculture, Ecosystems and Environment,* 116,15-26.

Smaling, E.M.A, Nandwa, S.M. & Janssen, B.H. (1997). Soil fertility is at stake. In: Buresh RJ, Sanchez PA & Calhoun F (Eds). *Replenishing soil fertility in Africa.* Special publication No. 51 Madison, Wisconsin: American Society of Agronomy and Soil Science society of America. pg 47-61.

Smaling, E.M.A, Nandwa, S.M., Prestele, H., Roetter, R. & Muchena, F.N. (1992). Yield response of maize to fertilizers and manure under different agro-ecological conditions in Kenya. *Soil and Water,* 41,241-252.

Smaling, E.M.A. & Braun, A.R. (1996). Soil fertility research in sub-Saharan Africa: new dimensions, new challenges, *Comm. Soil. Sci.,* 7,365-386.

Smaling, E.M.A. (Ed.), (1998). Nutrient balances as indicators of productivity and sustainability in Sub- Saharan African agriculture. *Agriculture, Ecosystems and Environment,* 71,1-3).

Smaling, E.M.A., (1993). *An ecological framework for integrated nutrient management with special reference to Kenya.* Ph.D. thesis, Wageningen Agricultural University, Wageningen.

Snapp, S.S., Rohrbach, D.D., Simtowe, F. & Freeman, H.A., (2002). Sustainable soil management options for Malawi: can smallholder farmers grow more legumes? *Agriculture, Ecosystems and Environment,* 91,159-174.

Speirs, M. & Olsen, O. (1992) *Indigenous Integrated Farming Systems in the Sahel.* World Bank Technical Paper No. 179, Africa Technical Department Series. World Bank, Washington, DC.

Stoorvogel J.J. & Smaling, E.M.A. (1990). *Assessment of soil nutrient depletion in Sub-Saharan Africa: 1983–2000.* Winand Staring Centre Report 28. Winand Staring Centre Wageningen, The Netherlands. pp. 137

Stoorvogel, J.J., Smaling, E.M.A. & Janssen, B.H. (1993). Calculating soil nutrient balances at different scale. I. Supra-national scale. *Fertilizer Research*, 35,227-235.

Tittonell, P., Vanlauwe, B., Leffelaar, P.A., Rowe, E. & Giller, K.E. (2005). Exploring diversity in soil fertility management of smallholder farms in western Kenya. I. Heterogeneity at region and farm scale. *Agriculture, Ecosystems and Environment.* 110,149-165.

TSBF-CIAT (2003). Defines strategic direction for 2003-2007. *The Comminutor*: Newsletter of the TSBF Institute of CIAT. Nairobi: Kenya.

UN Millenium Project, (2005). *Investing in development: a practical plan to achieve the Millenium Development Goals*. UN Development Programme, New York.

Van den Bosch, H., Jager, A.D. & Vlaming, J. (1998). Monitoring nutrient flows and economic performance in African farming systems (NUTMON) II. Tool development. *Agriculture, Ecosystems and Environment*, 71,49-62.

Van der Pol, F. (1992). *Soil mining. An unseen contributor to farm income in southern Mali*. Bull. 325. Royal Tropical Institute, Amsterdam, 48 pp.

van Noordwijk, M. (1996). Mulch and shade model for optimum alley-cropping design depending on soil fertility. In: Ong, C.K., Huxley, P. (Eds.), *Tree–crop Interactions: A Physiological Approach*. CAB International, Wallingford, UK, pp. 51–72.

Vanlauwe, B. & Giller, K.E. (2006). Popular myths around soil fertility management in sub-Saharan Africa, *Agriculture, Ecosystems and Environment*, 116,34-46.

Versteeg, M.N., Amadji, F., Eteka, A., Gogan, A., and Koudokpon, V., (1998). Farmers adaptability of Mucuna fallowing and Agroforestry technologies in the coastal savannah of Benin. *Agricultural Systems*, 56,269-287.

Vlaming, J., Bosch, H. van de, Wijk, M.S. van, Jager, A. de, Bannink, A. & Keulen, H. (2001). *Monitoring Nutrient Flows and Economic Performance in Tropical Farming Systems (NUTMON)*. NUTMON Tool Box, Wageningen.

Vlek, P.L.G., (1990). The role of fertilizers in sustaining agriculture in sub-Saharan Africa. Fertilizer Res. 26, 327-339.

Winpenny, J., Malga, A., Teme, B., Coulibaly, B., Diarra, L., Kergna. A. & Tigana, K. (1995). *Structural Adjustment and Sustainable Development in Mali*: Draft Working Paper, ODI/WWF.

Wivstad, M., (1997). *Green-Manure Crops as a source of Nitrogen in Cropping Systems*. Department of Crop Production Science, SUAS, Uppsala, Sweden.

Wopereis, M.C.S., Tamelokpo, A., Ezui K., Gnakpenou, D., Fofana, B. & Breman, H. (2005) Mineral fertilizer management of maize on farmer fields differing in organic inputs in the West African savanna. *Field Crops Research*, 96,355–362.

World Agroforestry Centre (2008). Addressing Africa's Food Crisis from the Ground Up. 10.07.2011. Available
http://www.cgiar.org/pdf/news_icraf_food_crisis_statement-final.pdf.

World Bank, (2007). *Fertilizer Use in African Agriculture: Lessons Learned and Good Practice Guidelines*. The world Bank. Washington DC. Pp 146.

Young, A., (1986). *The potential of agroforestry for soil conservation!. Erosion control*. ICRAF Working Paper 42, Nairobi, 68 pp.

7

Liquid Cattle Manure Application to Soil and Its Effect on Crop Growth, Yield, Composition, and on Soil Properties

Theodora Matsi
Soil Science Laboratory, School of Agriculture,
Aristotle University of Thessaloniki, Thessaloniki,
Greece

1. Introduction

Soil application of liquid cattle manure (LCM) (excrements plus urine, occasionally containing bedding material) can enhance plant growth and increase crop yield (Beauchamp, 1986; Culley et al., 1981; Evans et al., 1977; Kaffka & Kanneganti, 1996; Lithourgidis et al., 2007; Matsi et al., 2003; Motavalli et al., 1989; Sutton et al., 1986; Zhang et al., 2006; Zebarth et al., 1996). In most of the cases, crop yield increases are accompanied by increases in plant macronutrients concentration and/or uptake (Culley et al., 1981; Lithourgidis et al., 2007; Matsi et al., 2003; Motavalli et al., 1989; Sutton et al., 1986).

The beneficial effect of LCM on crop growth, yield and macronutrients absorption is mainly due to the improvement of soil fertility with respect to macronutrients, especially N. In general, the amounts of readily available N in manures, mainly in the form of NH_4-N, are lower than that of the inorganic fertilizers (Beauchamp, 1983; Jokela, 1992). However, LCM contains high amounts of immediately available N, due to its urine content. The quantity of this immediately available N can be almost half of the total N and this percentage is higher than that of the solid cattle manure (Beauchamp, 1986; Bechini & Marino, 2009; Sutton et al., 1986). A significant amount of the ammoniacal N in LCM can be lost, as NH_3 by volatilization, shortly after LCM application to soils (Beauchamp et al., 1982; de Jonge et al., 2004; Pain et al., 1989; Pfluke et al., 2011a, 2011b; Webb et al., 2010). However, this depends on the properties of the LCM and soil, where manure is applied, and also on the rate, manner, timing of application and weather conditions and interactions among these factors (Beauchamp et al., 1982; Mannheim et al., 1995; Mattila et al., 2003; Pain et al., 1990; Reijs et al., 2007; Rochette et al., 2006; Sorensen, 1998; Thompson et al., 1990a, 1990b). Soil incorporation of LCM, mainly by using injection techniques, seems to reduce NH_3 volatilization drastically and such techniques are now applicable not only to arable soils but also to grasslands, no-tillage and forest systems (Maguire et al., 2011a, 2011b). Due to the high quantity of immediately available N, LCM can be as efficient as inorganic fertilizers in satisfying plant needs in respect to N, when it is applied at equivalent rates (Kaffka & Kanneganti, 1996; Lithourgidis et al., 2007; Matsi et al., 2003; Zhang et al., 2006). In addition to N, LCM can increase soil available P and K, upon its use as an organic fertilizer

(Beauchamp, 1983; Culley et al., 1981; Lithourgidis et al., 2007; Matsi et al., 2003; Pratt & Laag, 1981; Randall et al., 2000; Sutton et al., 1979, 1986; Zhang et al., 2006).

Apart from macronutrients, LCM contains also micronutrients, essential for plant growth. Therefore, it can serve directly as a source of micronutrients, upon its use as basal dressing for crops, increasing micronutrients plant uptake and probably concentration (Brock et al., 2006; Nikoli & Matsi, 2011). In addition, an indirect effect of LCM on the availability of the soil native micronutrients cannot be excluded. Application of the LCM to soil for a long period and/or at high rates can increase the soil organic matter especially the dissolved fraction (Antil et al., 2005a; Culley et al., 1981; Nikoli & Matsi, 2011), since a considerable part of the organic matter of manure (around 20 %) exists in its liquid phase (Japenga et al., 1992). Consequently, soil application of LCM can enhance solubilization of metal micronutrients through their complexation with the dissolved organic matter and consequently increase availability to plants (Japenga et al., 1992). Also, after use of LCM as a fertilizer for many years and/or at high rates, a possible improvement of the soil structure, due to organic matter increase, cannot be excluded (Olesen et al., 1997; Mellek et al., 2010).

Apart from the beneficial effects of using LCM as an organic fertilizer, certain adverse effects might be involved for plants, soils and the environment upon LCM application to soil; such as increasing salinity in the soil profile, NO_3-N leaching to the underground water and P accumulation in the top soil (with its subsequent translocation to surface water reservoirs) (Beauchamp, 1983, 1986; Comfort et al., 1987; Culley et al., 1981; Daliparthy et al., 1994; Evans et al., 1977; Heathwaite et al., 1998; Lithourgidis et al., 2007; Motavalli et al., 1985; Phillips et al., 1981; Pratt & Laag, 1981; Sutton et al., 1979, 1986; Vellidis et al., 1996). These adverse effects are mainly connected to LCM application for long periods and/or at high rates and such applications should be avoided. In addition, in certain cases, micronutrient phytotoxicities, especially of Cu and Zn are possible; these phytotoxicities, however, are associated mainly with LCM enriched with solutions used for cattle hoof baths (Bolan et al., 2003; Jokela et al., 2010; McBride & Spiers, 2001).

The objectives of this chapter are to compile and evaluate existing research and knowledge concerning: a) composition of LCM and its application to soil, with emphasis on N, b) effect of LCM application on crop growth, yield and composition, c) beneficial effects of LCM application on soil properties, especially fertility, d) possible risks, of using LCM as a fertilizer, for plants, soils and the environment. In addition, the Greek experience with long-term use of LCM as a fertilizer is discussed.

2. Liquid cattle manure composition and application to soil, with emphasis on N

Liquid cattle manure composition depends on certain factors, such as the number and age of animals, the ration fed to animals, the inclusion of bedding material and cleaning water, the duration and conditions of storage and the kind of treatment prior to soil application (Marino et al., 2008; Reijs et al., 2007; Sorensen, 1998). Selected cases of LCM composition, reported in the literature, are presented in Table 1. Among the elements contained in LCM, N is of major concern, in respect of using the LCM as an organic fertilizer efficiently, without causing adverse effects to the environment.

Weight basis	pH†	Dry matter‡	Total-N	NH₄-N	Total-P	Total-K	Reference
				g kg⁻¹			
wet	7.1-7.8	57-92	3.3-4.0	1.6-1.8			(Amon et al., 2006)
wet	6.7-6.8	49-60	2.3-2.7				(Angers et al., 2006)
wet		60-78	2.4-3.1		0.4-0.6	1.6-2.4	(Beauchamp, 1983)
wet		68-71	2.6-3.8	1.5-2.0			(Beauchamp, 1986)
dry		49-117	26-75	9-37			(Bechini & Marino, 2009)
wet	6.4-7.5	18-80	1.9-3.3	0.88-1.8	0.07-0.82		(Bittman et al., 2011)
dry	7.2±0.1		30±4.8		8.0 ± 0.1	32 ± 0.5	(Briceno et al., 2008)
dry	8.1	135	4.9		4.5	1.6	(Brock et al., 2006)
dry			38	16	3		(Burger & Venterea, 2008)
wet		98-101	3.0-3.4	1.4-1.6	1.2-1.4	2.5-3.0	(Carter et al., 2010)
wet	7.2-7.2	20-63	1.4-6.1				(Chadwick & Pain, 1997)
wet		111±15	4.1±0.5				(Chadwick et al., 2000a)
wet	5.7-7.0	60-110	2.6-3.2	0.93-1.6	0.39-0.53	2.0-3.1	(Comfort et al., 1987)
dry		88	29		7	24	(Culley et al., 1981)
dry	7.5±0.2	70 ± 4	53±0.8	25 ± 2			(de Jonge et al., 2004)
dry	7.6-8.1	104-107	73-79	41-56	23-40	13-59	(Evans et al., 1977)
wet	7.0-7.7	36-85	2.1-3.2	1.3-1.6			(Hansen et al., 2003)
dry	6.1-9.1	35-229	12-40	3.5-46	4.1-18		(He et al., 2004)
dry		190-239	21-29	6.7-13			(Jokela, 1992)
dry		71-88	49-53	19-22	9.5-11	36-48	(Kaffka & Kanneganti, 1996)
dry			4.3-31		3.7-8.4	45-17	(Lund et al., 1975)
wet	7.5±0.3	95 ± 32	3.8±1.0	1.5 ± 0.5	0.65±0.24	2.6 ± 0.8	(Marino et al., 2008)
wet	7.8-7.8	77-83	3.0-3.1	1.2-1.3	0.65-0.71	2.2-2.8	(Matsi et al., 2003)
wet	7.3-7.9	108-136	2.7-3.9	1.3-1.6	0.3-0.4	2.6-4.5	(Misselbrook et al., 1995)
wet	7.0-7.4	26-48	1.8-4.4	0.70-1.2			(Misselbrook et al., 1996)
dry	6.8-8.0	11-92		6.0-15			(Misselbrook et al., 2002)
wet		71-80	3.2-3.4				(Pain et al., 1986)
dry	7.6	103	46	27	12	60	(Pain et al., 1989)
wet	7.1-8.4		2.3-3.7	1.1-1.8			(Pain et al., 1990)
wet	6.2-6.6	56-62	2.9-3.4	1.7-2.4			(Paul & Beauchamp, 1993)
dry		82-131	33-78	13-48			(Reijs et al., 2007)
dry	7.2	101	31		7.1		(Siddique & Robinson, 2003)
wet	7.4-8.0	69-74	3.4-3.5	1.9-2.3			(Sorensen, 2004)
wet	4.9-6.5	64-91	2.4-3.5	0.6-1.7	0.4-0.7	1.7-3.5	(Sutton et al., 1979)
wet		45±19	1.8±0.6	0.7 ± 0.4	0.5±0.1	1.4±0.4	(Sutton et al., 1986)
wet	7.5-8.3	64-65	2.6-3.7	1.1-1.5			(Thompson et al., 1990a)
wet	8.3-8.4	65-86	3.2-3.9	1.4-1.8			(Thompson et al., 1990b)
wet		24-110	1.5-4.6	0.23-1.7			(Unwin et al., 1986)
wet		80-162	2.8-5.6	0.72-2.4			(Whitehead et al., 1989)
dry		35-72	55-61		11-14		(Withers et al., 2001)
wet			1.2		0.6	2.3	(Zhang et al., 2006)

† pH was measured directly in LCM or in a suspension with water.
‡ In all cases dry matter is expressed on wet weight basis.

Table 1. Some properties of liquid cattle manures (LCM), reported in the literature, with emphasis on the plant macronutrient concentrations (Values represent: mean values, mean values and their standard deviations, or ranges of mean values).

Liquid cattle manure contains high amounts of immediately plant available N, in the form of NH_4^+ (Bechini & Marino, 2009; Sorensen, 2004) as it is shown in Table 1, due to its urine content. For example, Bechini & Marino (2009) and Sorensen (2004) found that NH_4-N content of the LCMs studied ranged from 33 to 55 % and 50 to 60 % of the total N, respectively usually, LCM contains higher levels of immediately plant available N than the solid cattle manure (Beauchamp, 1986; Sutton et al., 1986). Beauchamp (1986) found that on average the NH_4-N was 53 % of the total N in LCM and 9 % in solid beef cattle manure. However, a high amount of this NH_4-N can be lost through NH_3 volatilization or immobilized after LCM addition to soil (Burger & Venterea, 2008; Pain et al., 1990; Sorensen & Jensen, 1995). The organically bound N in LCM is expected to mineralize slowly and provide less plant available N than solid cattle manure (Burger & Venterea, 2008). Chadwick et al. (2000a) reported that from fifty manures analyzed, in order to characterize their N fractions and assess their potential organic N supply, the percentage of N mineralized was the lowest for a dairy cow slurry (< 2 %), whereas for a beef cattle manure was about 6 %.

Nitrogen in LCM is subjected to certain changes during the storage, i.e. mineralization, microbial immobilization, NH_3 volatilization, nitrification and denitrification (Whitehead & Raistrick, 1993a). During storage, N transformations depend on the properties of the LCM, such as dry matter content, C/N ratio and pH, the inclusion of bedding material in the LCM and the period and conditions of storage (Amon et al., 2006; Sorensen, 1998; Whitehead & Raistrick, 1993a). Nitrogen losses consist mainly of emissions of NH_3 and N_2O. In order to reduce N gases but also other gases emissions, Amon et al. (2006) evaluated different treatments of dairy cattle slurry during storage, such as slurry separation, anaerobic digestion, slurry aeration and straw cover, in comparison to no treatment. They reported that the anaerobic digestion of the slurry during storage reduced greenhouse gasses emissions, without increasing NH_3 emissions compared to the untreated slurry, whereas all other slurry treatments increased NH_3 emissions. The acidification of cattle slurry to pH 5 and the addition of nitrification inhibitors have also been suggested for the reduction of N losses, prior to slurry application to soil (Pain et al., 1990).

After LCM application to soil, its N can be subjected to the same changes just mentioned, but also to others after its mineralization, i.e. retention to clay minerals in exchangeable form, fixation by clay, uptake by plants, leaching (Bechini & Marino, 2009). Nitrogen losses after soil application of LCM consist of N_2O and NH_3 emissions and NO_3^- leaching. However, the main losses of LCM N seemed to occur through NH_3 volatilization (Carter et al., 2010), which is expected to be higher in LCM than in solid cattle manure, because of the higher urine and consequently urea content of the former (Whitehead & Raistrick, 1993b). Ammonia volatilization usually occurs within a short period, after LCM surface application to soil, ranging from a few hours to a few days, with the greatest NH_3 emissions occurring within a few hours after application (Beauchamp et al., 1982; de Jonge et al., 2004; Pain et al., 1989; Pfluke et al., 2011a, 2011b; Webb et al., 2010). At this stage, N transformations and losses depend not only on the properties, storage conditions and treatments during storage of the LCM, but also depend on soil characteristics, the rate, manner and timing of LCM application to soils, the weather conditions and the interactions of these factors (Beauchamp et al., 1982; Mannheim et al., 1995; Mattila et al., 2003; Pain et al., 1990; Reijs et al., 2007; Rochette et al., 2006; Sorensen, 1998; Thompson et al., 1990a, 1990b).

Among the properties of the LCM, that most affects N transformations, after application to soils, is the C/N ratio. Slurries with a low C/N ratio are expected to promote N mineralization, whereas slurries with high C/N ratios are expected to promote N immobilization (Chadwick et al., 2000a). Also, Whitehead et al. (1989) reported that the water-insoluble material of cattle slurries immobilized N that would otherwise have been plant available from the whole slurries and/or the soil. This was attributed to the higher C/N ratio of the water-insoluble fraction of the cattle slurries compared to the whole slurries. In addition, they concluded that the fine particle size fractions of the water-insoluble material of the slurries had the greatest effect on N mineralization-immobilization. Another property that affects NH_3 volatilization from LCM applied on the soil surface is the dry matter content of the LCM (Braschkat et al., 1997; Dell et al., 2011; Misselbrook et al., 2005). It seems that as the dry matter content of LCM increases the viscosity of the material also increases. The result is that smaller amounts of NH_4^+ in LCM can infiltrate and be retained by soil components and in this way be preserved from loss, through NH_3 volatilization (Braschkat et al., 1997). The strong effect of dry matter content on NH_3 losses after LCM addition to soils was also reported for grasslands and arable soils by Pain et al. (1989) and Sommer & Olesen (1991). Finally, Thompson et al. (1990b) found an inverse relationship between cattle slurry application rate and the proportion of NH_4-N volatilized, after cattle slurry applied to grassland.

Soil properties, that affect LCM N transformations, after its soil application, seem to be pH, texture and water regime, along with land use, cultivation practices and weather conditions. Ammonia volatilization is expected to be reduced after LCM application to acid soils compared to calcareous soils (Bechini & Marino, 2009). Thompson et al. (1990a) reported that the total loss of NH_4-N, through NH_3 volatilization, from cattle slurry applied to grassland was approximately 1.5 times that from slurry applied to bare soil. Sommer & Ersboll (1994) reported that harrowing the soil before cattle slurry application reduced NH_3 volatilization, whereas de Jonge et al. (2004) found no significant effect. Bechini & Marino (2009) reported that nitrification and NO_3^- production was extremely rapid after LCM application on unsaturated soils, regardless of soil texture. Loro et al. (1997) and Lowrance et al. (1998) reported that denitrification and N_2O production after soil application of LCM was positively correlated with the soil water content and the cumulative production of N_2O was found to be higher for the solid than the liquid cattle manure (Loro et al., 1997). Rochette et al. (2008) found no differences between LCM and solid cattle manure in respect to N_2O emissions and reported that the N_2O emissions were affected by soil texture in conjunction to weather conditions. Soil texture also may influence the LCM NH_4-N immobilization; it was found that the net immobilization of N due to soil application of cattle slurry was increased with increasing soil clay content (Sorensen & Jensen, 1995). In addition, the same soil property along with the weather conditions seems to strongly affect LCM decomposition after its application to soil (Bechini & Marino, 2009; Rochette et al., 2006). Also after LCM application to soil, NH_3 volatilization tends to be increased with temperature and pH and suppressed temporarily by rainfall (Beauchamp et al., 1982; Sommer & Olesen, 1991) and N_2O emissions were found to be lower in the dry than in wet season (Chadwick et al., 2000b).

In order to reduce NH_3 emissions, but also nutrient losses in runoff, and increase N utilization by plants after LCM addition to soils, soil incorporation of the LCM is needed and for this purpose various application (mainly injection) techniques are proposed in the literature. Numerous researchers (Beauchamp, 1983; Beauchamp et al., 1982; Carter et al, 2010; Dell et al., 2011; Hansen et al., 2003; Maguire et al., 2011a, 2011b; Mannheim et al., 1995; Mattila & Joki-Tokola, 2003; Mattila et al., 2003; Misselbrook et al., 2002; Pfluke et al., 2011a, 2011b; Powell et al., 2011; Ross et al., 1979; Webb et al., 2005, 2010) agree that soil incorporation of LCM as soon as possible after its application or LCM surface banding or injection are preferable than the conventional surface broadcast application, for arable land but also for grassland, no-tillage and forage systems, although in the latter cases, using injection techniques, there is the possibility of grass sward damage and soil disturbance (Mattila et al., 2003). However, Laws et al. (2002) reported that shallow disc injection and, in particular, trailing shoe application of cattle slurry to grassland improved silage quality and reduced herbage contamination, without damaging it (except of the case of LCM injection on tall swards) compared with the conventional surface broadcasting. They suggested that shallow injection should be used on short swards wherever is possible, preferably after cutting, whereas slurry can be applied by trailing shoes on taller swards. In addition, Maguire et al. (2011a, 2011b) reported many techniques that facilitate the incorporation of liquid manures into the soil with restricted or minor soil disturbance, such as shallow disk injection, chisel injection, aeration infiltration and pressure injection.

Certain cases of increasing N_2O emissions (Comfort et al., 1990; Dell et al., 2011; Flessa & Beese, 2000; Thompson et al., 1987; Webb et al., 2010) due to the use of reduced-NH_3 emission application techniques (mainly injection) are reported in the literature. However, such increases are probably not inevitable and N_2O emissions can be reduced by slurry injection to such depths that will increase the diffusion path to soil surface sufficiently, leading to the emission of most nitrified N as N_2 (Webb et al., 2010). Furthermore, Powell et al. (2011) and Pfluke et al. (2011b) reported that injection of dairy slurry reduced not only NH_3 emissions but also NO_3^- leaching compared to surface broadcast application or surface broadcast application followed by partial incorporation of the slurry. On the other hand, it is reported that soil incorporation of LCM can promote LCM NH_4-N immobilization compared to surface banding (Sorensen, 2004).

A general rule that could be followed in order to reduce the overall N losses from LCM applied to soil and to increase plant efficiency utilization of LCM N, is the application of the LCM as close as possible to the period of maximum crop uptake (Bechini & Marino, 2009).

3. Effect of liquid cattle manure application on growth, yield and composition of crops

The beneficial effect of using LCM as an organic fertilizer on the yield of various crops has been proven by means of field experiments. Selected experiments are presented in Table 2 and discussed in this section. As one can see from Table 2, there is a large variety of LCM application rates, even within the same crop. This could be attributed to differences in composition of the cattle slurries used, but also to the different application approaches regarding the expected amount of LCM N that would be available for plant uptake during the growing season.

Generally, the application rates of the LCM are chosen on the basis of its N content. However, for reasons mentioned in the previous section, it is difficult to determine precisely the amount of the initial available NH_4-N of the LCM that will be lost from the soil or immobilized, or the percentage of the initial organically bound N of the LCM that will be mineralized and become available for plant uptake, or even the extent of the LCM influence on the transformations of the soil native N, during the growing season. Because crop availability of N in LCM is expected to be lower than that from inorganic fertilizers (Beauchamp, 1983; Jokela, 1992), greater LCM N rates were applied in comparison to N inorganic fertilizers (Beauchamp, 1983; Evans et al., 1977; Sutton et al., 1986; Zebarth et al., 1996). Such LCM application rates resulted in increased crop yields at levels similar to the inorganic fertilizers. However, the same was evident when LCM was applied at rates equivalent to the recommended inorganic N fertilization for crops, based on LCM total N content (Kaffka & Kanneganti, 1996; Lithourgidis et al., 2007; Matsi et al., 2003; Zhang et al., 2006). In addition, there are cases of applying cattle slurry at rates equivalent to the recommended inorganic N fertilizers, based on LCM initial plant available N content (Patni & Culley, 1989; Randall et al., 2000), or on this plus the expected amount of organically bound N that would be mineralized during the growing season (Beauchamp, 1986; Griffin et al., 2002). In both cases, the obtained crop yields, N uptake and recovery, as well as other plant parameters gave lower and more variable responses compared to inorganic fertilizers.

Increased crop yields upon LCM application to soil are usually accompanied by increases in plant uptake of macronutrients. Many researchers reported that N, P and K uptake of different plant species were increased, upon repeated annual applications of LCM for certain years, at levels similar or higher than the inorganic fertilization (Culley et al., 1981; Lithourgidis et al., 2007; Matsi et al., 2003; Motavalli et al., 1989). However, Motavalli et al. (1989) reported that although the N, P and K uptake by corn plants was increased with increasing application rate of injected dairy cow slurry, at levels similar to inorganic N, P and K fertilization, crop recoveries of fertilizer N, P and K were generally higher than those of slurry total N, P and K. In addition, Paul & Beauchamp (1993) found that N recovery by the harvested portion of the corn (grain + stover) was higher for urea than for LCM treatments. This was attributable to several possible causes, including: a) lower availability of organically bound nutrients in LCM, b) higher quantities of nutrients applied with the LCM, c) differences in nutrient placement and d) greater loss of nutrients from the LCM treatments (Motavalli et al., 1989).

The beneficial effect of LCM application to soil on macronutrient concentrations in plant tissues is not apparent (Evans et al., 1977; Lithourgidis et al., 2007; Matsi et al., 2003; Parsons et al., 2007; Sutton et al., 1986). Sutton et al. (1979, 1986) reported that LCM did not consistently increase corn leaf N and P and Matsi et al. (2003) reported that N, P and K in the aboveground biomass of wheat remained unchanged upon LCM or inorganic fertilization application, whereas Evans et al. (1977) found that LCM, relative to the unfertilized and inorganic fertilized treatments, increased the N, P and K concentrations in corn ear leaves, grain and stover. The same was reported by Lithourgidis et al. (2007) for the three macronutrients in the aboveground biomass of corn at the R3 growth stage. This was probably due to the different application period and rates of LCM.

The beneficial effect of soil application of the LCM on micronutrient concentrations in plant tissues and uptake by plant species is ambiguous. In any case, increases of these plant parameters are expected after repeated annual applications of LCM to soil for many years and/or at high rates (Evans et al., 1977; Nikoli & Matsi, 2011).

LCM application rates	Plant species	Years	Reference
67-269 kg ha^{-1} yr^{-1} as N 70-560 kg ha^{-1} yr^{-1} as N	Corn (*Zea mays* L.)	3 3	(Beauchamp, 1983)
100-600 kg ha^{-1} yr^{-1} as N	Corn	3	(Beauchamp, 1986)
200-400 kg ha^{-1} yr^{-1} as N	Tall fescue (*Festuca arundinacea* Schreber)	6	(Bittman et al., 2011)
25-50 m^3 ha^{-1}, 3-4 times per yr	Orchardgrass (*Dactylis glomerata* L.) Reed canarygrass (*Phalaris arundinacea* L.)	2 2	(Carter et al., 2010)
53-138 Mg ha^{-1} yr^{-1}	Corn	3	(Comfort et al., 1987)
224-879 kg ha^{-1} yr^{-1} as N	Corn	5	(Culley et al., 1981)
94-199 m^3 ha^{-1} yr^{-1}	Potato (*Solanum tuberosum* L.)	3	(Curless et al., 2005)
112 & 336 kg ha^{-1} yr^{-1} as N	Alfalfa (*Medicago sativa* L.)	2	(Daliparthy et al., 1994)
224 metric tons ha^{-1} yr^{-1}	Corn	2	(Evans et al., 1977)
84 & 56 kg ha^{-1} as N	Mixed forage species	6	(Griffin et al., 2002)
240 kg ha^{-1} yr^{-1} as N	Corn	3	(Jokela, 1992)
150-450 kg ha^{-1} yr^{-1} as N	Orchardgrass	2	(Kaffka & Kanneganti, 1996)
80 Mg ha^{-1} yr^{-1}	Corn	4	(Lithourgidis et al., 2007)
45-135 metric tons ha^{-1} yr^{-1}	Coastal bermudagrass (*Cynodon dactylon* L.)	3	(Lund et al., 1975)
100-300 kg ha^{-1} yr^{-1} as N	Corn	3	(McGonigle & Beauchamp, 2004)
40 Mg ha^{-1} yr^{-1}	Winter wheat (*Triticum aestivum* L.)	4	(Matsi et al., 2003)
33-62 Mg ha^{-1} yr^{-1}	Meadow fescue (*Festuca pratensis* Huds), timothy [*Phleum pretense* (L.) Trabud] Timothy	3 3	(Mattila et al., 2003)
53-138 Mg ha^{-1} yr^{-1}	Corn	2	(Motavalli et al., 1989)
60 kg ha^{-1} yr^{-1} as N	Rye grass (*Lolium perenne* L.), white clover (*Trifolium repens* L.)	2	(Misselbrook et al., 1996)
80-160 kg ha^{-1} yr^{-1} as N	Herbage	3 (27 sites)	(Pain et al., 1986)
32.1-64.3 Mg ha^{-1}	Corn, wheat, soybean [*Glysine max* (L.) Merr]	2	(Parsons et al., 2007)
90 Mg ha^{-1} yr^{-1}	Corn	3	(Patni & Culley, 1989)
100-300 kg ha^{-1} yr^{-1} as N	Corn	3	(Paul & Beauchamp, 1993)
75 m^3 ha^{-1} yr^{-1}	Oat (*Avena sativa* L.), corn, winter rye (*Secale cereale* L.)	4	(Powell et al., 2011)
21 & 42 metric tons ha^{-1} yr^{-1}	Barley (*Hordeum vulgare* L.), sudangrass (*Sorghum sudanese* L.)	4	(Pratt & Laag, 1981)
154-224 kg ha^{-1} yr^{-1} as N	Corn	4	(Randall et al., 2000)
112-336 Mg ha^{-1} yr^{-1}	Corn Alfalfa, orchardgrass	3 2	(Sutton et al., 1979)
112-336 Mg ha^{-1} yr^{-1}	Corn	6	(Sutton et al., 1986)
75 m^3 ha^{-1}	Rye grass	3 (7 trials)	(Unwin et al., 1986)
175-525 kg ha^{-1} yr^{-1} as N	Corn	2	(Zebarth et al., 1996)
100 & 200 kg ha^{-1} as N	Smooth bromegrass (*Bromus inermis* Leyss) Oat	3 3	(Zhang et al., 2006)

Table 2. Field experiments: liquid cattle manure (LCM) application rates, the plant species studied and the duration of the experiment.

Because of the higher levels of the readily available N in LCM than in solid cattle manure (Beauchamp, 1986; Sutton et al., 1986), LCM seems to a be more effective organic fertilizer than solid cattle manure, when applied at equivalent N rates (Beauchamp, 1986; Kaffka & Kanneganti, 1996; Lund et al. 1975; Paul & Beauchamp, 1993; Zhang et al., 2006). Beauchamp (1986) reported that crop yield responses were higher in the LCM than the solid cattle manure treatments and the same is reported by Lund et al. (1975), Kaffka & Kanneganti (1996) and Zhang et al. (2006). In addition, Kaffka & Kanneganti (1996) found that N uptake by plants grown in plots that had received LCM was higher than the plots that had received solid cattle manure and Paul & Beauchamp (1993) found that N recovery by the harvested portion of the corn (grain + stover) was higher for LCM than for solid beef manure treatments. However, Evans et al. (1977) found that upon both manures application, crop yields increased significantly compared to control, at levels similar to the inorganic fertilization, and this could be attributed to the heavy application rates of both manures but also to the different application rates. Also, Evans et al. (1977) reported a residual beneficial effect of both manures on crop yield for two years, whereas Zhang et al. (2006) reported the same effect but only for LCM, although the opposite was expected, since solid cattle manure contains more organically bound N than LCM, which could be available for plant uptake after its mineralization for a longer period. In addition, Sutton et al. (1986) found that corn yields were increased the following year, after LCM application for five years at high rates. Surprisingly, Beauchamp (1987) reported that corn response to residual N from animal manures, including LCM and solid beef manure, after two years from application, was lower than that obtained for urea, the first year following the two years of application. However, the second year, following the two years of application, there was only a small response to residual N from any of the sources, organic or inorganic.

4. Effect of liquid cattle manure application on soil properties

The beneficial effect of LCM on crop yields has been connected to the improvement of soil fertility mainly, after LCM application to soils. In addition, a possible improvement of soil physical properties, through the increase of soil organic matter due to LCM application, cannot be excluded in the cases of long term and/or heavy applications. However, there are certain risks involved for plants, soils and the environment following LCM application to soils, such phytotoxicity of micronutrients, nutrient losses from soil by leaching and/or in the runoff and increase of soil salinity to unacceptable levels. All these aspects are discussed in this section.

Upon the use of LCM as an organic fertilizer for crops, soil availability of the plant macronutrients N, P and K is expected to be increased and maintained at desirable levels, when LCM is applied at optimal rates (Beauchamp, 1983; Culley et al., 1981; Lithourgidis et al., 2007; Matsi et al., 2003; Pratt & Laag, 1981; Randall et al., 2000; Sutton et al., 1979, 1986; Zhang et al., 2006). On the other hand, there are certain risks of plant macronutrients accumulation in the soil and their subsequent losses to the underground water or to surface water reservoirs (Misselbrook, et al., 1995; Soupir et al., 2006). These risks are more pronounced in the case of LCM application at high rates and/or for a long period. Such risks are mainly the NO_3^- loss below the root zone due to leaching and P accumulation in the top soil and its subsequent loss in the runoff (Culley et al., 1981; Evans et al., 1977; Pratt & Laag, 1981; Sutton et al., 1979, 1986; Vellidis et al., 1996), although also NO_3^- loss in the runoff and P leaching cannot be excluded. However, it is uncertain if these risks are greater in the case of LCM application than when applying inorganic fertilizers (Beauchamp, 1983, 1986; Comfort et al., 1987; Daliparthy et al., 1994; Heathwaite et al., 1998; Lithourgidis et al., 2007; Motavalli et al., 1985, Phillips et al., 1981; Randall et al., 2000).

In addition to gaseous N losses following LCM application to soils, reported in the second section of this chapter, N can be lost as NH_4-N associated with suspended soil particles in the runoff (Smith et al., 2001a), but the main loss is through NO_3^- leaching. The LCM N (the inorganic NH_4-N or the organically bound N, after its mineralization) can be readily transformed to NO_3-N, which is highly soluble and thus it is susceptible to leaching. Indeed, many researchers found elevated concentrations of NO_3^- in the soil profile upon LCM application. However, these increased concentrations were at similar or lower levels than those caused by the inorganic fertilizers, especially urea, applied at rates equivalent or even lower than the LCM N (Beauchamp, 1983, 1986; Comfort et al., 1987; Daliparthy et al., 1994; Jokela, 1992; Lithourgidis et al., 2007; Motavalli et al., 1985; Phillips et al., 1981; Randall et al., 2000). For example, Phillips et al. (1981) reported that NO_3-N concentration in the tile-drain effluent from silage corn receiving LCM at a rate of 897 kg N ha^{-1} was no greater than that from 134 kg N ha^{-1} applied as inorganic fertilizer. Beauchamp (1983) found that 560 kg N ha^{-1} as LCM resulted in less soil NO_3-N than 208 kg N ha^{-1} as urea and Beauchamp (1986) reported that application of LCM at a rate of 600 kg N ha^{-1} did not increase soil NO_3-N levels above those from urea or the lower LCM application rates.

As far as P concerns, the major problem seems to be P build-up in the plow layer and loss in the runoff following LCM application to soil (Smith et al., 2001b; Soupir et al., 2006), because usually P is strongly associated to soil particles or exist in the form of insoluble substances in the soil and thus it moves down the soil profile with difficulty. However, the P leaching cannot be excluded (Pratt & Laag, 1981; Tarkalson & Leytem, 2009), since appreciable amounts of water soluble P can be found in the LCM (Kleinman et al., 2005). Again, P accumulation in the upper soil layer was found after LCM applications for many years and/or at high rates (Culley et al, 1981; Pratt & Laag, 1981; Sutton et al., 1986). Furthermore, it is questionable if the risk of P build-up following soil LCM application is greater than that from inorganic fertilizers (Withers et al., 2001). Phosphorus in LCM treated soils was found to be less available than in soils treated with triple superphosphate (Withers et al., 2001); however, Siddique & Robinson (2003) and Tarkalson & Leytem (2009) reported that P availability and mobility in LCM treated soils were higher than in soils treated with potassium di-hydrogen phosphate or mono-ammonium phosphate, respectively.

The effect of soil application of the LCM on the availability of plant micronutrients has not been investigated adequately in the literature, probably because this effect is inconsistent, even after repeated LCM applications for many years. However, the concentration of soil available micronutrients is likely to be increased after long-term repeated applications of LCM (Brock et al., 2006; Nikoli & Matsi, 2011). In certain cases, the risk of Cu and Zn phytotoxicities is possible upon soil application of enriched LCM with metals. The causes of such enrichments are the use of Cu and Zn feed additives to cattle and mainly the addition of hoof treatment solutions containing $CuSO_4$ or $ZnSO_4$ to the manure storage (Bolan et al., 2003; Jokela et al., 2010; McBride & Spiers, 2001). In order to clarify this risk, Brock et al. (2006) studied the accumulation, depth distribution and bioavailability of Cu and Zn in 109 fields, amended with LCM for 5 to 40 years. They found increased soil total Cu and Zn concentrations in the plow layer, but Cu and Zn soil available concentrations were low and the same was evident for Cu and Zn concentrations in the leachates collected from soil cores (0-50 cm). They concluded that there was no evidence that Cu and Zn accumulation in the plow layer had reached toxicity thresholds, even after 40 years of LCM application.

Increases of soil total organic C and N resulting from cattle manures application to soils are mainly connected to the addition of solid cattle manure (Chang et al., 1991; Eghball, 2002) than LCM, due to the higher dry matter content of the former in comparison to the latter (Sutton et al., 1986). Consequently, since LCM contains low amounts of dry matter and thus low amounts of organic matter, the beneficial effect of LCM application on soil total organic C and N becomes apparent after many years of continuous application and/or at high rates (Antil et al., 2005a, 2005b; Culley et al., 1981; Nikoli & Matsi, 2011). Culley et al. (1981) found that soil total organic C increased significantly upon LCM application at high rates, for 5 years. Nikoli & Matsi (2011) reported that soil total and dissolved organic C increased significantly after nine years of LCM addition to soil, at rates equivalent to the recommended inorganic fertilization for crops. At that time, no significant increase of soil total N was evident. Significant increases of both total organic C and N in the top soil were also reported, after addition of cattle slurry in fallow and cropped plots, for 28 and 38 years, respectively (Antil et al., 2005a, 2005b). On the other hand, Mellek et al. (2010) observed a tendency for increases in soil total organic C due to LCM application to a no-tillage soil for only two years and Briceno et al. (2008) reported that LCM application at rates of 100-300 m³ ha⁻¹ to a soil with high initial content of organic matter, although it did not increase total organic C, resulted in increased dissolved organic C immediately after addition. Angers et al. (2006), who studied the dynamics of soil dissolved organic C following application of LCM to a loamy and a clay soil, reported that their results were inconsistent and the overall, temporal variations in soil dissolved organic C content were large and greater than the fluctuations directly attributable to LCM addition.

The impact of LCM application on soil physical properties has not been adequately investigated in the literature. However, because improved soil properties are strongly connected to increased soil organic matter content, in addition to other soil properties that also influence soil structure, the improvement of soil physical properties due to LCM application is expected after long-term continuous applications of LCM and/or at high rates. Olesen et al. (1997) reported that water holding capacity of two soils differed in texture increased after addition of LCM at rates of 15-20 % and Mellek et al. (2010) found that application of LCM in a no-tillage soil for two years improved soil structure by changing physical properties, such as bulk density, macroporosity, aggregates mean weight diameter, saturated hydraulic conductivity and water infiltration rate.

Although the beneficial effect of LCM application on soil properties is adequately established in the literature, there are cases of questioning this effect. Jokela et al. (2009) reported that LCM application to soil at a rate of 110 m³ ha⁻¹ yr⁻¹, for four years did not improve soil quality. This was attributed to removal of the large particle-size solids from the LCM prior its use to the field, as well as to the fact that the experimental field was in no-tillage production with various crop rotations and had received periodic application of manure for twenty years before the experiment.

The risk of increased soil salinity at unacceptable levels and consequently the risk of possible plant injury are possible after repeated heavy applications of LCM (Culley et al.,

1981; Evans et al., 1977; Sutton et al., 1979, 1986). However, it is uncertain if this risk is greater than that due to the use of inorganic fertilizers. Evans et al. (1977) and Sutton et al. (1979) reported that soil salinity increased significantly, upon application of LCM at rates up to 336 metric tons ha^{-1} yr^{-1}, but remained at levels below the critical limit, even at the highest application rate. Lithourgidis et al. (2007) found increased salinity in the soil profile after eight years of LCM application to soil, at rates equivalent to the recommended inorganic fertilization for crops, but at levels acceptable for most crops and similar to the levels caused by the inorganic fertilization. Pratt et al. (1977), who studied salts leaching as a function of application rates of solid and liquid cattle manures and irrigation systems, reported that large amounts of K, Ca and Mg were accumulated in the soil, but there was a net loss of Na. The percentage of leached cations coming from manures declined as the application rate of manures increased. They suggested that, under most irrigation systems, addition of manures to fine-textured soils can result in a reduction of salts leaching to underground water compared to coarse-textured soils, due to their lower infiltration rate.

As far as the soil pH concerns, LCM application to soils at high rates or at rates equivalent to the recommended inorganic fertilization for crops for many years is not expected to affect it (Briceno et al., 2008; Nikoli & Matsi, 2011). Briceno et al. (2008) reported that soil pH increased immediately on addition of LCM at high rates to soils but returned to values similar to control and Nikoli & Matsi (2011) found that soil pH remained unchanged after nine annual applications of LCM, at rates equivalent to the recommended inorganic fertilization for crops.

5. The Greek experience of using liquid cattle manure as a fertilizer

The effect of LCM application to soil on wheat and corn and soil properties was studied in comparison to commercial inorganic fertilizers (both applied at equivalent N-P recommended rates), by means of a field experiment, in Northern Greece (Dordas et al., 2008; Lithourgidis et al., 2007; Matsi et al., 2003; Nikoli & Matsi, 2011).

The experiment was established in a field of the Farm of Aristotle University of Thessaloniki (22°, 59', 6.17'' north latitude and 40°, 32', 9.32'' east longitude), the fall of 1996. Field soil was a calcareous loam (Typic Xerorthent) (Matsi et al., 2003) and cultivated with winter wheat for four years, remained uncultivated for one year and then cultivated with corn until 2011. The size of the experimental plots was 60 m^2 and the experimental design was completely randomized blocks with six replications.

The fertilization treatments (Lithourgidis et al, 2007; Matsi et al., 2003) were established in the same plots each year and were: a) Soil incorporation of LCM, before sowing; b) application of the recommended for each crop N-P inorganic fertilization, as a single basal dressing, before sowing; c) application of the recommended for each crop N-P inorganic fertilization, but with split application of the N inorganic fertilizer, half of the amount as basal dressing before sowing and the other half at a specific growth stage of the crop (at tillering for wheat, broadcast and at the V8 growth stage for corn, as side dressing); d) no organic or inorganic fertilization (control).

The LCM was collected in an open tank, occasionally agitated during storage and diluted with water to obtain density of almost 1 g mL^{-1} prior its use for the field experiments. Analysis of the LCM composition was performed for three consecutive years prior to 1996 and repeated during the first two years of the wheat experiment. The results showed that

LCM properties were almost constant over the years, in respect to pH, dry matter content and N and P concentrations. Potassium concentrations were affected by the ration composition (unpublished data). Consequently, the mean values of total N and P contents of manure (Matsi et al., 2003)(Table 1) were taken as the basis for LCM application rates, for both experiments with wheat and corn and LCM was applied at 40 and 80 Mg ha[-1] yr[-1] (wet weight basis), for wheat and corn, respectively. The recommended N-P inorganic fertilization consisted of 120 kg N ha[-1] yr[-1] and 26 kg P ha[-1] yr[-1], for wheat and of 260 kg N ha[-1] yr[-1] and 57 kg P ha[-1] yr[-1], for corn (with single or split application of the N fertilizer). Fertilizers (inorganic or LCM) were applied a few days before wheat or corn sowing and incorporated into the soil as soon as possible (Lithourgidis et al, 2007; Matsi et al., 2003).

Each year, plant samples were collected at two specific growth stages of the crops and analyzed for nutrients; the first growth stage was the heading and the R3 growth stage, for wheat and corn, respectively and the second growth stage was the harvest for both crops. In addition, samples from the soil surface or deeper layers were collected and analyzed (Lithourgidis et al, 2007; Matsi et al., 2003; Nikoli & Matsi, 2011).

The results of the 4-yr (1996-2000) field experiment with winter wheat (Matsi et al., 2003) showed that application of the LCM to soil did not affect seed germination and N, P and K concentrations in plant tissues. However, upon LCM addition to soil, aboveground biomass of wheat, grain yield and plant uptake of the three macronutrients increased significantly compared to control, at levels similar to the inorganic fertilization. Similar increases were obtained for the available NO_3-N, P and K concentrations in the surface soil layer. After four years of LCM application to soil, there was no evidence of N, P and K build-up in soil, whereas soil salinity, organic C and total N levels remained unchanged.

The same beneficial effects of the LCM, as for wheat, were evident for corn in the first four years (2002-2005) of the corn experiment (Lithourgidis et al, 2007). Moreover, concentrations of the macronutrients N, P and K in the aboveground corn biomass, at the R3 growth stage increased significantly relative to control, at levels similar or higher that the inorganic fertilization treatments, in the years 2004 and 2005. As far as the soil properties, combining the results of the first four years corn experiment with those of the wheat experiment showed that, annual LCM application for eight years maintained the amounts of the available NO_3-N, P and K in the surface soil layer, at desirable levels, each year of the application period. Upon LCM addition to soil, available NO_3-N in the soil profile (0-90 cm) increased significantly compared to control, at levels similar to the commercial fertilizers. The same was evident for soil salinity, but in all cases salinity levels were acceptable for most crops. After eight years of annual LCM application to soil, soil organic C and total N remained unchanged.

During the same period (2002-2005), Dordas et al. (2008) measured dry matter accumulation and partitioning at silking and harvest, yield components, morphological characteristics, chlorophyll content and N uptake and partitioning and calculated N remobilization and use efficiency in corn. They reported that upon LCM application to soil, all properties increased significantly compared to control and were at levels similar to the inorganic fertilization treatments, in the years 2004 and 2005.

The beneficial effect of soil application of LCM on the availability of micronutrients was apparent after seven years of repeated LCM applications (Nikoli & Matsi, 2011). Although, the Cu, Zn, Fe, Mn and B concentrations in corn aboveground biomass, collected at the R3 growth stage, were not affected by fertilization, the uptake of micronutrients by corn grown on manured plots increased significantly compared to control and were at levels similar to inorganic fertilizer treatments, in the years 2005 and 2006. In addition, by 2007, i.e. after nine years of LCM addition to soil, the soil available micronutrients increased significantly and this increase was accompanied by increases in soil total and dissolved organic C.

Measurements of the plant and soil parameters performed after 2007 revealed that the beneficial impact of soil application of LCM on macro- and micronutrient concentrations in corn aboveground biomass, plant uptake and soil availability and on soil total and dissolved organic C was persistent (unpublished data). Also soil total N was increased, upon soil LCM application of LCM, but the C/N ratio remained unchanged (unpublished data). In addition, in 2009, i.e. eleven years of repeated annual LCM additions to soil, available NO_3-N and salinity in the soil profile (0-90 cm) were found to be significantly increased compared to control, but at levels similar or lower than the commercial fertilizers. In all cases, soil salinity levels were acceptable for most crops (unpublished data).

6. Conclusions

The use of liquid cattle manure (LCM) as an organic fertilizer for crops is based on the fact that LCM is a valuable source of plant nutrients, especially of N; since it contains high amounts of readily plant available NH_4-N, due of its urine content. Large percentages of this ammoniacal N can be lost during the storage of the LCM, but especially after LCM application to soils through NH_3 volatilization. For this reason, soil incorporation of the LCM is suggested as soon as possible after land application or even at the moment of application. For this purpose, many techniques and equipment have been developed recently for use on arable soils but also in grassland, no-tillage and forest systems.

The application rates of LCM to soil, which are based usually on its N content, can be variable, depending on the expected LCM N that will be available for plant uptake during the growing season. The quantification of LCM N transformations and losses after LCM application to soils seems to be a black box, even today. The problem is complex, since in addition to LCM properties, soil properties along with the weather conditions regulate these transformations and losses. When the LCM is applied to soils at rates higher than the recommended (in respect to N) inorganic fertilization rates for crops, it can enhance crop growth, yield and macronutrients uptake and maintain soil fertility at desirable levels. However, the same has been proven for soil application of the LCM at rates equivalent to the recommended inorganic fertilizers rates for crops, based on the total N content of the LCM.

The beneficial effect of soil application of the LCM on micronutrients availability in soil and plant uptake can become apparent only after a long-term continuous use of LCM as a fertilizer. The same is true and for its effect on soil organic matter and physical properties. This is probably the reason for the few relevant studies on these topics in the scientific literature.

The risk of increased soil salinity, NO_3^- leaching and P build-up in the top soil at unacceptable levels due to LCM application are connected mainly to repeated heavy application rates, which should be avoided anyway. However, in any case it is uncertain if this risk is greater from LCM than from inorganic fertilizers use. The risk of Cu and Zn phytotoxicites is connected to the use of LCM enriched with these metals, but again this fact has not been adequately established.

Overall, it can be concluded that soil incorporation of liquid cattle manure (LCM) at rates equivalent to the recommended inorganic fertilization rates for crops can enhance crop growth, yield and nutrient uptake and maintain soil fertility at desirable levels, without causing adverse effects to plants, soils and the environment.

7. References

Amon, B., Kryvoruchko, V., Amon, T. & Zechmeister-Boltenstern, S. (2006). Methane, nitrous oxide and ammonia emissions during storage and after application of dairy cattle slurry and influence of slurry treatment. *Agriculture, Ecosystems and Environment*, Vol.112, No.2-3, (February 2006), pp. 153-162, ISSN 0167-8809.

Angers, D.A., Chantigy, M.H., Rochette, P. & Gagnon, B. (2006). Dynamics of soil water-extractable organic C following application of dairy cattle manures. *Canadian Journal of Soil Science*, Vol. 86, No.5, (November 2006), pp. 851-858, ISSN 1918-1841.

Antil, R.S., Gerzabek, M.H., Haberhauer, G. & Eder, G. (2005a). Long-term effects of cropped vs. fallow and fertilizer amendments on soil organic matter. I. Organic carbon. *Journal of Plant Nutrition and Soil Science*, Vol.168, No.1, (February 2005), pp. 108-116, ISSN 1522-2624.

Antil, R.S., Gerzabek, M.H., Haberhauer, G. & Eder, G. (2005b). Long-term effects of cropped vs. fallow and fertilizer amendments on soil organic matter. II. Nitrogen. *Journal of Plant Nutrition and Soil Science*, Vol.168, No.2, (April 2005), pp. 212-216, ISSN 1522-2624.

Beauchamp, E.G. (1983). Response of corn to nitrogen in preplant and sidedress applications of liquid dairy cattle manure. *Canadian Journal of Soil Science*, Vol.63, No.2, (May 1983), pp. 377-386, ISSN 1918-1841.

Beauchamp, E.G. (1986). Availability of nitrogen from three manures to corn in the field. *Canadian Journal of Soil Science*, Vol.66, No.4, (November 1986), pp. 713-720, ISSN 1918-1841.

Beauchamp, E.G. (1987). Corn response to residual N from urea and manures applied in previous years. *Canadian Journal of Soil Science*, Vol.67, No.4, (November 1987), pp. 931-942, ISSN 1918-1841.

Beauchamp, E.G., Kidd, G.E. & Thurtell, G. (1982). Ammonia volatilization from liquid dairy cattle manure in the field. *Canadian Journal of Soil Science*, Vol.62, No.1, (February 1982), pp. 11-19, ISSN 1918-1841.

Bechini, L. & Marino, P. (2009). Short-term nitrogen fertilizing value of liquid dairy manures is mainly due to ammonium. *Soil Science Society of America Journal*, Vol.73, No.6, (November-December 2009), pp. 2159-2169, ISSN 1435-0661.

Bittman, S, Hunt, D.E., Kowalenko, C.G., Chantigny, M., Buckley, K. & Bounaix, F. (2011). Removing solids improves responses of grass to surface-banded dairy manure slurry: A multiyear study. *Journal of Environmental Quality*, Vol.40, No.2, (March 2011), pp. 393-401, ISSN 1537-2537.

Bolan, N.S., Khan, M.A., Donaldson, J., Adriano, D.C. & Matthew, C. (2003). Distribution and bioavailability of copper in farm effluent. *The Science of the Total Environment*, Vol.309, No.1-3, (June 2003), pp. 225-236, ISSN 0048-9697.

Braschkat, J., Mannheim, T. & Marschner, H. (1997). Estimation of ammonia losses after application of liquid cattle manure on grassland. *Journal of Plant Nutrition and Soil Science*, Vol.160, No.2, (April 1997), pp. 117-123, ISSN 1522-2624.

Briceno, G., Demanet, R., de la Luz Mora, M. & Palma, G. (2008). Effect of liquid cow manure on Andisol properties and atrazine adsorption. *Journal of Environmental Quality*, Vol.37, No.4, (August 2008), pp. 1519-1526, ISSN 1537-2537.

Brock, E.H., Ketterings, Q.M. & McBride, M. (2006). Copper and zinc accumulation in poultry and dairy manure-amended fields. *Soil Science*, Vol.171, No.5, (May 2006), pp. 388-399, ISSN 1538-9243.

Burger, M. & Venterea, R.T. (2008). Nitrogen immobilization and mineralization kinetics of cattle, hog, and turkey manure applied to soil. *Soil Science Society of America Journal*, Vol.72, No.6, (November-December 2008), pp. 1570-1579, ISSN 1435-0661.

Carter, J.E., Jokela, W.E. & Bosworth, S.C. (2010). Grass forage response to broadcast or surface-banded liquid dairy manure and nitrogen fertilizer. *Agronomy Journal*, Vol.102, No.4, (July-August 2010), pp. 1123-1131, ISSN 1435-0645.

Chadwick, D.R., John, F., Pain, B.F., Chambers, B.J. & Williams, J. (2000a). Plant uptake of nitrogen from the organic nitrogen fraction of animal manures: a laboratory experiment. *Journal of Agricultural Science, Cambridge*, Vol.134, No.2, (March 2000), pp. 159-168, ISSN 1469-5146.

Chadwick, D.R. & Pain, B.F. (1997). Methane fluxes following slurry applications to grassland soils: laboratory experiments. *Agriculture, Ecosystems and Environment*, Vol.63, No.1, (May 1997), pp. 51-60, ISSN 0167-8809.

Chadwick, D.R., Pain, B.F. & Brookman, S.K.E. (2000b). Nitrous oxide and methane emissions following application of animal manures to grassland. *Journal of Environmental Quality*, Vol.29, No.1, (January-February 2000), pp. 277-287, ISSN 1537-2537.

Chang, C., Sommerfeldt, T.G. & Entz, T. (1991). Soil chemistry after eleven applications of cattle manure. *Journal of Environmental Quality*, Vol.20, No.2, (April-June 1991), pp. 475-480, ISSN 1537-2537.

Comfort, S.D., Kelling, K.A., Keeney, D.R. & Converse, J.C. (1990). Nitrous oxide production from injected liquid dairy manure. *Soil Science Society of America Journal*, Vol.54, No.2, (March-April 1990), pp. 421-427, ISSN 1435-0661.

Comfort, S.D., Motavalli, P.P., Kelling, K.A. & Converse, J.C. (1987). Soil profile N, P, and K changes from injected liquid dairy manure or broadcast fertilizer. *Transactions of the ASAE*, Vol.30, No.5, (September-October 1987), pp. 1364-1369, ISSN 0001-2351.

Culley, J.L.B., Phillips, P.A., Hore, F.R. & Patni, N.K. (1981). Soil chemical properties and removal of nutrients by corn resulting from different rates and timing of liquid dairy manure applications. *Canadian Journal of Soil Science*, Vol.61, No.1, (February 1981), pp. 39-46, ISSN 1918-1841.

Curless, M.A., Kelling, K.A. & Speth, P.E. (2005). Nitrogen and phosphorus availability from liquid dairy manure to potatoes. *American Journal of Potato Research*, Vol.82, No.4, (July-August 2005), pp. 287-297, ISSN 1099-209X.

Daliparthy, J., Herbert, S.J. & Veneman, P.L.M. (1994). Dairy manure applications on alfalfa: Crop response, soil nitrate, and nitrate in soil water. *Agronomy Journal*, Vol.86, No.6, (November-December 1994), pp. 927-933, ISSN 1435-0645.

de Jonge, L.W., Sommer, S.G., Jacobsen, O.H. & Djurhuus, J. (2004). Infiltration of slurry liquid and ammonia volatilization from pig and cattle slurry applied to harrowed and stubble soils. *Soil Science*, Vol.169, No.10, (October 2004), pp. 729-736, ISSN 1538-9243.

Dell, C.J., Meisinger, J.J. & Beegle, D.B. (2011). Subsurface application of manures slurries for conservation tillage and pasture soils and their impact on the nitrogen balance. *Journal of Environmental Quality*, Vol.40, No.2, (March 2011), pp. 352-361, ISSN 1537-2537.

Dordas, C.A., Lithourgidis, A.S., Matsi, Th. & Barbayiannis, N. (2008). Application of liquid cattle manure and inorganic fertilizers affect dry matter, nitrogen accumulation, and partitioning in maize. *Nutrient Cycling in Agroecosystems*, Vol.80, No.3, (March 2008), pp. 283-296, ISSN 1573-0867.

Eghball, B. (2002). Soil properties as influenced by phosphorus- and nitrogen-based manure and compost applications. *Agronomy Journal*, Vol.94, No.1, (January 2002), pp. 128-135, ISSN 1435-0645.

Evans, S.D., Goodrich, P.R., Munter, R.C. & Smith, R.E. (1977). Effects of solid and liquid beef manure and liquid hog manure on soil characteristics and on growth, yield, and composition of corn. *Journal of Environmental Quality*, Vol.6, No.4, (October-December 1997), pp. 361-368, ISSN 1537-2537.

Flessa, H. & Beese, F. (2000). Laboratory estimates of trace gas emissions following surface application and injection of liquid cattle slurry. *Journal of Environmental Quality*, Vol.29, No.1, (January-February 2000), pp. 262-268, ISSN 1537-2537.

Griffin, T., Gilberson, E. & Wiedenhoeft, M. (2002). Yield response of long-term grassland swards and nutrient cycling under different nutrient sources and management regimes. *Grass and Forage Science*, Vol.57, No.3, (September 2002), 268-278, ISSN 1365-2494.

Hansen, M.N., Sommer, S.G. & Madsen, N.P. (2003). Reduction of ammonia emission by shallow slurry injection: Injection efficiency and additional energy demand. *Journal of Environmental Quality*, Vol.32, No.3, (May 2003), pp. 1099-1104, ISSN 1537-2537.

He, Z., Griffin, T.S. & Honeycutt, C.W. (2004). Phosphorus distribution in dairy manures. *Journal of Environmental Quality*, Vol.33, No.4, (July 2004), pp. 1528-1534, ISSN 1537-2537.

Heathwaite, A.L., Griffiths, P. & Parkinson, R.J. (1998). Nitrogen and phosphorus in runoff from grassland with buffer strips following application of fertilizers and manures. *Soil Use and Management*, Vol.14, No.3, (September 1998), pp. 142-148, ISSN 1475-2743.

Japenga, J., Dalenberg, J.W., Wiersma, D., Scheltens, S.D., Hesterberg, D. & Salomons, W. (1992). Effect of liquid animal manure application on the solubilization of heavy metals from soil. *International Journal of Environmental Analytical Chemistry*, Vol.46, No.1-3, (1992), pp. 25-39, ISSN 1029-0397.

Jokela, W.E. (1992). Nitrogen fertilizer and dairy manure effects on corn yield and soil nitrate. *Soil Science Society of America Journal*, Vol.56, No.1, (January-February 1992), pp. 148-154, ISSN 1435-0661.

Jokela, W.E., Grabber, J.H., Karlen, D.L., Balser, T.C. & Palmquist, D.E. (2009). Cover crop and liquid manure effects on soil quality indicators in a corn silage system. *Agronomy Journal*, Vol.101, No.4, (July-August 2009), pp. 727-737, ISSN 1435-0645.

Jokela, W.E., Tilley, J.P. & Ross, D.S. (2010). Manure nutrient content on Vermont dairy farms: Long-term trends and relationships. *Communications in Soil Science and Plant Analysis*, Vol.41, No.5, (2010), pp. 623-637, ISSN 1532-2416.

Kaffka, S.R. & Kanneganti, V.R. (1996). Orchardgrass response to different types, rates and application patterns of dairy manure. *Field Crops Research*, Vol.47, No.1, (July 1996), pp. 43-52, ISSN 0378-4290.

Kleinman, P.J.A., Wolf, A.M., Sharpley, A.N., Beegle, D.B. & Saporito, L.S. (2005). Survey of water-extractable phosphorus in livestock manures. *Soil Science Society of America Journal*, Vol.69, No.3, (May-June 2005), pp. 701-708, ISSN 1435-0661.

Laws, J.A., Smith, K.A., Jackson, D.R. & Pain, B.F. (2002). Effects of slurry application method and timing on grass silage quality. *Journal of Agricultural Science*, Vol.139, No.4, (December 2002), pp. 371-384, ISSN 1916-9760.

Lithourgidis, A.S., Matsi, T., Barbayiannis, N. & Dordas, C.A. (2007). Effect of liquid cattle manure on corn yield, composition and soil properties. *Agronomy Journal*, Vol.99, No.4, (July-August 2007), pp. 592-596, ISSN 1435-0645.

Loro, P.J., Bergstrom, D.W. & Beauchamp, E.G. (1997). Intensity and duration of denitrification following application of manure and fertilizer to soil. *Journal of Environmental Quality*, Vol.26, No.3, (May 1997), pp. 706-713, ISSN 1537-2537.

Lund, Z.F., Doss, B.D. & Lowry, F.E. (1975). Dairy cattle manure-Its effect on yield and quality of coastal bermudagrass. *Journal of Environmental Quality*, Vol.4, No.3, (July-September 1975), pp. 358-362, ISSN 1537-2537.

Lowrance, R, Johnson, J.C., Newton, G.L. & Williams, R.G. (1998). Denitrification from soils of a year-round forage production system fertilized with liquid cattle manure. *Journal of Environmental Quality*, Vol.27, No.6, (November-December 1998), pp. 1504-1511, ISSN 1537-2537.

McBride, M.B. & Spiers, G. (2001). Trace element content of selected fertilizers and dairy manures as determined by ICP-MS. *Communications in Soil Science and Plant Analysis*, Vol.23, No.1-2, (2001), pp. 139-156, ISSN 1532-2416.

McGonigle, T.P. & Beauchamp, E.G. (2004). Relation of yield corn (*Zea mays* L.) to nitrogen in shoot and soil during the early-season following manure application to field plots. *Canadian Journal of Soil Science*, Vol.84, No.4, (November 2004), pp. 481-490, ISSN 1918-1841.

Maguire, R.O., Kleinman, P.J.A. & Beegle, D.B. (2011a). Novel manure management technologies in no-till and forage systems: Introduction to the special series. *Journal of Environmental Quality*, Vol.40, No.2, (March 2011), pp. 287-291, ISSN 1537-2537.

Maguire, R.O., Kleinman, P.J.A., Dell, C.J., Beegle, D.B., Brandt, R.C., McGrath, J.M. & Ketterings, Q.M. (2011b). Manure application technology in reduced tillage and forage systems: A review. *Journal of Environmental Quality*, Vol.40, No.2, (March 2011), pp. 292-301, ISSN 1537-2537.

Mannheim, T., Braschakat, J. & Marschner, H. (1995). Reduction of ammonia emissions after application of liquid cattle manure on arable soil and grassland: Comparison of wide spread application, application in narrow bands and injection. *Journal of Plant Nutrition and Soil Science*, Vol.158, No.6, (December 1995), pp. 535-542, ISSN 1522-2624.

Marino, P., de Ferrari, G. & Bechini, L. (2008). Description of a sample of liquid dairy manures and relationships between analytical variables. *Biosystems Engineering*, Vol.100, No.2, (June 2008), pp. 256-265, ISSN 1537-5110.

Matsi, T., Lithourgidis, A.S. & Gagianas, A.A. (2003). Effects of injected liquid cattle manure on growth and yield of winter wheat and soil characteristics. *Agronomy Journal*, Vol.95, No.3, (May-June 2003), pp. 592-596, ISSN 1435-0645.

Mattila, P.K. & Joki-Tokola, E. (2003). Effect of treatment and application technique of cattle slurry on its utilization by ley: I. Slurry properties and ammonia volatilization. *Nutrient Cycling in Agroecosystems*, Vol.65, No.3, (March 2003), pp. 221-230, ISSN 1573-0867.

Mattila, P.K., Joki-Tokola, E. & Tanni, R. (2003). Effect of treatment and application technique of cattle slurry on its utilization by ley: II. Recovery of nitrogen and composition of herbage yield. *Nutrient Cycling in Agroecosystems*, Vol.65, No.3, (March 2003), pp. 231-242, ISSN 1573-0867.

Mellek, J.E., Jeferson, D., da Silva, V.L., Favaretto, N., Pauletti, V., Vezzani, F.M. & de Souza, J.L.M. (2010). Dairy liquid manure and no tillage: Physical and hydraulic properties and carbon stocks in a Cambisol of Southern Brazil. *Soil and Tillage Research*, Vol.110, No.1, (September 2010), pp. 69-76, ISSN 0167-1987.

Misselbrook, T.H., Laws, J.A. & Pain, B.F. (1996). Surface application and shallow injection of cattle slurry on grasslands: nitrogen losses, herbage yields and nitrogen recoveries. *Grass and Forage Science*, Vol.51, No.3, (September 1996), pp. 270-277, ISSN 1365-2494.

Misselbrook, T.H., Pain, B.F., Stone, A.C. & Scholefield, D. (1995). Nutrient runoff following application of livestock wastes to grassland. *Environmental Pollution*, Vol.88, No.1, (1995), pp. 51-56, ISSN 0269-7491.

Misselbrook, T.H., Smith, K.A., Johnson, R.A. & Pain, B.F. (2002). Slurry application techniques to reduce ammonia emissions: Results of some UK field-scale experiments. *Biosystems Engineering*, Vol.81, No.3, (March 2002), pp. 313-321, ISSN 1537-5110.

Misselbrook, T.H., Scholefield, D. & Parkinson, R. (2005). Using time domain reflectometry to characterize cattle and pig slurry infiltration into the soil. *Soil Use and Management*, Vol.21, No.2, (June 2005), pp. 167-172, ISSN 1475-2743.

Motavalli, P.P., Comfort, S.D., Kelling, K.A. & Converse, J.C. (1985). Changes in soil profile N, P, and K from injected liquid dairy manure or fertilizer, *Agricultural waste utilization and management, Proceedings of the 5th International Symposium on Agricultural Wastes*, pp. 200-210, ISBN 0916150763, Chicago, IL, USA, December, 16-17, 1985.

Motavalli, P.P., Kelling, K.A. & Converse, J.C. (1989). First-year nutrient availability from injected dairy manure. *Journal of Environmental Quality*, Vol.18, No.2, (April-June 1989), pp. 180-185, ISSN 1537-2537.

Nikoli, Th. & Matsi, Th. (2011). Influence of liquid cattle manure on micronutrients content and uptake by corn and their availability in a calcareous soil. *Agronomy Journal*, Vol.103, No.1, (January-February 2011), pp. 113-118, ISSN 1435-0645.

Olesen, T, Moldrup, P. & Henriksen, K. (1997). Modeling diffusion and reaction in soils. 6. Ion diffusion and water characteristics in organic manure-amended soil. *Soil Science*, Vol.162, No.6, (June 1997), pp. 399-409, ISSN 1538-9243.

Pain, B.F., Smith, K.A. & Dyer, C.J. (1986). Factors affecting the response of cut grass to the nitrogen content of dairy cow slurry. *Agricultural Wastes*, Vol.17, No.3, (1986), pp. 189-202, ISSN 09608524.

Pain, B.F., Phillips, V.R., Clarkson, C.R. & Klarenbeek, J.V. (1989). Loss of nitrogen through ammonia volatilization during and following the applications of pig or cattle slurry to grassland. *Journal of the Science of Food and Agriculture*, Vol.47, No.1, (1989), pp. 1-12, ISSN 1097-0010.

Pain, B.F., Thompson, R.B., Rees, Y.J. & Skinner, J.H. (1990). Reducing gaseous losses of nitrogen from cattle slurry applied to grassland by the use of additives. *Journal of the Science of Food and Agriculture*, Vol.50, No.2, (1990), pp. 141-153, ISSN 1097-0010.

Parsons, K.J., Zheljazkov, V.D., MacLeod, J. & Caldwell, C.D. (2007). Soil and tissue phosphorus, potassium, calcium, and sulfur as affected by dairy manure application in a no-till corn, wheat, and soybean rotation. *Agronomy Journal*, Vol.99, No.5, (September 2007), pp. 1306-1316, ISSN 1435-0645.

Patni, N.K. & Culley, J.L.B. (1989). Corn silage yield, shallow groundwater quality and soil properties under different methods and times of manure application. *Transactions of the ASAE*, Vol.32, No.6, (November-December 1989), pp. 2123-2129, ISSN 0001-2351.

Paul, J.W. & Beauchamp, E.G. (1993). Nitrogen availability for corn in soils amended with urea, cattle slurry, and solid and composted wastes. *Canadian Journal of Soil Science*, Vol.73, No.2, (May 1993), pp. 253-266, ISSN 1918-1841.

Phillips, P.A., Culley, J.L.B., Hore, F.R., & Patni N.K. (1981). Pollution potential and corn yields from selected rates and timing of liquid manure applications. *Transactions of the ASAE*, Vol. 24, No. 1, (1981), pp. 139-144, ISSN 0001-2351.

Pfluke, P.D., Jokela, W.E. & Misselbrook, T.H. (2011a). Dairy slurry application method impacts ammonia emission and nitrate leaching in no-till corn silage. *Journal of Environmental Quality*, Vol.40, No.2, (March 2011), pp. 383-392, ISSN 1537-2537.

Pfluke, P.D., Jokela, W.E. & Bosworth, S.C. (2011b). Ammonia volatilization from surface-banded and broadcast application of liquid dairy manure on grass forage. *Journal of Environmental Quality*, Vol.40, No.2, (March 2011), pp. 374-382, ISSN 1537-2537.

Powell, J.M., Jokela, W.E. & Misselbrook, T.H. (2011). Dairy slurry application method impacts ammonia emission and nitrate leaching in no-till corn silage. *Journal of Environmental Quality*, Vol.40, No.2, (March 2011), pp. 383-392, ISSN 1537-2537.

Pratt, P.F., Davis, S. & Laag, A.E. (1977). Manure management in an irrigation basin relative to salt leachate to ground water. *Journal of Environmental Quality*, Vol.6, No.4, (October-December 1977), pp. 397-402, ISSN 1537-2537.

Pratt, P.F. & Laag, A.E. (1981). Effect of manure and irrigation on sodium bicarbonate-extractable phosphorus. *Soil Science Society of America Journal*, Vol.45, No.5, (September-October 1981), pp. 887-888, ISSN 1435-0661.

Randall, G.W., Iragavarapu, T.K. & Schmitt, M.A. (2000). Nutrient losses in surface drainage water from dairy manure and urea applied for corn. *Journal of Environmental Quality*, Vol.29, No.4, (July-August 2000), pp. 1244-1252, ISSN 1537-2537.

Reijs, J.W., Sonneveld, M.P.W., Sorensen, P., Schils, R.L.M., Groot, J.C.J. & Lantinga, E.A. (2007). Effects of different diets on utilization of nitrogen from cattle slurry applied to grassland on a sandy soil in The Netherlands. *Agriculture, Ecosystems and Environment*, Vol.118, No.1-4, (January 2007), pp. 65-79, ISSN 0167-8809.

Rochette, P., Angers, D.A., Chantigny, M.H., Gagnon, B. & Bertrand, N. (2006). In situ mineralization of dairy cattle manures as determined using soil-surface carbon dioxide fluxes. *Soil Science Society of America Journal*, Vol.70, No.3, (May 2006), pp. 744-752, ISSN 1435-0661.

Rochette, P., Angers, D.A., Chantigny, M.H., Gagnon, B. & Bertrand, N. (2008). N₂O fluxes in soils of contrasting textures fertilized with liquid and solid dairy cattle manures. *Canadian Journal of Soil Science*, Vol.88, No.2, (May 2008), pp. 175-187, ISSN 1918-1841.

Ross, I.J., Sizemore, S., Bowden, J.P. & Haan, C.T. (1979). Quality of runoff from land receiving surface application and injection of liquid dairy manure. *Transactions of the ASAE*, Vol.22, No.5, (1979), pp. 1058-1062, ISSN 0001-2351.

Siddique, M.T. & Robinson, J.S. (2003). Phosphorus sorption and availability in soils amended with animal manures and sewage sludge. *Journal of Environmental Quality*, Vol.32, No.3, (May 2003), pp. 1114-1121, ISSN 1537-2537.

Smith, K.A., Jackson, D.R. & Pepper, T.J. (2001a). Nutrient losses by surface run-off following the application of organic manures to arable land. 1. Nitrogen. *Environmental Pollution*, Vol. 112, No.1, (2001), pp. 41-51, ISSN 0269-7491.

Smith, K.A., Jackson, D.R. & Withers, P.J.A. (2001b). Nutrient losses by surface run-off following the application of organic manures to arable land. 2. Phosphorus. *Environmental Pollution*, Vol. 112, No.1, (2001), pp. 53-60, ISSN 0269-7491.

Sommer, S.G. & Ersboll, A.K. (1994). Soil tillage effects on ammonia volatilization from surface-applied or injected animal slurry. *Journal of Environmental Quality*, Vol.23, No.3, (May-June 1994), pp. 493-498, ISSN 1537-2537.

Sommer, S.G. & Olesen, J.E. (1991). Effects of dry matter content and temperature on ammonia loss from surface applied cattle slurry. *Journal of Environmental Quality*, Vol.20, No.3, (July-September 1991), pp. 679-683, ISSN 1537-2537.

Sorensen, P. (1998). Effects of storage time and straw content of cattle slurry on the mineralization of nitrogen and carbon in soil. *Biology and Fertility of Soils*, Vol.27, No.1, (May 1998), pp. 85-91, ISSN 1432-0789.

Sorensen, P. (2004). Immobilization, remineralisation and residual effects in subsequent crops of dairy cattle slurry nitrogen compared to mineral fertilizer nitrogen. *Plant and Soil*, Vol.267, No.1-2, (December 2004), pp. 285-296, ISSN 1573-5036.

Sorensen, P. & Jensen, E.S. (1995). Mineralization-immobilization and plant uptake of nitrogen as influenced by the spatial distribution of cattle slurry in soils of different texture. *Plant and Soil*, Vol.173, No.2, (June 1995), pp. 283-291, ISSN 1573-5036.

Soupir, M.L., Mostaghimi, S. & Yagow, E.R. (2006). Nutrient transport from livestock manure applied to pastureland using phosphorus-based management strategies. *Journal of Environmental Quality*, Vol.35, No.4, (July 2006), pp. 1269-1278, ISSN 1537-2537.

Sutton, A.L., Nelson, D.W., Kelly, D.T. & Hill, D.L. (1986). Comparison of solid vs. liquid dairy manure applications on corn yield and soil composition. *Journal of Environmental Quality*, Vol.15, No.4, (October-December 1986), pp. 370-375, ISSN 1537-2537.

Sutton, A.L., Nelson, D.W., Moeller, N.J. & Hill, D.L. (1979). Applying liquid dairy waste to silt loam soils cropped to corn and alfalfa-orchard grass. *Journal of Environmental Quality*, Vol.8, No.4, (October-December 1979), pp. 515-520, ISSN 1537-2537.

Tarkalson, D.D. & Leytem, A.B. (2009). Phosphorus mobility in soil columns treated with dairy manures and commercial fertilizer. *Soil Science*, Vol.174, No.2, (February 2009), pp. 73-80, ISSN 1538-9243.

Thompson, R.B., Ryden, J.C. & Lockyer, D.R. (1987). Fate of nitrogen in cattle slurry following surface application or injection to grassland. *Journal of Soil Science*, Vol.8, No.4, (December 1987), pp. 689-700, ISSN 1365-2389.

Thompson, R.B., Pain, B.F. & Lockyer, D.R. (1990a). Ammonia volatilization from cattle slurry following surface application to grassland. I. Influence of mechanical separation, changes in chemical composition during volatilization and the presence of the grass sward. *Plant and Soil*, Vol.125, No.1, (June 1990), pp. 109-117, ISSN 1573-5036.

Thompson, R.B., Pain, B.F. & Rees, Y.J. (1990b). Ammonia volatilization from cattle slurry following surface application to grassland. II. Influence of application rate, wind speed and applying slurry in narrow bands. *Plant and Soil*, Vol.125, No.1, (June 1990), pp. 119-128, ISSN 1573-5036.

Unwin, R.J., Pain, B.F. & Whinham, W.N. (1986). The effect of rate and time of application of nitrogen in cow slurry on grass cut for silage. *Agricultural Wastes*, Vol.15, No.4, (1986), pp. 253-268, ISSN 09608524.

Vellidis, G., Hubbard, K., Davis, J.G., Lowrance, R., Williams, R.G., Johnson, J.C., Jr. & Newton, G.L. (1996). Nutrient concentrations in the soil solution and shallow groundwater of a liquid dairy manure land application site. *Transactions of the ASAE*, Vol.39, No.4, (July-August 1996), pp. 1357-1365, ISSN 0001-2351.

Webb, J., Menzi, H., Pain, B.F., Misselbrook, T.H., Dammgen, U., Hendriks, H. & Dohler, H. (2005). Managing ammonia emissions from livestock production in Europe. *Environmental Pollution*, Vol.135, No.3, (June 2005), pp. 399-406, ISSN 0269-7491.

Webb, J., Pain, B., Bittman, S. & Morgan, J. (2010). The impacts of manure application methods on emissions of ammonia, nitrous oxide and on crop response-A review. *Agriculture, Ecosystems and Environment*, Vol.137, No.1-2, (April 2010), pp. 39-46, ISSN 0167-8809.

Withers, P.J.A., Clay, S.D. & Breeze, V.G. (2001). Phosphorus transfer in runoff following application of fertilizer, manure, and sewage sludge. *Journal of Environmental Quality*, Vol.30, No.1, (January-February 2001), pp. 180-188, ISSN 1537-2537.

Whitehead, D.C., Bristow, A.W. & Pain, B.F. (1989). The influence of some cattle and pig slurries on the uptake of nitrogen by ryegrass in relation to fractionation of the slurry N. *Plant and Soil*, Vol.117, No.1, (June 1989), pp. 111-120, ISSN 1573-5036.

Whitehead, D.C. & Raistrick, N. (1993a). Nitrogen in the excreta of dairy cattle: changes during short-term storage. *Journal of Agricultural Science, Cambridge*, Vol.121, No.1, (August 1993), pp. 73-81, ISSN 1469-5146.

Whitehead, D.C. & Raistrick, N. (1993b). The volatilization of ammonia from cattle urine applied to soils as influenced by soil properties. *Plant and Soil*, Vol.148, No.1, (January 1993), pp. 43-51, ISSN 1573-5036.

Zebarth, B.J., Paul, J.W., Schmidt, O. & McDougall, R. (1996). Influence of the time and rate of liquid-manure application on yield and nitrogen utilization of silage corn in south coastal British Columbia. *Canadian Journal of Soil Science*, Vol.76, No.2, (May 1996), pp. 153-164, ISSN 1918-1841.

Zhang, M., Gavlak, R., Mitchell, A. & Sparrow, S. (2006). Solid and liquid cattle manure application in a subarctic soil: Bromegrass and oat production and soil properties. *Agronomy Journal*, Vol.98, No.6, (November-December 2006), pp. 1551-1558, ISSN 1435-0645.

8

Indigenous Fertilizing Materials to Enhance Soil Productivity in Ghana

Roland Nuhu Issaka[1], Moro Mohammed Buri[1],
Satoshi Tobita[2], Satoshi Nakamura[2] and Eric Owusu-Adjei[1]
[1]CSIR-Soil Research Institute, Academy Post Office, Kwadaso-Kumasi
[2]Japan International Research Center for Agricultural Sciences,
Ohwashi, Tsukuba,
[1]Ghana
[2]Japan

1. Introduction

Ghana is divided into six ecological zones namely; Sudan Savannah, Guinea Savannah, Forest Savannah Transition, Semi-Decideous Rainforest, High Rainforest and Coastal Savannah (Figure 1). The Guinea Savannah zone covers the whole of Upper West and Northern regions. It also occupies parts of Uppr East region and the northern part of Brong Ahafo and Volta regions. This zone has a single rainfall season lasting from May to October. Annual rainfall is about 1000 mm. The Sudan Savannah occupies the north-eastern part of Upper East region with an annual rainfall of between 500 – 700 mm. The Forest Savannah Transition lies within the middle portion of Brong Ahafo region, the northern part of both Ashanti and Eastern regions and the western part of Volta region. This zone has a bimodal rainfall with an annual rainfall of about 1200 mm. The Semi-Decideous zone cut across the northern part of Western region through southern Brong Ahafo, Ashanti and Eastern regions. It also occupies the estern part of Volta region and most parts of th Central region. It also has a bimodal rainfall with an annual rainfall of 1400mm.

Most parts of Western region is within the High Rainfall zone. A small part of Central region also falls within this zone. Annual rainfall is over 2000 mm with a bimodal partern. The Coastal Savannah stretches from Central region through Greater Accra to the Votal region. It has only one rainy season of about 600 mm.

1.1 Agriculture productivity

Table 1 shows major crops and the respective areas on which these crops were cultivated in 2007. Due to inappropriate farming practices actual yield per unit area is less than 30% of achievable yield. The trend is similar for most crops. After several interventions including the introduction of improved varieties yield per unit area for most crops is still very low. Soil fertility has been identified as a major factor militating against crop yield. Mineral fertilizers to boost crop production are expensive and sometimes unavailable.

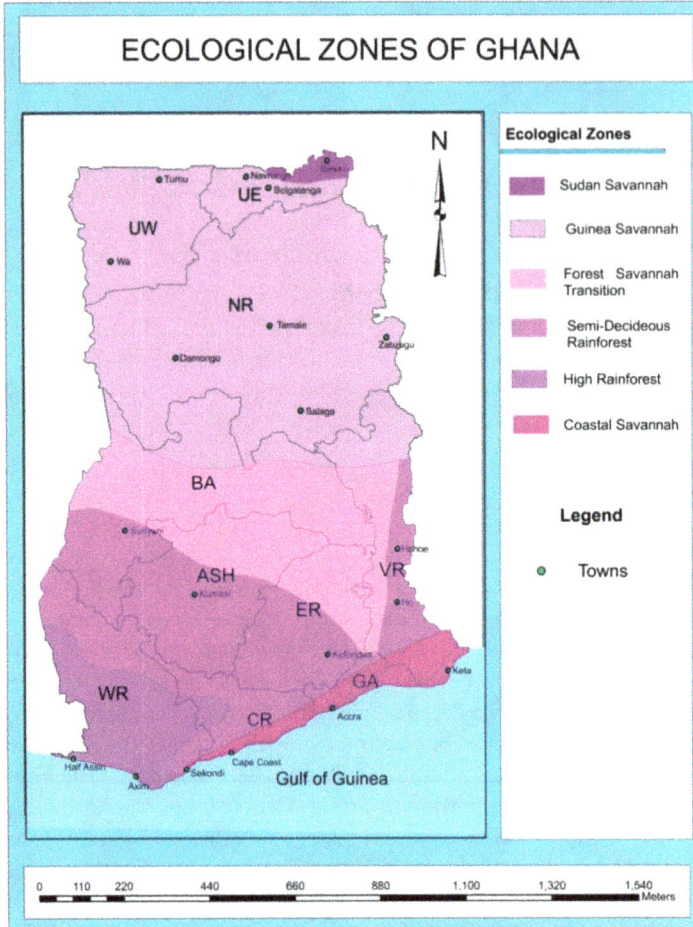

Fig. 1. Ecological zones of Ghana

Crop	Area ('000 ha)	Average Yield (Mg/ha)	Achievable Yield (Mg/ha)	Potential Yield Gap (Mg/ha)
Cassava	886	13.8	48.7	34.9
Maize	954	1.7	6.0	4.3
Rice	162	2.4	6.5	4.1
Yam	379	15.3	49.0	33.7
Plantain	325	11.0	20.0	9.0
Cocoa	1,600	0.4	1.0	0.6

MoFA (2009).

Table 1. Cultivated Area, Average Yield and Potential Yield of Selected Crops in Ghana

2. Status of soil fertility in Ghana

2.1 Major soils

The Ghanaian classification (Brammer, 1962) was equated to the World Reference Base (WRB) classification, ISSS/ISRIC/FAO (1998) by Adjei-Gyapong and Asiamah (2000) as follows:

Savanna Ochrosols (WRB: Lixisols/Luvisols) – These soils occur in northern Ghana and parts of the coastal savanna. They are highly weathered and moderately to strongly acid in the surface soil. Organic matter is low (<15 gkg^{-1} soil). Soil fertility is generally low.

Forest Ochrosols (WRB: Acrisols/Alfisols/Lixisols/Ferralsols/Nitisols/Plinthosols) - These soils occur within the forest zone and parts of the forest-savanna transition. They are deep and highly weathered and are generally moderate to strongly acid in the surface soil. These soils have high organic matter content in the top horizon which may contribute significantly to the phosphorus pool, exchangeable bases and Nitrogen levels.

Forest Oxysols (WRB: Ferralsols/Acrisols) - These occur in the high rainfall zone (south-west of Western Region). These soils are deep and highly weathered but strongly acid (pH<5.0). Organic matter is very high with high potential in N and P supply. P fixation is very high due to the presence of large amounts of Al and Fe oxides.

Groundwater Laterites (WRB: Plinthosol/Planosol) - These soils occur mostly in northern Ghana. They are shallow to plinthite and low in organic matter. Soil fertility is generally poor. They have high P fixation due to the presence of abundant iron concretions.

Parameter	Mean	Range	St. Dev.
pH (water)	4.6	3.7 - 7.4	0.5
Total C (g kg^{-1})	6.10	0.6 - 19	3.0
Total N (g kg^{-1})	0.65	0.1 - 1.6	0.3
C:N ratio	9.3	5.0-14.3	1.4
Available P (mg kg^{-1})	1.5	Tr - 5.4	0.9
Exchangeable K {cmol (+) kg^{-1}}	0.22	0.04 - 1.1	0.17
Exchangeable Ca {(+) kg^{-1}}	2.10	0.53 - 15	1.9
Exchangeable Mg {(+) kg^{-1}}	1.00	0.27 - 5.87	0.27
Exchangeable Na {(+) kg^{-1}}	0.12	0.1 - 0.72	0.11
Exchangeable. Acidity {(+) kg^{-1}}	1.00	0.05 - 1.80	0.48
Clay content (g kg^{-1})	66	40 - 241	39
Silt content (g kg^{-1})	607	347 - 810	107

Number of samples: 90; Source: Buri et al. 2000; Topsoil (0-20 cm)

Table 2a. Mean soil fertility characteristics of lowlands within the Guinea Savannah.

Characteristics of lowland soils: Most lowlands within the Guinea savannah and Semi-deciduous rainforest are mainly inland valleys and river flood plains. Rectilinear valleys occur within the Savannah agro-ecological zone while convex valleys are common within the Forest agro-ecological zone. Concave valleys, however, occur in both zones. Major soil types in these two zones are basically Gleysols and to a lesser extent, Fluvisols. *Volta* and *Lima* series are prominent within the savanna while *Oda, Kakum* and *Temang* series are prominent in the forest zone.

Soil fertility levels as observed for selected parameters are low across locations, particularly within the Savanna zone (Buri et al. 2009). Available phosphorus (P) is the most deficient nutrient in both zones. Soils of the Savanna were also observed to be quite acidic. Exchangeable Cations (K, Ca, Mg, Na) are quite moderate across locations within the Forest agro-ecology but relatively low for the Savannah, particularly Ca. Both total carbon and nitrogen levels, even though low, were comparatively higher for the forest than the savanna zone. To increase yield levels under these conditions the fertility levels of these soils must be improved.

Parameter	Mean	Range	St. Dev.
pH (water)	5.7	4.1 – 7.6	0.89
Organic C (g kg^{-1})	12	3.6 – 36.5	0.58
Total N (g kg^{-1})	1.1	0.30 – 3.20	0.05
C: N ratio	11	4.9 – 14.2	1.26
Available P (mg kg^{-1})	4.9	0.1 – 28.5	5.36
Exchangeable K {(+) kg^{-1}}	0.42	0.03 – 1.28	0.25
Exchangeable Ca {(+) kg^{-1}}	7.5	1.1 – 26.0	5.1
Exchangeable Mg {(+) kg^{-1}}	4.1	0.3 – 12.3	2.6
Exchangeable Na {(+) kg^{-1}}	0.32	0.04 – 1.74	0.26
Exchangeable. Acidity {(+) kg^{-1}}	0.31	0.04 – 1.15	0.29
Clay content (g kg^{-1})	127	41 - 301	8.2
Silt content (g kg^{-1})	502	187 - 770	45.8

Number of samples: 122; Source: Buri et al. 2009. Topsoil 0-20 cm

Table 2b. Mean soil fertility characteristics of lowlands within the Semi-deciduous rainforest

Characteristics of upland soils: Nutrient levels for upland soils are characteristically very low (Tables 3 a, b and c). The soils are generally acidic. Soil pH values for the high rainforest zones are strongly acidic with very low exchangeable cations. Prolonged weathering and leaching under high rainfall regime has resulted in soils with very low pH regimes. In the Semi-deciduous rainforest the soils are relatively richer. Nutrient levels, however, suggest the need for improvement for any profitable production levels to be achieved. In the Savannah zone, nutrient levels are similar to the high rainfall zone. Soil pH values are, however, higher in the Savannah zone. While organic matter and nitrogen levels vary between ecologies, available P is a problem throughout the country. Low total P coupled with high fixation are major factors affecting P availability.

Parameter	Mean	Range
pH (water)	4.1	3.4-5.4
Total C (g kg^{-1})	13.0	3.9-24.3
Total N (g kg^{-1})	1.9	0.7-3.5
Available P (mg kg^{-1})	2.8	0.4-13.6
Exchangeable K {(+) kg^{-1}}	0.13	0.04-0.33
Exchangeable Mg {(+) kg^{-1}}	1.0	0.3-3.3
Exchangeable Ca {(+) kg^{-1}}	1.9	0.5-5.1
Exchangeable. Acidity {(+) kg^{-1}}	4.1	2.1-5.6
ECEC {(+) kg^{-1}}	3.1	0.9-8.1
Base saturation (%)	78	38-98
Sand (%)	-	-
Silt (%)	-	-
Clay (%)	6.1	2.1-18.1

Mean of 30 samples (0-20 cm)

Table 3a. Mean soil fertility characteristics of lowlands within the High rainforest (Western region)

Parameter	Mean	Range
pH (water)	5.5	4.6-6.6
Total C (g kg^{-1})	18.0	13.9-22.3
Total N (g kg^{-1})	1.3	0.8-2.5
Available P (mg kg^{-1})	6.8	2.4-18.6
Exchangeable K {(+) kg^{-1}}	0.18	0.05-033
Exchangeable Mg {(+) kg^{-1}}	2.5	1.5-4.6
Exchangeable Ca {(+) kg^{-1}}	4.9	2.0-10.1
Exchangeable. Acidity {(+) kg^{-1}}	1.5	0.2-3.6
ECEC {(+) kg^{-1}}	8.5	3.8-14.5
Base saturation (%)	85	75-95
Sand (%)	88	80-92
Silt (%)	15	5-21
Clay (%)	9	3-15

Mean of 40 samples (0-20 cm)

Table 3b. Mean soil fertility characteristics of lowlands within the Semi-deciduous rainforest

The inability of farmers to buy adequate amounts of mineral fertilizers to improve their crop yields is a major factor affecting food security. Use of locally available materials for soil improvement is an option that must be fully exploited. Use of these materials (manures, dungs, crop residue, mineral deposits) will significantly improve on the soils ability to sustain higher crop yield.

Parameter	Mean	Range
pH (water)	5.4	4.6-6.6
Total C (g kg^{-1})	12.0	2.9-18.3
Total N (g kg^{-1})	0.9	0.8-1.5
Available P (mg kg^{-1})	4.8	0.4-11.6
Exchangeable K {(+) kg^{-1}}	0.11	0.03-0.23
Exchangeable Mg {(+) kg^{-1}}	1.5	0.8-2.6
Exchangeable Ca {(+) kg^{-1}}	2.9	1.0-5.1
Exchangeable. Acidity {(+) kg^{-1}}	1.1	0.5-2.6
ECEC {(+) kg^{-1}}	5.1	2.8-8.5
Base saturation (%)	88	65-98
Sand (%)	92	82-92
Silt (%)	11	4-21
Clay (%)	7	2-12

Mean of 40 samples (0-20 cm)

Table 3c. Mean soil fertility characteristics of lowlands within the Savannah agro-ecological zone

Table 4 shows the effect of some of these materials (poultry manure, cow dung and rice husk) on rice yield. Sole application of these materials or in combination with mineral fertilizer increased rice yield over the control. In some instances sole application of some of these materials was as good as applying the recommended rate of mineral fertilizer. This clearly shows that the productivity of these poor soils can be improved through the use of locally available fertilizing materials.

Treatments	Potrikrom	Biemso No. 1	Biemso No. 2
Absolute Control	1.7a	2.6a	1.5a
90-60-60 (N-P$_2$O$_5$-K$_2$O) kg/ha (Urea as N source)	7.0d	7.5d	4.0ef
90-60-60 (N-P$_2$O$_5$-K$_2$O) kg/ha (SA as N source)	6.6d	7.3d	3.9def
Poultry Manure (7.0 t/ha)	6.0d	6.4c	3.3bc
PM (3.5 t/ha) +45-30-30 (N-P$_2$O$_5$-K$_2$O) kg/ha (Urea as N source)	6.3d	7.3d	3.7cde
Cattle Manure (7.0 t/ha)	4.5c	6.3c	3.4cd
CM (3.5 t/ha) +45-30-30 (N-P$_2$O$_5$-K$_2$O) kg/ha (Urea as N source)	4.9c	6.2c	3.7cde
Rice Husk (7.0 t/ha)	3.3b	5.5b	2.8b
RH (3.5 t/ha) +45-30-30 (N-P$_2$O$_5$-K$_2$O) kg/ha (Urea as N source)	4.4c	6.3c	3.3c

Source: Buri et al. 2004

Table 4. Effect of organic amendments and mineral fertilizer on paddy yield (t/ha)

3. Fertilizing potential of indigenous materials

Organic waste is produced wherever there is human habitation. The main forms of organic waste are household food waste, agricultural waste, human and animal waste. The economies of most developing countries dictates that materials and resources must be used to their fullest potential, leading to a culture of reuse, repair and recycling. In many developing countries there exists a whole sector of recyclers, scavengers and collectors, whose business is to salvage 'waste' material and reclaim it for further use. Where large quantities of waste are created, usually in the major cities, there are inadequate facilities for dealing with it. Much of this waste is either left to rot in the streets, or is collected and dumped on open land near the city limits. There are few environmental controls in these countries to prevent such practices. In addition, mineral deposits can also be exploited for several purposes including agriculture, particularly soil fertility improvement.

Types of organic waste include: (a) Domestic or household waste (cooked or uncooked food scraps), (b) Agricultural residue (e.g. stover of crops, rice husks, etc.). (c) Commercially produced organic waste (waste generated from schools, hotels, restaurants etc.), (d) Human faecal residue and (e) Animal residue (dung and manures).

3.1 Plant origin

Large amount of plant waste are annually generated in the form of plant materials. These include maize stover, rice straw, rice husk, millet/sorghum resulting from annual production of these crops. Large amounts of saw dust and wood shavings also come from the wood industry (mostly from timber firms and to a lesser extent from commercial carpenters). Maize stover is normally left on the field after harvest. On the other hand rice is generally brought to a particular spot on the farm for threshing. This results in huge amount of rice straw being put at particular spots on the farm. Rice husk is generated during milling. Large amount of rice husk (about 33 % of paddy weight) is seen in hips close to rice mills in villages or towns. These materials can be used in various forms (direct application, composting, charring or ash) to increase the productivity of the soil.

Rice straw and husk: An estimation of of rice straw and rice husk produced in Ghana in 2007 show that over 366,000 Mt of rice straw and 63,000 Mt of rice husk were produced as waste. Large amount of rice straw (> 120,000 Mt) was produced in the Northern region followed by the Volta (Figure 2A). Upper East region, Eastern and Western regions also produced substantial quantities of rice straw. The trend for rice husk is similar to rice straw but of lower quantities.

Nutrient equivalent of rice straw and husk is presented in Figure 2B. About 2528 Mt of N, 990 Mt of P_2O_5, 5,459 Mt of K_2O (Table 2) is potentially available in these materials. Large amount of calcium and magnesium are also potentially available. Over 50% of these nutrients are potentially available in the large amounts of rice straw and husk produced in the Northern region. These materials can be used as amendments to improve the productivity of lowland soils, for higher grain yields of rice. Particularly the very low organic matter and P status of lowland soils in Northern Ghana (savanna zone) can be improved through effective management of these materials. Total amount of rice straw and husk generated in 2007 is presented in Table 5. These materials can be used to improve both physical and chemical properties of the soil.

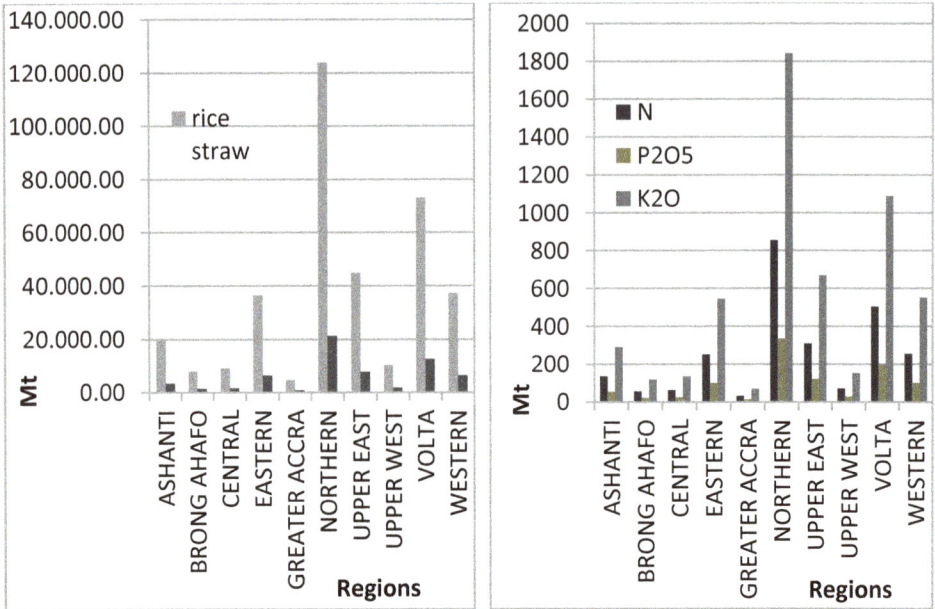

Source: Issaka et al., 2010

Fig. 2. Quantity of rice straw and husk (A) and nutrient equivalent (B) for the various regions

Organic source	Quantity (Mt)	N (Mt)	P_2O_5 (Mt)	K_2O (Mt)	CaO (Mt)	MgO (Mt)
Straw	366,975.2	1,834.9	587.2	5,137.7	-	-
Husk	63015.9	693.2	403.3	321.4	352.9	894.8
Total	429,991.1	2,528.1	990.5	5,459.1	352.9	894.8

Table 5. Estimated quantities of rice straw and husk

Saw dust: Generally timber firms are located in the Brong Ahafo, Ashanti and Eastern regions. Some can also be found in the Central and Western regions. Large hips of saw dust can be located on the out skirts of towns or cities where these factories are located. Saw dust constitute a good source of fertilizing material when properly managed.

3.2 Animal origin

Animal waste that can be classified as fertilizing materials include, poultry manure, cow dung, manures from sheep, goats and pigs. Human excreta is another source of fertilizing material. The amount of cow dung produce annually is far larger than poultry manure (Figure 3). Cattle are mostly reared in Northern Ghana (Northern, Upper East and Upper West regions) and hence large amounts of cow dung are obtained from these areas. The three regions together produce over 1.2 million Mt of cow dung annually. On the other hand poultry rearing is more concentrated in Greater Accra and Ashanti regions and like cow dung each of the regions produce some amount of it.

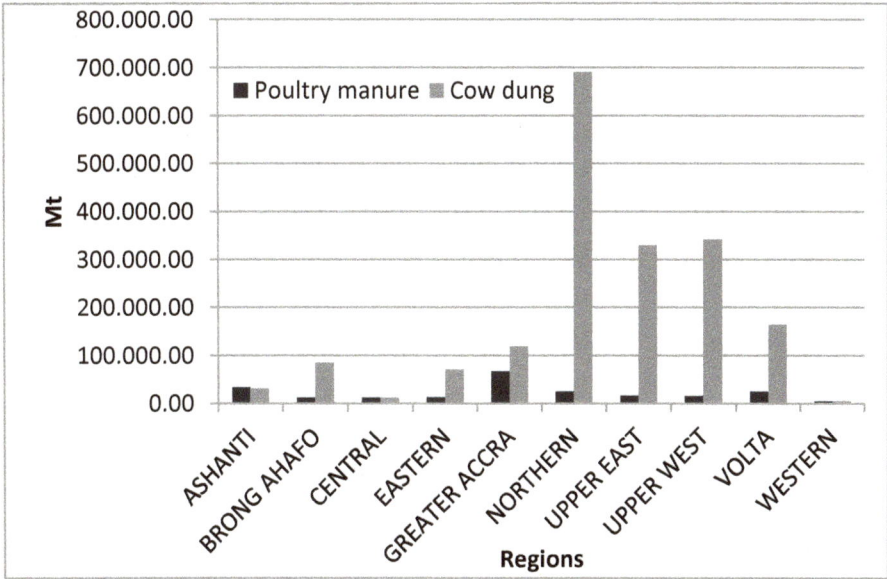

Fig. 3. Quantities of poultry manure and cow dung for the various regions

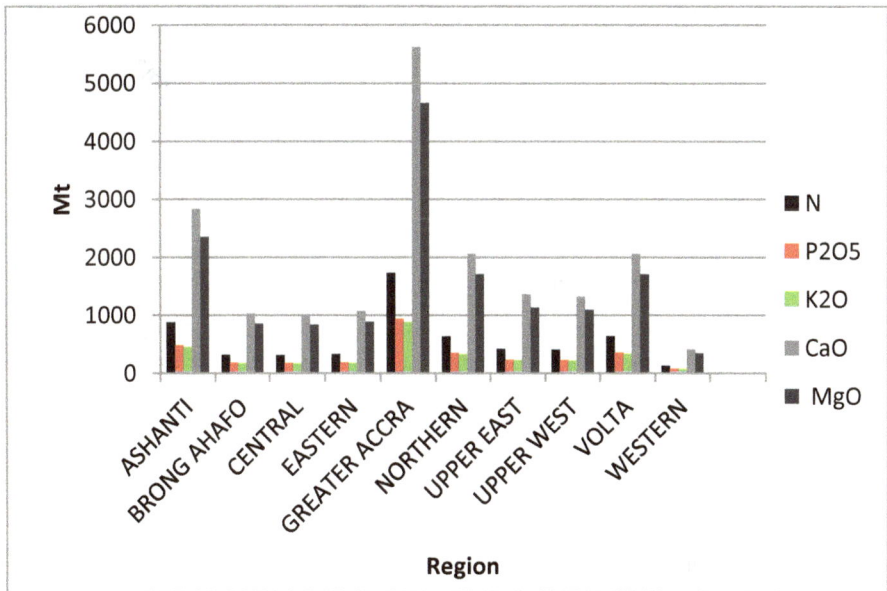

Fig. 4. Nutrient equivalent of poultry manure for the various regions

Nutrient equivalent of poultry manure is presented in Figure 4. Calcium and magnesium content of poultry manure is high. Addition of calcium enriched food materials into poultry feeds especially layers is the most possible reason. Poultry manure is a very good material that may even improve the pH of the soil when applied in large quantities. The manure contains all the major nutrients. These materials can be used to improve the productivity of soils in the forest and coastal Savanna zones.

Figure 5 shows nutrient equivalent of cow dung. Cow dung contains all the major nutrients while the urine is very rich in N and K. Provision of beddings for the cattle results in an improved material since the urine is absorbed by the beddings. Cow dung will increase crop yield significantly especially the very poor soils in the Savanna zone.

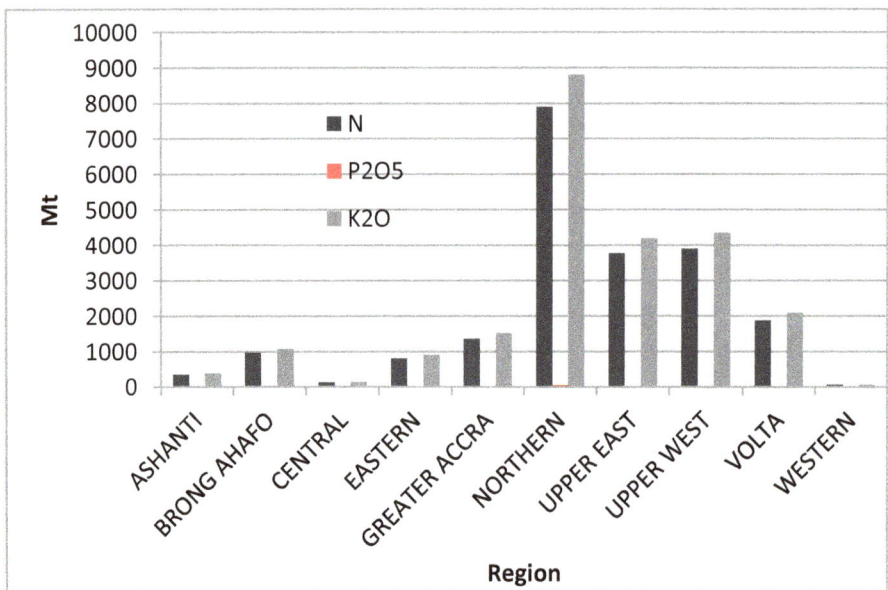

Fig. 5. Nutrient equivalent of cow dung and urine for the various regions

Sheep and goats are reared throughout the country. Northern and Volta regions rear a lot of sheep and goats while in the Upper West region a lot of sheep are reared. Generally these animals are reared throughout the country. Large amount of sheep and goat manures are produced annually (Figure 6). Quantity of manure produced is directly related to number of animals in these regions. Sheep and goat manure contain high amount of nitrogen. Manure produced by these animals is normally mixed with the urine resulting in the relatively high amount of nitrogen in their manure.

Generally about 3.2 million Mt of manure was produced by animals in 2007 (Table 6). This amount of manure can reduce the amount of money spent on mineral fertilizer if most of it is used in crop production.

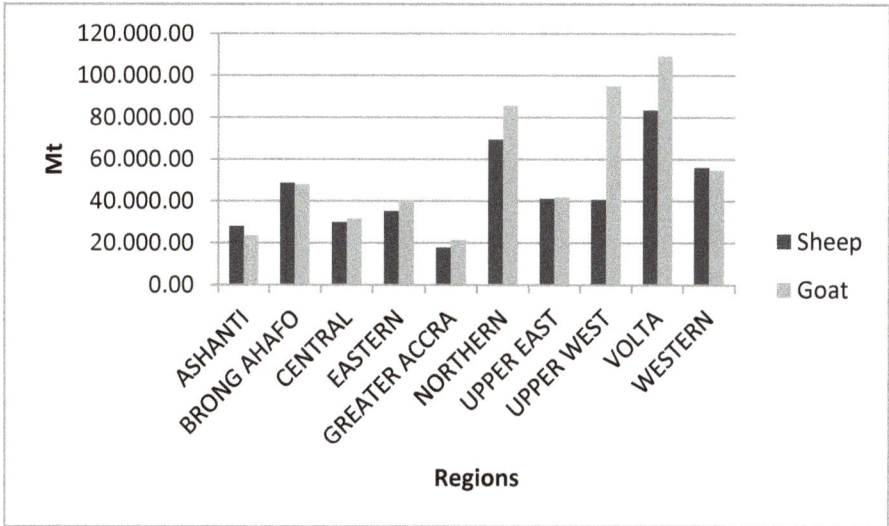

Fig. 6. Quantities of sheep and goat manure for the various regions

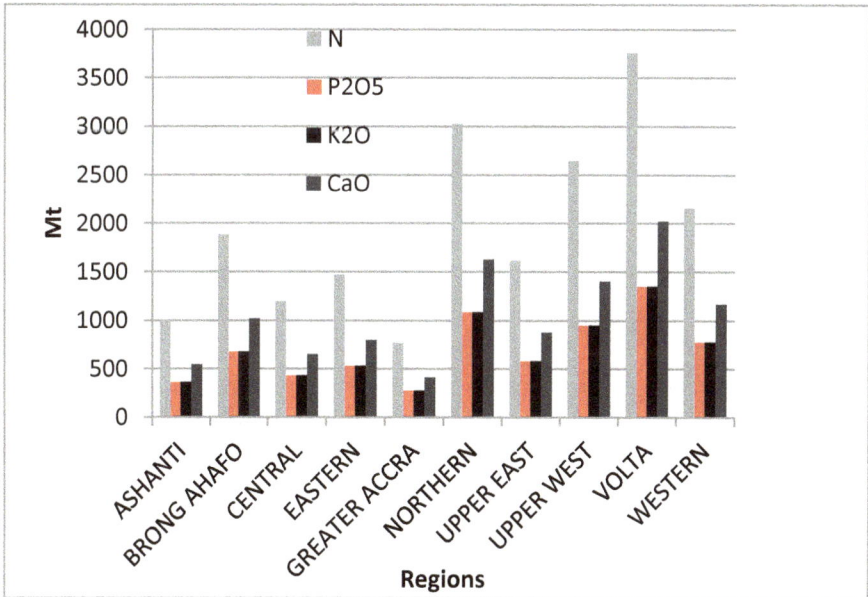

Fig. 7. Combined nutrient equivalent of sheep and goat manure

Type	Q'ty of manure/yr Mt	Nutrient content (Mt)				
		N	P_2O_5	K_2O	CaO	MgO
Poultry	222,228.0	5,777.9	3,111.2	2,911.2	18,778.3	15,556.0
Cattle	1,848,798.0	20,706.5	31,614.4	8,874.2	14,420.6	127,012.4
Sheep	449,388.0	8,763.1	3,145.7	3,145.7	4,988.2	-
Goat	551,354.4	10,751.4	3,859.5	3,859.5	5,513.5	
Pig	118,281.9	2,696.8	2,152.7	2,129.1	378.5	-
TOTAL	3,190,050.3	48,695.7	43,883.5	20,919.7	44,079.1	142,568.4

Table 6a. Quantity and fertilizer equivalent of animal manures

Region	Q'ty of urine/yr Mt	N	P_2O_5	K_2O
Cattle	1,746,087.0	21,127.65	174.61	23,572.17
Sheep	855.0	12.569	0.428	16.758
Goat	1049.0	15.42	0.525	20.560
Pig	294.6	1.119	0.295	2.917
TOTAL	1,748,285.6	21,156.8	175.9	23,612.4

Table 6b. Urine and total nutrient content

A study of quantity of human excreta produced in parts of Accra and Kumasi show that large amount of the material is available and a very good source of amendment for soil improvement (Table 7). Treatment and management of human excreta is generally poor and require some attention.

Location	Quantity Disposed/yr* (Mt)	Nitrogen content/yr (Mt)	P_2O_5 content/yr (Mt)
Accra	292,000	586.04	17.23
Kumasi	54,750	109.88	3.23
Total	346,750	695.93	20.46

Issaka et al 2010

Table 7. Potential available nutrients from human excreta in parts of Accra and Kumasi in 2007

4. Agro-minerals and rocks

Large deposit of limestone and dolomite are found in several parts of the country (Table 8). Mining of some of these minerals will support the agriculture sector of the economy. The very low levels of basic cations need to be significantly raised for the application of Nitrogen, Phosphorus and Potassium fertilizers to be effective and efficient. Basic slag is a good source of P (Table 9) and can be harnessed to improve available P in our soils for better crop production.

Deposit	Type	Deposit (million tons)	CaO and MgO contents (%)	Lime-Potential quantity (million tons)
Akuse region	Shells	1.3	53.5 (0.65)	0.7
Tano	Limestone	1000		
Nauli area	Limestone	400	48.7-52 (0.57-1.84)	194.8
Nauli deposit	Limestone	23	48.9 (1.04)	11.3
Bongo Da	Limestone	15	40.0-49.3 (1.2-4.2)	6.0
Bongo Da	Dolomite	23-30	30.8-36 (12.1-19.0)	7.1
Bupei Upper horizon	Limestone	6		
Bupei Lower horizon	Dolomite	144	29.3-31 (17.5-19.2)	42.2
Oterpolu	Limestone	8-10	38.3 (3.65)	3.1
Po river	Limestone	1.5		
Daboya	Dolomite	0.7		
Total				>265.2

Adapted from Kesse (1985, 1988) cited by Owusu-Bennoah, 1997. Content of MgO in parenthesis.

Table 8. Lime resources of Ghana

Steel Industries	Capacity Mt Steel/yr	Actual Production Mt Steel/yr	Basic Slag Produced/yr* Mt	Quantity of P_2O_5 Produced/yr Mt
Tema Steel Complex	30,000	4500	675	60.75
Ferro-aibric Limited	15000	9000	1350	121.50
Total	45,000	13,500	2,025	182.2

*Adapted from: Owusu-Bennoah, 1997. * Basic slag is 15% of steel produced and contains 9% P2O5*

Table 9. Annual production of basic slag in Ghana

5. Uses of indigenous materials in the various ecological zones

Plant Material: Even though there are many sources of plant materials emphasis will be on rice straw, rice husk and saw dust.

Rice straw is generated on rice farms and is usually burnt during or before land preparation. About 40 percent of the nitrogen (N), 30 - 35 percent of the phosphorus (P), 80 - 85 percent of the potassium (K), and 40 - 50 percent of the sulfur (S) taken up by rice remains in vegetative plant parts at crop maturity (Dobermann and Fairhurst, 2002). Burning causes almost complete N loss, P losses of about 25%, K losses of 20%, and S losses of 50 – 60%. The amount of nutrients lost depends on the method used to burn the straw (Dobermann and Fairhurst, 2002). More losses occur when straw is burnt under windy conditions since most of the ash will be blown away. Thus, huge amount of nutrients are lost annually due to poor management of rice straw. Proper management of this material will greatly improve soil fertility especially in the savanna areas where organic matter is very low. Rice husk may be used as poultry and pig feed but is hardly used as amendment hence all the nutrients in the materials is lost. In Ghana saw dust is normally treated as waste. The material may be used as landfills, deposited in the out skirt of the towns/cities and burnt or left to decompose. Generally saw dust decomposes very slowly and may pose a problem when applied directly on the field.

Animal manure: Poultry manure is on high demand and farmers are unable to get enough of it. Formerly huge piles of poultry manure could be seen by the side of roads close to poultry farms. Farmers could go for the material free of charge. The story is now completely opposite. Farmers have to pay in advance (including cost of transportation) before the manure is conveyed to the field. The effect of poultry manure in the farming industry is only significant in the forest zone (especially Ashanti and Eastern regions) and the coastal savanna zone (GT Accra).

Cow dung is a material highly cherished by rural dwellers in the Upper East region. Cow dung is used for crop production, as a binding agent for plastering houses, for cooking (fuel) and to a lesser extent for trapping termites to feed chicks. Unlike poultry manure the material is not sold and therefore not easy to obtain. In the Upper West region cow dung is not valued. Farmers consider the material to be too heavy and bulky to carry to the field (farms are generally far from home). The story is mixed in the Northern region were farms are also located far from the homes. However, some areas in this region have started making good use of this material. This is in contrast to the Upper East region where most farms are close to the house. In the forest zone cow dung is scarcely used for crop production. Some of it is, however, used in crop production in the coastal savanna zone.

Sheep and goat manures are used widely in the Upper East region. Both the dung and urine are normally mixed resulting in high nitrogen content of the manure. Inclusion of litter (grasses or plant residue) results in the production of farm yard manure. In other parts of the savanna zone the manure is scarcely used. Some amount of it is used in the Volta and Western regions.

Human excreta are used in parts of the Northern region. Farmers may hire trucks to collect the effluent which is deposited on their fields in its raw form. This practice is rather unhygienic. In parts of Kumasi, near Kwame Nkrumah University of Science and Technology, human excreta is allowed to decompose. Farmers are then given the option to collect the decomposed material to their fields. Patronage is however low.

6. Management of indigenous materials

Plant materials (saw dust, rice straw, rice husk etc) can be decomposed, charred or even ashed so as to make the material more user friendly and the nutrients more available to plants. Within the farm it should be possible to decompose rice straw before application.

Rice husk which is normally out of the farmers reach can be carbonized (material becomes easier to transport in this form) before application. The material can also be ashed (this may be environmentally unfriendly) before transporting to field. Ashed rice husk or straw has a pH of over 11.0 and hence can be used to improve the rather acidic soils that are common in Ghana. Decomposed saw dust is generally a very good material. It takes more than 6 months to get the material well decomposed. The material, however, can easily be charred or ashed.

7. Conclusion and recommendations

A wealth of indigenous materials are available in Ghana. These materials can be used to significantly improve the fertility of the soil. Some of these materials include rice straw and husk; saw dust, poultry manure, cow dung, goat and sheep manure, and human excreta. It is strongly recommended that materials of plant origin should be composted or charred before usage. Human excreta need to be composted before it is used.

8. Acknowledgements

The authors are grateful for the fund received from the Ministry of Agriculture, Forestry and Fisheries (MAFF), Japan, through the Japan International Research Center for the Agricultural Sciences (JIRCAS), as an out source contract for the study on "Improvement of Soil Fertility with Use of Indigenous Resources in Rice Systems of the Sub-Sahara Africa" with Dr.Satoshi Tobita as Project Leader. We also wish to thank technicians and all staff at Soil Research Institute laboratory for the analysis of both soil and plant materials.

9. References

Adjei-Gyapong, T. and Asiamah, R. D. 2000. The interim Ghana soil classification system and its relation with the World Reference Base for Soil Resources. FAO Report on Soil Resources No. 98.

Brammer, H. 1962. Soils: *In* Agriculture and Land use in Ghana. J. B. Wills ed., Oxford University Press London, Accra, New York. pp 88-126.

ISSS/ISRIC?FAO. 1998. *World Reference Base for Soil Resources.* World Soil Resource Report 84, FAO, Rome.

Issaka, R. N., Buri, M. M. Adjei, O. E. and Ababio, F. O. 2010. Characterization of indigenous resources in Ghana: Locally available soil amendments and fertilizers for improving and sustaining rice production. CSIR-SRI/CR/RNI/2010/01.

Buri, M. M., Iassaka, R. N., Fujii , H. and Wakatsuki, T. 2009. Comparison of Soil Nutrient status of some Rice growing Environments in the major Agro-ecological zones of Ghana. International Journal of Food, Agriculture & Environment Vol. 8 (1): 384-388

Buri, M. M., Issaka, R. N. T. Wakatsuki and E. Otoo 2004 Soil organic amendments and mineral fertilizers: Options for sustainable lowland rice production in the Forest agro-ecology of Ghana. Agriculture and Food Science Journal of Ghana. Vol., 3, 237-248

Dobermann. A. and Fairhurst, T. H   2002. Rice Straw Management.   Better Crops
      International Vol. 16, Special Supplement.

MOFA 2009 Agriculture in Ghana. Facts and Figures. SRID, Ministry of Food and
      Agriculture

Part 3

Integrated Nutrient Management: Case Studies from Central Europe, South America and Africa

9

Mineral and Organic Fertilization to Improve Soil Fertility and Increase Biomass Production and N Utilization by Cereals

Tamas Kismanyoky and Zoltan Toth
University of Pannonia Georgikon Faculty
Hungary

1. Introduction

Long term field experiments play a significant role in agronomy research, since they are sources of information from which we can learn a lot about the factors that influence soil fertility and its sustainability. In most cases, effects and interactions of experimental factors and treatments can be understood only from long-term data across different soil and climatic conditions (Kismanyoky & Toth, 2007).

Agronomic experiments are like conversations with the plant, the soil and the studied ecosystem in general. The results of long-term field experiments are valid for a given territorial unit and a given time interval but we want to use this information as the scientific basis for general recommendations over a larger area and for a longer period (Várallyay, 2009). Such recommendations are often communicated to individual farmers by agronomic advisory services and extension agencies.

Soil processes run very slowly and are often only measurable and/or quantifiable after decades. Long-term field experiments are an indispensable aid to our knowledge of predominantly practical solutions for sustainable land use. A large proportion of current agronomic problems can be clarified exclusively using long-term experiments (Körschens, 2005).

Several long-term field experiments were set up to study the effects and interactions of organic and mineral fertilizers on cereal production in Hungary. Adequate plant nutrition, especially nitrogen sufficiency, is a major preoccupation in agronomy due to the influence of nitrogen (N) on crop quantity or quality (Szentpétery et al., 2005; Berzsenyi & Lap, 2005; Tanács et al., 2005; Staugaitis et al., 2007; Rühlmann, 2006; Tajnsek et al., 2005; Hoffmann et al., 2006a). Since the N storage capacity of the soil is limited and fertilizers are expensive, N fertilizers should be applied precisely. There are several forms of N in the soil, but only nitrate and ammonium ions are absorbed in appreciable quantities by the roots. The soil NO_3^--N concentration plays an important role as a mineral nutrient for plants as well as for microorganisms involved in decomposition, nutrient cycling and other soil biochemical processes . On the other hand, excess NO_3 represents an environmental hazard due to its solubility, which causes it to leach through the soil profile (Bircsák et al., 2005; Kádár &

Németh, 2004; Tóth, 2006; Sethi et al., 2005). Since most of the soil N stock is stored by the soil organic matter (SOM), SOM management has also a great importance. In addition, humic substances are well known to influence the physical condition of the soil, resulting in several effects on the physical, chemical and biological processes of the soil. SOM stocks are also a sink and a source of the atmospheric CO_2, which is an important issue too (Reuter et al., 2007). In the upper soil layer, SOM is constantly being created and decomposed, but its pool is relatively stable (Jenny, 1941). Several authors reported the important role of mineral and organic fertilizers in maintaining or increasing SOM content (Uhlen, 1991; Teesalu et al., 2006; Rühlmann & Ruppel, 2005; Hoffmann et al., 2006b).

The international working group for soil fertility, under the leadership of Prof. Dr. Boguslawski, formed the IOSDV network with 10 participating European countries in 1984 (Boguslawski, 1995). Most of the field trials established under the auspices of this network continue to the present. In this paper, some results of the international mineral and organic nitrogen fertilization trial located in Keszthely (Hungary) are reported. The long-term field experiment testing mineral nitrogen fertilizer in combination with different forms of organic manure has provided specific results concerning the interactions of fertilizer and manure treatments on cereal production.

The present chapter gives a concise report of the most important results of the productivity and N use efficiency of cereals in the eighth crop rotation period (2005-2007). We report soil fertility data from the 24 year long-term experimental period (eight crop rotation cycles). We will discuss the effects of mineral and organic fertilization on biomass production (grain and straw), nutrient balances, N uptake of maize, winter wheat and winter barley, as well as on some indices of N utilization (N harvest index %, N use efficiency, agronomic efficiency, apparent recovery efficiency) and soil fertility. The figures, results and conclusions introduced in the chapter are based on experimental results.

2. Material and methods

The study was conducted in the international mineral and organic nitrogen fertilization trial (IOSDV) located in Keszthely in the west part of Hungary (46°44' N, 17°13' E, 112 m above sea level). The experiment was set up in the autumn of 1983 and the first harvest year was 1984.

The soil class at the study site was a Ramann-type brown forest soil (Eutric Cambisol) containing 410 g kg^{-1} sand, 320 g kg^{-1} silt, and 270 g kg^{-1} clay. The natural available phosphorus (P) content of this sandy loam soil was low (ammonium-lactate soluble [AL] P_2O_5: 60-80 mg kg^{-1}), the potassium (K) content medium (AL-K_2O: 140-160 mg kg^{-1}) and the humus (H) content fairly low (16-17 g kg^{-1}), with a pH_{KCl} value of 7.1. The bulk density of undisturbed soil is 1.53 g cm^{-3}. The 100 year average annual precipitation was 683 mm, but the distribution was often unequal (most rainfall occurs in June –79.0 mm - and low rainfall can be observed in January – 34.5 mm). The long-term annual mean temperature was 10.8 ºC.

The factorial experiment has a strip-plot design with three replications. The size of the subplots is 48 m^2. The factorial treatments were mineral N fertilization (5 rates) combined with organic fertilizers (3 levels). Treatments were applied in a field with three-year cereal crop rotation (maize, winter wheat, winter barley) system. The mineral N fertilizer rates were 0, 70, 140, 210 and 280 kg N ha⁻¹ in case of maize, 0, 50, 100, 150 and 200 kg N ha⁻¹ for winter wheat and 0, 40, 80, 120, 160 kg N ha⁻¹ for winter barley. N rates are referred to in the text as N_0, N_1, N_2, N_3 and N_4. Supplemental P and K fertilizers at rates of 100 kg ha⁻¹ P_2O_5 and K_2O were applied on all the experimental plots (even on the N control plots). The organic fertilizer treatments were applied as a complementary fertilization with the mineral NPK fertilizers having 3 different variants: (I) no organic fertilizer application (control), (II) farmyard manure (FYM) application (35 t ha⁻¹, in every third years before maize), (III) straw (St) incorporation (completed with 10 kg mineral N for each t straw ha⁻¹). After winter barley on the "St" plots an extra green manure (GM) was applied (*Raphanus sativus* var. Oleiformis) as a 2nd crop sowing on barley stubble.

In this paper, the productivity and N use efficiency of winter wheat, maize and winter barley was analysed in the eighth crop rotation period (2005-2007) Yield data was collected by threshing the grain collected a 2 m wide swath in the center of each plot and expressed on a dry matter (DM) basis converted into t ha⁻¹ units. The indices of N use efficiency were calculated according to Huggins & Pan (1993) (1-2) and Crosswell & Godwin (1984) (3-4) as follows:

$$N \text{ harvest index } [\%] = \frac{N \text{ accumulated in grain}}{N \text{ accumulated in total biomass}} \times 100 \qquad (1)$$

$$N \text{ use efficiency (NUE)} = \frac{\text{Grain yield } [kg\ ha^{-1}]}{N \text{ absorbed at maturity } [kg\ ha^{-1}]} \qquad (2)$$

Agronomic efficiency (AE) =

$$\frac{\text{Grain yield of fertilized crop } [kg\ ha^{-1}] - \text{Grain yield of unfertilized crop } [kg\ ha^{-1}]}{\text{Quantity of N applied } [kg\ ha^{-1}]} \qquad (3)$$

Apparent recovery efficiency of N (AREN) =

$$\frac{N \text{ uptake of fertilized crop } [kg\ ha^{-1}] - N \text{ uptake of unfertilized crop } [kg\ ha^{-1}]}{\text{Quantity of N applied } [kg\ ha^{-1}]} \qquad (4)$$

Soil fertility parameters were evaluated each year after harvest on topsoil (0-20 cm depth) collected from each plot. One composit soil sample was collected per plot. Soil tillage as all the soil and pest managent practices were applied as usual in convential farming systems. The soil organic carbon (SOC) and humus content (H) was quantified by Tyurin's method, the total N content was determined by Kjeldahl's method, while the soluble P_2O_5 and K_2O content of soil was quantified from Ammonium-lactate (AL) extraction by spectrophotometric and flamephotometric method, respectively (Ballenegger and Di Gléria 1962).

The statistical significance of the experimental factors and treatments (increasing rates of N fertilizer, organic fertilizer and the mineral N x organic fertilizer interaction) was tested by regression analysis and analysis of variance as two factor experiment on a P<0.05 level. SPSS statistical software was used for tests.

3. Results and discussion

3.1 Yields of the crops in the rotation as a function of different rates and forms of fertilizers

The relationship between N fertilization and yield can be described by a quadratic equation. The yield of winter wheat varied between 1.5-5 t ha^{-1} depending on the different manure treatments when averaged over the rotation cycle (Figure 1, Table 1). The maximal grain yield was obtained when 150 kg ha^{-1} N was applied.

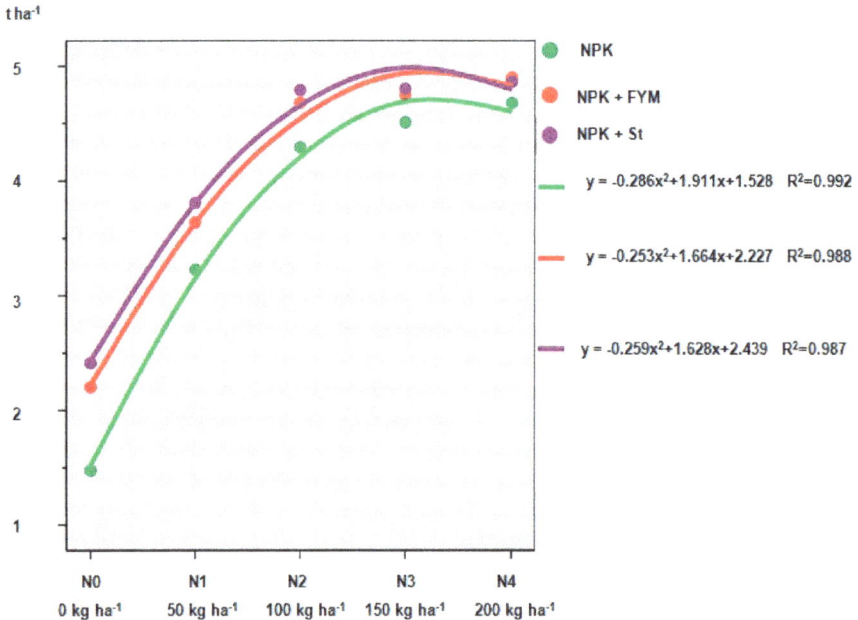

Fig. 1. Grain yield of winter wheat as a function of increasing rates of N fertilizer in the 8th rotation (2005-2007) at the long-term IOSDV trial, Keszthely, Hungary.

The average yield was lower when plots received NPK fertilizer alone (no organic fertilizer) compared to plots that received FYM or straw applied to complement the mineral fertilizer inputs. The N_0 plots yielded only ca. 2 t ha^{-1} when averaged over the two decades, whereas N_0 plots with complementary organic fertilizer application N_0 plots yielded 1000 kg ha^{-1} more. The positive effect of organic fertilizer compared to the NPK fertilizer alone was significant in all of the N fertilizer treatments (Table 1).

The maximal biomass production in grain and straw was measured in the N_3 treatment, when 150 kg ha^{-1} N fertilizer was applied. However, the complementary organic fertilizer application resulted in significantly higher biomass production in the N_3 treatment. The ratio between the grain and straw yields was closely to 1:1 (in DM values).

The grain yield of maize varied between 5-10 t ha^{-1} depending on the different N fertilizer and manure treatments (Figure 2). The maximal grain yield was obtained when 210 kg ha^{-1} N was applied.

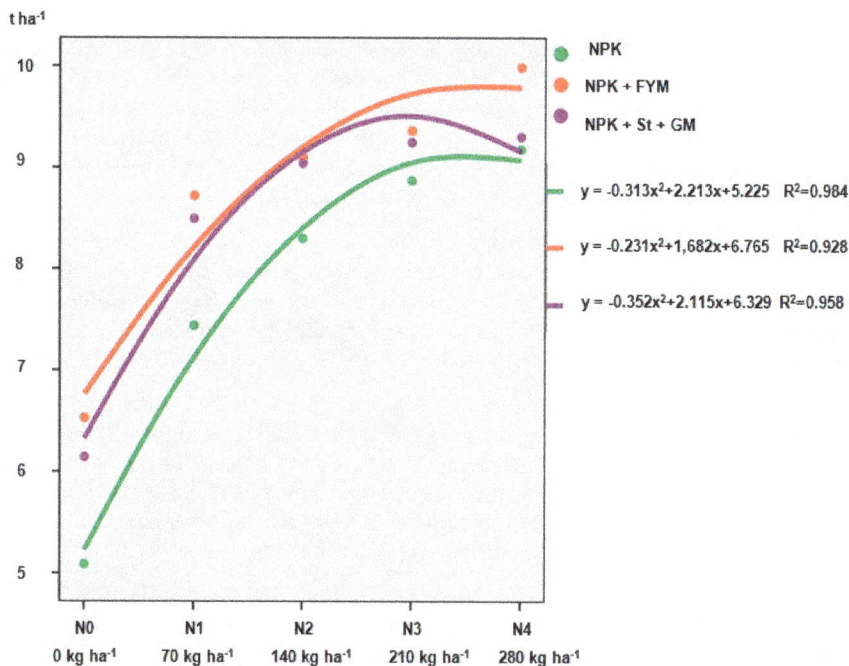

Fig. 2. Grain yield of maize as a function of increasing rates of N fertilizer in the 8th rotation (2005-2007) at the long-term IOSDV trial, Keszthely, Hungary.

Maize		Winter wheat		Winter barley			
Grain D.M. t ha⁻¹	Stalk D.M. t ha⁻¹	Grain D.M. t ha⁻¹	Straw D.M. t ha⁻¹	Grain D.M. t ha⁻¹	Straw D.M. t ha⁻¹		Treatments
4.37	3.03	1.48	1.33	1.68	1.56	N_0	NPK (I)
7.12	5.74	4.29	5.17	3.95	4.23	N_2	
7.88	7.38	4.68	5.27	4.44	4.70	N_4	
6.46	*5.38*	*3.48*	*3.92*	*3.36*	*3.50*	*4.35*	*Mean*
5.73	4.27	2.20	2.24	2.78	2.88	N_0	NPK+FYM (II)
8.34	6.66	4.68	5.49	4.40	4.65	N_2	
8.31	8.46	4.90	5.60	4.70	4.90	N_4	
7.46	*6.46*	*3.93*	*4.44*	*3.96*	*4.14*	*5.07*	*Mean*
5.71	4.28	2.42	2.16	2.68	2.97	N_0	NPK+St (+Gm) (III)
8.29	6.98	4.79	5.26	4.27	4.61	N_2	
8.05	8.45	4.87	5.45	4.70	4.28	N_4	
7.35	*6.57*	*4.03*	*4.29*	*3.88*	*3.95*	*5.01*	*Mean*
1.01	*1.04*	*0.40*	*0.36*	*0.58*	*0.51*	$LSD_{5\%}$ *between combinations*	

Table 1. Yields of crops from selected treatments in the 8th rotation (2005-2007) at the long-term IOSDV trial, Keszthely, Hungary.

Average yield level was significantly lower with NPK fertilizer alone than when organic fertilizers were applied to complement mineral fertilizers. On the N_0 plots, the complementary organic fertilizer application resulted in 1.0-1.5 tha⁻¹ extra yield. No significant difference in yield was detected between the NPK+FYM and the NPK+St+GM treatments.. The yield of the N_0 plots was significantly lower than each of the fertilized ones. The yield increase tended to diminish with higher N rates and the interactive effect of mineral N x organic fertilizer applications was also smaller at the high N rates.

In case of stalk/straw yield the tendencies are similar, but the values were registered within a greater interval (lower minimum and higher maximum values) than in case of grain. In addition the high N rates resulted in higher stalk yield increase than in case of grain. The extra yields resulting from complementary organic fertilizer application was proportional to the additional nutrient input from organic fertilizers.

The ratio between the grain and stalk yields was close to 1:1 (in DM values) at the highest fertilizer rates, while at the lower rates, the ratio of the grain yield was higher. In the case of the N_0 plots, this ratio was 1.37. The increase in grain yield due to organic fertilizer was higher at the lower rates of mineral N treatments and reached a plateau, but stalk/straw yield showed a continual linear increase over the range of mineral N x organic fertilizer combinations in this study. At higher N fertilizer rates, the efficiency of N fertilization decreased.

In the crop rotation, the winter barley followed the winter wheat. The effects of N fertilizer on barley were similar to those observed for wheat (Figure 3). The maximal grain yield was obtained when 120 kg ha⁻¹ N was applied.

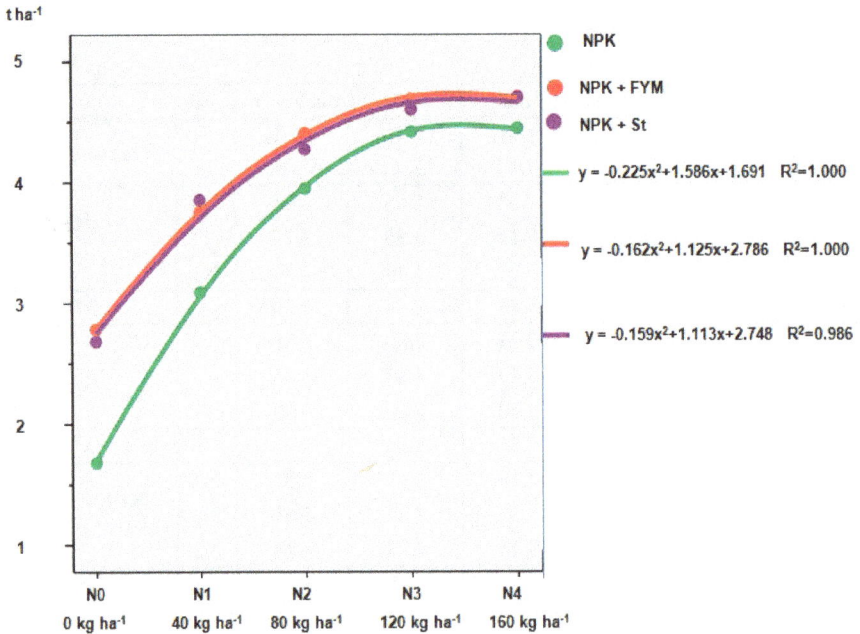

Fig. 3. Grain yield of winter barley as a function of increasing rates of N fertilizer in the 8th rotation (2005-2007) at the long-term IOSDV trial, Keszthely, Hungary.

The effect of FYM as a 3rd year effect in the crop rotation resulted in about 1000 kg ha^{-1} extra yield compared to the N0 plots. Straw incorporation (with complementary N application) resulted in the same yield increase as achieved with FYM. This positive effect of organic fertilizer was measured at higher N rates too. The biggest grain yield was harvested at N_4 treatment (160 kg ha^{-1} N). The grain:straw ratio was nearly 1:1 in alltreatments.

3.2 Nutrient content of the crops in the rotation as a function of different rates and forms of fertilizers

Table 2 demonstrates the nutrient content of wheat at harvest time. The N content of wheat grain was 23.80 g kg^{-1} while in case of straw it was merely 4.47 g kg^{-1} when averaged over the factorial mineral N x organic fertilizer combinations. The results are consistent with other experimental results (Sarkadi, 1975; Huggins & Pan, 1993; Bischoff, 1995; Berecz & Kismányoky, 2005).

More than 80% of the N uptake was found in the grain and less than 20% remained the straw at harvest, with some deviation depending on the rate of N supply. The N content (concentration – g kg^{-1}) in the grain and straw increased significantly with greater N fertilizer inputs compared to the N0 treatments.

Treatments		N g kg^{-1}		P_2O_5 g kg^{-1}		K_2O g kg^{-1}	
		Grain	Straw	Grain	Straw	Grain	Straw
NPK	N0	21.70	3.33	5.44	1.93	4.24	11.93
	N2	20.50	3.47	4.88	1.58	4.01	10.81
	N4	27.20	4.28	5.23	1.23	4.09	14.35
NPK+FYM	N0	22.80	5.40	5.17	2.12	3.88	13.26
	N2	23.20	3.86	5.12	1.69	4.16	13.53
	N4	25.60	5.09	5.94	1.53	3.82	15.63
NPK+St	N0	23.30	3.61	5.37	1.79	4.36	12.43
	N2	23.60	4.51	5.08	1.63	4.15	12.60
	N4	26.80	6.15	5.05	1.56	3.94	15.66
LSD$_{5\%}$		1.44	1.12	0.45	0.28	-	3.64
Probability		**	***	*	***	n.s.	***

Table 2. Nutrient content of winter wheat biomass from selected treatments in the 8th rotation (2005-2007) at the long-term IOSDV trial, Keszthely, Hungary.

The P_2O_5 values were fairly similar (constant) in each treatment, regardless of the N fertilizer inputs (4.88-5.94 g kg^{-1} P in grain and 1.23-2.12 % P in straw). The K_2O content of grain was rather stable (3-4 g kg^{-1} K) while that of straw increased in parallel with rising N rates (10-15 g kg^{-1} K). This is why the K ratio between the grain and straw was 1:3 in the N_0 treatment and increased to 1:4 in the N_4 treatment.

The nutrient content of maize biomass is presented in Table 3. The N content was 9-14 g kg^{-1} in the grain, while that of stalk was 3-9 g kg^{-1}. Greater N supply increased the N content (concentration – g kg^{-1}) in both grain and stalk.

Treatments		N g kg⁻¹		P₂O₅ g kg⁻¹		K₂O g kg⁻¹	
		Grain	Stalk	Grain	Stalk	Grain	Stalk
NPK	N0	9.31	3.86	2.62	2.10	2.48	9.42
	N2	10.79	5.41	2.71	1.89	2.41	11.60
	N4	11.40	7.32	2.74	2.27	2.45	10.80
NPK+FYM	N0	12.65	6.70	2.95	1.81	2.76	11.00
	N2	12.70	7.50	3.05	1.92	2.67	6.36
	N4	13.60	8.33	3.15	1.89	2.83	12.56
NPK+St+Gm	N0	12.66	5.67	2.98	170	2.70	6.63
	N2	12.86	7.35	2.89	1.69	2.49	10.90
	N4	14.36	9.04	3.33	2.08	2.93	11.57
LSD₅%		1.90	2.38	0.52	-	-	3.64
Probability		***	**	*	n.s.	n.s.	*

Table 3. Nutrient content of maize biomass from selected treatments in the 8th rotation (2005-2007) at the long-term IOSDV trial, Keszthely, Hungary.

The P and K level in the grain and straw of maize were lower than in wheat biomass and N fertilization resulted in no consistent change.

The nutrient content of winter barley biomass is given in Table 4. The N content of grain ranged from 17-21 g kg⁻¹ while it was 03-05 g kg⁻¹ in straw depending on the different rates of N fertilizers. In general, N content of grain was lower while that of straw was higher than values for winter wheat. The P_2O_5 content (concentraqtion - g kg⁻¹) of grain and straw were not affected by N fertilization and remained constant around 5 g kg⁻¹ and 2 g kg⁻¹, respectively.

The K_2O content of winter barley was also fairly constant - grain contained 7 – 8 g kg⁻¹ K and straw had 7 – 12 g kg⁻¹ K..

Treatments		N g kg⁻¹		P₂O₅ g kg⁻¹		K₂O g kg⁻¹	
		Grain	Straw	Grain	Straw	Grain	Straw
NPK	N0	17.66	3.87	5.67	2.26	6.97	7.96
	N2	18.96	3.68	6.67	2.27	8.62	11.05
	N4	19.63	5.39	5.53	2.40	7.13	11.32
NPK+FYM	N0	15.00	3.22	5.49	2.22	7.78	8.82
	N2	16.63	4.54	5.59	2.44	8.29	11.00
	N4	20.56	5.16	6.02	2.32	7.29	11.90
NPK+St	N0	12.76	5.39	4.94	2.32	7.14	9.60
	N2	21.16	5.16	6.94	2.38	8.51	11.00
	N4	20.03	5.67	5.67	2.30	7.16	12.76
LSD₅%		4.14	1.26	n.s.	n.s	1.07	2.13
Probability		**	**	+	+	***	***

Table 4. Nutrient content of w. barley biomass from selected treatments in the 8th rotation (2005-2007) at the long-term IOSDV trial, Keszthely, Hungary.

3.3 Nutrient balance and soil test results

The nutrient balance was calculated from the yield results (grain and total aboveground biomass) the absorbed nutrients (NPK) in crops. Tables 5, 6, and 7 show the nutrient balances of winter wheat, maize and winter barley from the IOSDV experiment.

The nutrient balance of winter wheat is presented in Table 5. Without organic fertilizer (I), the yield of crops was smaller than in the treatments (II-III) that included complementary organic fertilizers. Without fertilization (N_0), the inherent soil fertility produced about 2 t ha^{-1} cereal grain when averaged over the three crops. The addition of FYM increased the yield by about 1000 kg ha^{-1} compare to the N_0 treatments in three crops. The biomass production (grain and straw) was maximal at the amount of 150 kg ha^{-1} N doses (N_3) in the I-II-III fertilizer systems alike where the N balances were at equilibrium or slightly positive.

Treatments	NPK (I)			NPK+FYM (II)			NPK+St (III)		
	N_0	N_2	N_4	N_0	N_2	N_4	N_0	N_2	N_4
Nutrient balance kg ha^{-1}									
N	-37.07	+8.34	+47.10	-61.38	+29.76	+46.03	-64.05	+36.79	+36.04
P_2O_5	+89.25	+70.89	+69.06	+83.88	+66.72	+67.24	+83.18	+67.04	+64.73
K_2O	+77.82	+27.52	+5.05	+62.52	+6.41	-6.16	+62.70	+13.92	-4.48
Soil test results									
H g kg^{-1}	19.1	19.9	19.8	23.9	24.0	22.5	23.0	22.0	23.1
N_{total} g kg^{-1}	1.3	1.2	1.2	1.6	1.7	1.6	1.4	1.3	1.4
AL-P_2O_5 mg kg^{-1}	361	394	320	660	713	749	331	331	275
AL-K_2O mg kg^{-1}	243	198	209	463	332	357	260	247	170

LSD$_{5\%}$ H A×B: n.s., LSD$_{5\%}$ H Fert. Sys.: 1.8, LSD$_{5\%}$ H N rates: n.s., LSD$_{5\%}$ N total A×B: 0.3, LSD$_{5\%}$ AL-P_2O_5 A×B: 95, LSD$_{5\%}$ AL-K_2O A×B: 135

Table 5. Nutrient balance of winter wheat and soil test results in the 8th rotation (2005-2007) of the IOSDV long-term trial in Keszthely, Hungary.

The amount of N (kg ha^{-1}) removed by the plants in the N_0 plots indicates the N supplying ability of soil. The absorbed N was 37 kg ha^{-1} in case of wheat on the N_0 plots and the 2nd year effect of FYM resulted in 30 kg ha^{-1} above this, while organic fertilization with straw produced 29 kg ha^{-1} extra yield. It is well known that without N fertilizer input (N_0) soil N stocks are depleted, wich is also related to SOM cycle due to the lower amount of residues it is resulted in a negative H balance. The humus content of soil (concentration – g kg^{-1}) was increased by FYM application by some 5 g kg^{-1} (20 t ha^{-1} C_{org}). The effect of N fertilizer was negligible in this respect. The straw input (St) was valuable and increased the soil humus content by 2 to 3 g kg^{-1} as well.

The humus content (concentration - g kg^{-1}) and the humus pool (t ha^{-1}) increased significantly with organic fertilizer addition. It draws our attention to the fact that straw as a by-product is a very important C resource that has to be recycled by leaving straw residue in the field and tilling it into the soil, especially if there is no animal husbandry in the farm, in order to keep the C balance in the soil. This idea has also appeared in earlier publications on Hungarian soils (Kismányoky & Toth, 1997; Kismányoky & Toth, 2010).

The yield of maize (Table 1) varied between 3-8 t ha^{-1} depending on the amount of mineral N fertilizer applied. After farmyard manure (FYM) application and straw incorporation (St), 1000 kg ha^{-1} yield increase could be realized. In plots with the highest yield, the effect of straw was not so favourable (i.e., a yield plateau was reached), possibly because of the lack of required amount of K (negative K balance).

The highest yield was obtained at the equilibrium N nutrient balance (slightly positive). N supplying ability of soil was 50 kg ha^{-1} on the N_0 plots, while FYM resulted in 100 kg ha^{-1} N and straw+green manure resulted in 96 kg ha^{-1} N.

The P balance was positive in each treatment and there were significant P accumulation after farmyard manure (FYM) application. FYM and straw application resulted in increasing K_2O content in the soil. The K balance was positive in almost any case but at and more than 9 t ha^{-1} maize yield the amount of applied K fertilizer (100 kg ha^{-1} K_2O) proved to be insufficient and became negative.

Treatments	NPK (I)			NPK+FYM (II)			NPK+St+GM (III)		
	N_0	N_2	N_4	N_0	N_2	N_4	N_0	N_2	N_4
Nutrient balance kg ha^{-1}									
N	-52.37	+32.13	+136.15	-101.08	+5.48	+96.56	-96.54	-20.09	+88.03
P_2O_5	+82.11	+69.87	+64.47	+75.95	+61.79	+57.66	+76.23	+64.26	+56.25
K_2O	+60.63	+16,.27	+1.00	+37.23	+35.39	-31.01	+56.23	+3.28	-21.34
Soil test results									
H g kg^{-1}	18.2	18.7	17.6	21.4	19.7	22.4	19.6	19.5	20.3
N_{total} g kg^{-1}	1.2	1.3	1.4	1.7	1.7	1.7	1.5	1.7	1.7
AL-P_2O_5 mg kg^{-1}	392	404	382	646	559	598	537	492	404
AL-K_2O mg kg^{-1}	211	230	181	316	247	259	388	265	220

LSD$_{5\%}$ H A×B: n.s., LSD$_{5\%}$ H Fert. Sys.: 1.8, LSD$_{5\%}$ H N rates: n.s., LSD$_{5\%}$ N_{total} A×B: 0.3, LSD$_{5\%}$ AL-P_2O_5 A×B: 129, LSD$_{5\%}$ AL-K_2O A×B: 77

Table 6. Nutrient balance of maize and soil test results in the 8th rotation (2005-2007) of the IOSDV long-term trial in Keszthely, Hungary.

In case of winter barley (Table 7), the response to mineral and organic fertilization was similar to results from the wheat phase of the rotation. The 3rd year after effect of FYM and straw incorporation resulted in 1000 kg ha^{-1} extra yield in the crop rotation.

In the N_0 treatments 35 kg ha^{-1} N were absorbed by the barley plant. The 3rd year effect of FYM was 17 kg ha^{-1} N, while straw incorporation resulted in 15 kg ha^{-1} extra N absorption of barley in the crop rotation. The optimal N dose for winter barley was 120 kg ha^{-1}.

The P and K balance were positive in every treatment except at the highest N rate (N_4). The grain and straw removed 20-30 kg ha^{-1} P_2O_5 and 20-80 kg ha^{-1} K_2O from the soil annually. Plots receiving straw + GM exhibited notable K consumption (luxury uptake), supported by the high K_2O values in the laboratory analyses of harvested tissues.

Treatments	NPK (I)			NPK+FYM (II)			NPK+St (III)		
	N_0	N_2	N_4	N_0	N_2	N_4	N_0	N_2	N_4
Nutrient balance kg ha^{-1}									
N	-35.64	-10.04	+47.71	-52.40	-14.28	+38.05	-50.24	-34.11	+41.60
P_2O_5	+81.10	+64.10	+64.15	+78.33	+64.06	+60.34	+79.85	+59.42	+63.51
K_2O	+79.27	+26.89	+15.22	+52.94	+12.44	+11.96	+59.67	+19.47	+11.75
Soil test results									
H g kg^{-1}	19.76	19.40	18.53	22.03	23.73	22.19	22.16	21.40	23.53
N_{total} g kg^{-1}	1.31	1.31	1.24	1.30	1.36	1.35	1.33	1.38	1.39
AL-P_2O_5 $mgkg^{-1}$	539	399	332	583	553	521	403	361	327
AL-K_2O $mgkg^{-1}$	311	262	211	393	303	284	297	376	274

LSD$_{5\%}$ H A×B: 2.9, LSD$_{5\%}$ H Fert. Sys.: 1.7, LSD$_{5\%}$ H N rates: n.s., LSD$_{5\%}$ N_{total} A×B: n.s., LSD$_{5\%}$ AL-P_2O_5 A×B: 137, LSD$_{5\%}$ AL-K_2O A×B: n.s.

Table 7. Nutrient balance of winter barley and soil test results in the 8th rotation (2005-2007) of the IOSDV long-term trial in Keszthely, Hungary.

Balance	NPK			NPK+FYM			NPK+St		
	N_0	N_2	N_4	N_0	N_2	N_4	N_0	N_2	N_4
Average nutrient balance kg ha^{-1}									
N	-41.7	4.5	77.0	-71.8	6.9	60.2	-70.3	-6.1	55.3
P_2O_5	84.1	68.3	65.9	79.4	64.2	61.7	79.7	63.5	61.4
K_2O	72.6	23.5	7.1	50.8	18.1	-8.4	59.5	12.2	-4.7
Average soil test results									
H g kg^{-1}	19.00	19.34	18.63	22.43	22.47	22.35	21.58	20.96	22.32
N_{total} g kg^{-1}	1.27	1.26	1.26	1.53	1.56	1.55	1.42	1.46	1.47
AL-P_2O_5 mg kg^{-1}	430	399	344	630	608	623	423	394	335
AL-K_2O mg kg^{-1}	255	230	200	390	294	300	315	296	221

Table 8. Annual average nutrient balance of a winter wheat – maize – winter barley rotation grown from 2005-2007 in the IOSDV long-term trial in Keszthely, Hungary. Values were averaged (over 3 years) from the annual nutrient balances.

In Table 8 the nutrient balance in the complete crop rotation is demonstrated. The key points to note from this table are (1) the negative N balance in N_0 plots, which rely on the inherent soil N supply for crop production, (2) mineral N fertilization led to an equilibrium or positive N balance, with or without complementary organic fertilization, (3) the mineral P fertilization was sufficient for winter wheat, maize and winter barley production, evidenced by the positive P balance over the rotation, and (4) the mineral K fertilization was positive at most levels of mineral N input (N_0 to N_3) but a negative balance was observed in the N_4 plots that received complementary organic fertilizers, presumably due to the higher yields in these treatments.

3.4 N use indices of the crops in the rotation as a function of different rates and forms of fertilizers

Table 9, 10 and 11 shows the nitrogen use indices for each crop related to the effect of mineral N fertilization and organic fertilizer application.

Harvest index of N ($HI_{N\%}$) shows that wheat absorbed the highest proportion - more than 80 % - of the total above-ground N content in the grain, barley absorbed less than 80 %, while maize absorbed the lowest proportion – less than 70 % - when averaged over the different fertilizer treatments. The rest of the absorbed N was located in the vegetative above-ground biomass. When complementary organic fertilizers – both FYM and straw – were applied the $HI_{N\%}$ values were slightly reduced due to the dilution processes induced by the increase in biomass yield. With the rise in mineral N fertilizer rates (N_0-N_4) a definite reduction of $HI_{N\%}$ values was observed in case of maize, while in case of wheat and barley similar tendencies were not observed.

$HI_{N\%}$	NUE (kg ha⁻¹)	AE (kg ha⁻¹)	AREN (kg ha⁻¹)	Treatment	
86.5	49	-	-	N_0	Min. fert.
81.7	30	28.13	0.71	N_2	
87.8	31	16.01	0.58	N_4	
85.3	**37**	**22.07**	**0.64**		**Mean**
81.8	36	-	-	N_0	Min. fert. +FYM
83.7	36	16.53	0.68	N_2	
81.5	30	13.50	0.46	N_4	
82.3	**34**	**15.01**	**0.57**		**Mean**
86.5	34	-	-	N_0	Min. fert. +St
82.7	35	23.77	0.73	N_2	
79.6	30	12.26	0.50	N_4	
82.9	**33**	**18.01**	**0.61**		**Mean**
83.5	**34**	**18.37**	**0.61**		**Overall mean**

Table 9. N use indices of winter wheat from selected treatments in the 8th rotation (2005-2007) at the long-term IOSDV trial, Keszthely, Hungary.

HI$_{N\%}$	NUE (kg ha^{-1})	AE (kg ha^{-1})	AREN (kg ha^{-1})	Treatment	
77.7	83	-	-	N$_0$	Min. fert.
71.2	66	19.64	0.40	N$_2$	
62.5	55	12.54	0.33	N$_4$	
70.5	**68**	**16.09**	**0.36**		**Mean**
71.7	57	-	-	N$_0$	Min. fert. +FYM
68.0	54	18.65	0.39	N$_2$	
61.6	45	9.21	0.29	N$_4$	
67.1	**52**	**13.93**	**0.34**		**Mean**
74.9	59	-	-	N$_0$	Min. fert. +St+GM
67.5	53	18.43	0.44	N$_2$	
60.2	42	8.36	0.34	N$_4$	
67.5	**51**	**13.39**	**0.39**		**Mean**
68.4	**57**	**14.47**	**0.36**		**Main mean**

Table 10. N use indices of maize from selected treatments in the 8th rotation (2005-2007) at the long-term IOSDV trial, Keszthely, Hungary.

HI$_{N\%}$	NUE (kg ha^{-1})	AE (kg ha^{-1})	AREN (kg ha^{-1})	Treatment	
83.1	47	-	-	N$_0$	Min. fert.
82.8	44	28.35	0.68	N$_2$	
77.5	39	19.76	0.55	N$_4$	
81.1	**43**	**24.05**	**0.61**		**Mean**
81.8	55	-	-	N$_0$	Min. fert. +FYM
77.6	47	20.24	0.54	N$_2$	
79.3	39	13.71	0.51	N$_4$	
79.6	**46**	**16.97**	**0.52**		**Mean**
68.2	53	-	-	N$_0$	Min. fert. +St
79.2	37	19.83	0.80	N$_2$	
79.5	40	14.40	0.49	N$_4$	
75.6	**43**	**17.11**	**0.64**		**Mean**
78.8	**44**	**19.38**	**0.59**		**Main mean**

Table 11. N use indices of winter barley from selected treatments in the 8th rotation (2005-2007) at the long-term IOSDV trial, Keszthely, Hungary.

N use efficiency (NUE) shows the amount of grain yield (kg ha^{-1}) produced per each kg of absorbed N at maturity. Since winter wheat accumulated the highest proportion of N in the grain (the $HI_{N\%}$ values were the highest), the NUE values were the lowest in case of wheat (34,37 kg when averaged over the fertilizer treatments), while that of maize – which had the lowest $HI_{N\%}$ values – was the highest (57,03 kg). When complementary organic fertilizers were applied, NUE values were slightly reduced for wheat and maize. In the case of barley, the second year post effect of FYM application slightly increased N use efficiency. With the rise in mineral N fertilizer rates (N_0-N_4) due to the luxury consumption a definite reduction of NUE values was observed in case of maize, while in case of wheat and barley such definite tendencies were not observed.

Agronomic efficiency (AE) values show the amount of extra grain yield (kg ha^{-1}) production (yield increase) above the control per each kg of applied fertilizer N. The applied N fertilizer treatments resulted in higher AE values (higher extra yield) in case of wheat and barley than maize, when averaged over the fertilizer treatments. On the other hand it has to be mentioned that maize yielded much higher in the N control plots than the other cereals therefore the yield gain due to N fertilization was less. . When complementary organic fertilizers were applied, AE values were reduced in case of each crop. With the rise in mineral N fertilizer rates (N_0-N_4) a definite reduction of AE values can be observed for each crop.

Apparent recovery efficiency of N (AREN) describes the amount of N (kg ha^{-1}) included in the extra grain yield above the control per each kg of applied fertilizer N. The applied N fertilizer treatments resulted in higher AREN values in case of wheat and barley than maize, when averaged over the fertilizer treatments. When complementary organic fertilizers were applied AREN values were slightly reduced in case of wheat and barley. With the rise in mineral N fertilizer rates (N_0-N_4) a definite reduction of AE values were observed for each crop.

4. Conclusion

The nutrient absorption of crops in the N_0 treatments indicates the inherent (original) fertility of soil regarding the N status. In case of winter wheat and winter barley, the soil N supply was 30-40 kg ha^{-1} N while in case of maize, 50 kg ha^{-1} N were removed from the soil by the plants annually. Accordingly wheat and barley yielded 1.5-2.5 t ha^{-1}, while the yield of maize was 5.0-6.5 t ha^{-1} in the control plots.

The mineral N fertilizer doses increased the yield significantly until the 150 kg N ha^{-1} dose at wheat, the optimum point was 120 kg N ha^{-1} at barley and 210 kg N ha^{-1} for maize. Additional mineral N fertilizer applications did not produce significantly greater yields, as the response curves exhibited a quadratic relationship at higher mineral N rates. The N balance was always slightly positive above the equilibrium in this cases.

The joint effect of mineral and organic fertilizers were favourable. Supposedly there are positive interactions between the two sorts of fertilizers. This interaction exists on high level of nitrogen application (N_4) too, but not significantly. FYM and straw application positively influenced soil fertility. Organic fertilizers produced a 1000 kg ha^{-1} yield increase in each phase of the crop rotation. The 2nd and 3rd year post-application effect of FYM was similar for wheat and barley. The effect of FYM in the first year (applied directly before maize) was 47.8 kg ha^{-1} extra N. The second year post effect (wheat) was 24.3 kg ha^{-1} N, while the 3rd year post effect (barley) was 16.7 kg ha^{-1} N. The nutrient value of straw and mineral N fertilizer was nearly equivalent to the value of FYM.

The amount of nutrients provided by FYM application was proportional to the yield. The humus content (H g kg^{-1}) of soil after the 8th cycle of the crop rotation as follows: FYM plots showed an increase in humus content (H g kg^{-1}) of 3.4 %, while straw (+GM) had a 2.6 g kg^{-1} increase, compared to the plots without organic fertilizers. After wheat 21.9 g kg^{-1} H, at barley 21.4 g kg^{-1} H, after maize 19.4 g kg^{-1} H were found in the soil samples in the average of the N fertilizer treatments. The different doses of N fertilizer did not influence the soil organic matter content.

The N$_{total}$ values in the soil did not change as a result of the different amount of N fertilizer, but there were differences between organic fertilizer treatments (NPK - NPK+FYM - NPK+St+GM). Since soil samples were taken after harvest, supposedly the inorganic N forms had been utilized during the vegetation period.

K balance was positive in every treatment except at the high N fertilizer and yield level. Around and over 9 t ha^{-1} grain yield the K balance decreased, even turned to negative in case of maize. In consequence of organic fertilizer application, the K$_2$O content in the soil increased %. Compare to the initial 80 mg kg^{-1} value in soil, recently 200-300 mg kg^{-1} was measured. It means that in the course of 25 years 10 mg kg^{-1} increasing needed 30-40 kg ha^{-1} extra K$_2$O nutrient input above K balance.

P balance was positive in every treatment. The available (ammonium-lactate soluble) P$_2$O$_5$ content during the 25 years increased significantly. The soil AL-P$_2$O$_5$ content doubled after FYM. At the starting of the field experiment (1983) the soil P$_2$O$_5$ content was 12 mg kg^{-1} and recently it is 460 mg kg^{-1}. According to that 50 kg ha^{-1} extra P$_2$O$_5$ nutrient input above P balance equilibrium produces 10 mg kg^{-1} P$_2$O$_5$ increasing in the soil.

The proportional rates of mineral N fertilizer applied in the experiment allowed the calculation of some nitrogen use indices, which are in accordance with the results of the nutrient balances.

Harvest index of N (HI$_{N\%}$) was around 80 %for winter wheat and winter barley while in case of maize it was around 70 %. These values were decreased in parallel with the increase in N fertilizer rates.

Increasing N fertilizer rates generally decreased all the other N use indices as N use efficiency (NUE), agronomic efficiency (AE) and apparent recovery efficiency of N (AREN) as well, but on the other hand the yields (grain and straw) were increased significantly. Nitrogen use indices show the differences between the different crops regarding N utilization and productivity. It is remarkable that maize due to its high productivity (even on the control plots) performed the highest nitrogen use efficiency values, on the contrary HI$_{N\%}$, AE and AREN values were lower.

From the results it can be concluded that different rates and sorts of fertilizers influenced the productivity of crops as well as the efficiency of nutrients with interaction of the different characteristics of the certain crops. The different rates and types of fertilizers also have long-term effect on some important soil fertility parameters as well.

From the above mentioned results it can be concluded that optimal N fertilizer rates were 150, 210 and 120 kg ha^{-1} N in this order for winter wheat, maize and winter barley, respectively. The higher N rates do not increase the yield, even N surplus (strongly positive N balance) resulted in N losses and environmental hazard.

It also has to be emphasized that soil organic matter content can be sustained by organic fertilizer addition as well as recycling straw into the soil. It draws attention to the fact that straw as a by-product is a very important C resource that has to be recycled by leaving straw residue in the field and tilling it into the soil, especially if there is no animal husbandry in the farm, in order to keep the C balance in the soil.

5. Acknowledgment

This research was supported by the Hungarian Scientific Research Found (OTKA K 60314).

6. References

Berecz, K. & Kismányoky, T. (2005). Az ásványi és a szervestrágyázás néhány növénytermesztési és környezeti hatása. *Növénytermelés*, 34(3): 169-179.

Berzsenyi, Z. & Lap, DQ. (2005). Responses of maize (Zea mays L.) hybrids to sowing date, N fertiliser and plant density in different years. *Acta Agron. Hung.*, 53: 119-131.

Bircsák, E.; Csathó, P.; Radimszky, L.; Baczó, G.; Németh, T. & Németh, I. (2005). Residual effect of previous nitrogen application in two Hungarian long-term field trials. *Commun. Soil Sci. Pl. Anal.*, 36: 215-230.

Bischoff, R. (1995). The International organic long-term experiment (IOSDV) at Speyer. *Arch. Acker-Pfl. Boden.*, 39: 461-471.

Boguslawski, E.v. (1995). The combined effect of fertilizing with different forms of organic fertilizer in the IOSDV Rauischholzhausen (in German). *Arch. Acker-Pfl. Boden.*, 39: 403-411.

Crosswell, ET. & Godwin, DC. (1984). The efficiency of N fertilizers applied to cereals. In: *Advances in Plant nutrition.* PB. Tinker, (Ed.), pp. 1-55, NewYork

Győri, D. (1984). *Fertility of Soil* (In Hungarian). Mezőgazdasági Kiadó, ISBN 963-231-662-2, Budapest, Hungary

Hoffmann, S.; Debreczeni, K.; Hoffmann, B. & Nagy, E. (2006a). Grain yield and baking quality of wheat as affected by cropyear and plant nutrition. *Cereal Res. Commun.*, 34(1): 473-476.

Hoffmann, S.; Csitári, G.; Bankó, L. & Balázs, J. (2006b). Soil fertility characteristics due to different organic and mineral fertilization. *Cereal Res. Commun.*, 34(1): 203-205.

Huggins, DR. & Pan, WL. (1993). N efficiency component analysis. *Crop Sci.*, 85: 898-965.

Jenny, H. 1941. *Factors of Soil Formation.* McGraw-Hill Co. New York - London.

Kádár, I. & Németh, T. (2004). Leaching of NO₃-N and SO₄-S in soil after 28 years in a long-term field experiment (In hungarian). *Crop Production*, 53: 415-428.

Kismányoky, T. & Tóth, Z. (1997). Role of crop rotation and organic manure in sustainable land use. *Agrochemistry and Soil Science*, 4: 99-106.

Kismányoky, T. & Tóth, Z. (2010). Effect of mineral and organic fertilization on soil fertilityas well as on the biomass production and N fertilization of winter wheat in long-term cereal crop rotation. *Arch. Acker-Pfl. Boden.*, 56(4): 473-481.

Kismányoky, T. & Tóth, Z. (2007). Long-term field experiments as the base of site specific farming, Keszthely long-term field experiments. *Proceedings of the Joint International Conference on Long-term Experiments*, pp. 62-67. ISBN 978-963-473-054-5, Agricultural Research and Natural Resources. Debrecen – Nírlugos, Hungary. 31 May – 1 June, 2010

Koerschens, M. (2005). Global and regional importance of long-term field experiments. Arch. Acker-Pfl. Boden., 51(2): 111-117.

Reuter, G.; Böttcher, H.; Honermeier, B.; Kopp, H.; Lange, G.; Makowksi, N.; Mueller, K. & Teltschik, K. (2007). Diversity of humus stocks in the arable soils of the Late Pleistocene ground morains in Mecklenburg, West Pomerania, Germany. *Arch. Acker-Pfl. Boden.*, 53(2-3): 231-240.

Rühlmann, J. & Ruppel, S. (2005). Effects of organic amendments on soil carbon content and microbial biomass – results of the long-term box plot experiment in Grossbeeren. *Arch. Acker-Pfl. Boden.*, 51(2): 163-170.

Rühlmann, J. (2006). The Box Plot Experiment in Grossbeeren after six rotation: Effect of fertilization on crop yield. *Arch. Acker-Pfl. Boden.*, 52(3): 313-319.

Sarkadi, J. (1975). A műtrágyaigény becslésének módszerei. Mezőgazda Kiadó. Bp. pp. 250.

Sethi, RR.; Panda RK. & Singandhupe, RB. 2005. Study of nitrate movement in a sandy loam soil. *Arch. Acker-Pfl. Boden.*, 51(1): 41-50.

Staugaitis, G.; Vaisvila, Z.; Mazvila, J.; Arbaciauskas, J.; Adomaitis, T. & Fullen, MAA. (2007). Role of soil mineral nitrogen for agricultural crops: Nitrogen nutrition diagnostics in Lithuania. *Arch. Acker-Pfl. Boden.*, 53(2-3): 263-271.

Szentpétery, Zs.; Jolánkai, M.; Kleinheinc, Cs. & Szöllősi, G. (2005). Effect of Nitrogen Top-Dressing on Winter Wheat. *Cereal Res. Commun.*, 33: 619-626.

Tanács, L.; Matuz, J.; Gerő, L. & Petróczi, IM. (2005). Effect of NPK Fertilizers and Fungicides on the Quality of Bread Wheat in Different Years. *Cereal Res. Commun.*, 33: 627-634.

Tajnsek, A.; Santavec, I. & Ceh-Breznik, B. (2005). Using 'The Third Approximation of the Yield Law' for the determination of maximum yield and nitrogen fertilization of winter wheat. *Arch. Acker-Pfl. Boden.*, 51(5): 501-512.

Teesalu, T.; Kuldkepp, P.; Toomsoo, A. & Laidvee, T. (2006). Content of organic carbon and total nitrogen in Stagnic Albeluvisols depending on fertilization. *Arch. Acker-Pfl. Boden.*, 52(2): 193-200.

Tóth, Z.: (2006). Vertical distribution of soil NO₃⁻-N content in different cropping systems. *Cereal Res. Commun.*, 34(1): 339-341.

Uhlen, G. (1991). Long-term Effects of Fertilizers, Manure, Straw and Crop Rotation on Total-N and Total-C in Soil. *Acta Agricult. Scand.*, 41: 119-127.

Várallyay, Gy. (2009). Territorial and temporal extension of the results of long-term field experiments. In: *The importance of long-term field experiments in the improvement of crop production* (In Hungarian). *Proceedings of Jubilee Scientific Conference, 50 Years of Martonvásár Long-term Field Experiments.* Agricultural Research Institute of the Hungarian Academy of Sciences. Martonvásár, Hungary pp. 7-20. ISBN 978-963-8351-36-4

10

Constraints and Solutions to Maintain Soil Productivity: A Case Study from Central Europe

Witold Grzebisz and Jean Diatta
Department of Agricultural Chemistry and Environmental Biogeochemistry,
Poznan University of Life Sciences,
Poland

1. Introduction

Yield improvement in old farming areas has been recently recognized as a practice that will further increase crop production. One of the most important old farming areas is Europe (EU 27), which accounts only for 4% of World agricultural area, but at the end of the 20th century produced 14% of total cereals and 20% of meat (Olesen & Bindi, 2002). At present, Europe has a very diversified level of agriculture development. European agriculture is currently divided into eight agricultural regions based on natural and socio-economic factors. Two of them, i.e., the North Eastern region comprising Czech Republic, Slovak Republic, Poland and the South Eastern region, including Bulgaria, Hungary, Romania, Bosnia & Herzegovina, Croatia, Macedonia, Slovenia, Serbia and Montenegro, are of special interest to this chapter. Due to the "planned economy" experiment, which took place from the end of the 1940s to the end of the 1980s, these two regions are generally classified as Central Europe (CE). This macro-region is considered nowadays as one of the most important future producers of cereals (Rosegrant et al., 2001; Kelch & Osborne, 2001).

The current level of agriculture production in the CE is much lower than in the Western part of the continent. Yield potential of winter wheat, as the key crop in this region, excluding Bulgaria, is at the level of *ca* 10 t ha^{-1}. The "water limited" yield for most countries of the region shows the level of 6.5 t ha, but for Bulgaria and Romania it has been assessed at 4.5 t ha^{-1}. However, actual yields of all CE countries are much lower, representing only from 55 to 71% of attainable yields (Rabbinge & Diepen, 2000; Supit et al., 2010). Two main reasons for this deep disparity:

1. Natural climate and soil conditions;
2. Political transformations at the beginning of the 1990s.

The first group of factors is considered to be responsible for regional differences in volume of harvested yields, expressed as the yield gap (YG) (Dobermann & Casman, 2002). Its assessment requires insight into factors affecting crop growth and production. The realization of a plant yield potential, i.e., exploitation of a crop variety capacity under fixed climatic conditions, is undoubtedly related to the soil fertility level (Atkinson et al., 2005;

Rabbinge, 1993). The second reason of the YG, between Western and Central Europe countries is strictly rooted into the history of both parts of the continent during the 20th century. The general objective of agricultural growth in these two mega-regions was to reach food self-sufficiency after World War II. In Western Europe, due to the market-oriented economy system, this target was achieved by the end of the 1970s. In CE, where the "planned economy" ruled agricultural production for more than 40 years, this key objective was also reached. However, it was based on a system of state subsidies, prevailing as cheap prices of agricultural commodities such as fertilizers, pesticides, machinery. The sudden cut of subsidies at the beginning of the 1990s and huge increase of production prices forced farm managers to restructure the farm economy. The first step of its transformation into the market-oriented production system was to decrease the number of employees and reduce amount of applied fertilizers (Csatho & Radimszky, 2009). As a consequence, the naturally-occurring yield differences among these macro-regions increased unexpectedly, producing a new type of YG termed *the temporary yield gap (TYG)*, (Grzebisz et al., 2010).

The primary objective of this contribution is to outline key developments leading to improved yields of wheat, maize grain and sugar beet over a 23 year period, from 1986 to 2008, in the CE countries. Explanatory objectives for the present yield performance rely on the fertilizer nitrogen (N) index (partial factor productivity of fertilizer N, PFP_N). It was used (1) to calculate, for each of indicatory crop, indices of the yield gap expressed as the quantity of (a) virtually lost yield and/or (b) unassimilated fertilizer N, and (2) to make a long-term prognosis of crop yields, on the basis of the developed set of yield gap indices. The third part of this chapter describes some conceptual solutions, based on case studies aiming at improving soil productivity.

2. Natural constraints for crop production

The regionally diversified land productivity in Europe can be well assessed by using net primary productivity (NPP). The NPP of CE land, except the southern part of the Balkan Peninsula, ranges from 800 to 900 g m^{-2} y^{-1}, i.e., it is lower by 200 to 300 g m^{-2} y^{-1} than in Western Europe (Fronzek & Carter, 2007). Many reports discussing climate change stress the increasing sensitivity of key crops to year-to-year weather variability, including cereals (barley, wheat) grown in Central-Eastern Europe (Olesen et al., 2011; Falloon & Betts, 2010; Supit et al., 2010).

2.1 Climatic conditions

In agriculture, climate is considered as the driving factor for crop plant adaptations to prevailing weather conditions. Yield potential is strongly influenced by radiation, the sum of daily temperature and precipitation, and also on their distribution over the growth season. The first two factors, as a rule, are used for calculating yield potential, i.e., a realization of crop varieties' potential under undisturbed conditions of growth (Evans & Fisher, 1999). In real production conditions (country or regional scales) a third climatic constraint – water limitation – is the limiting factor of crop plant growth, as discussed by Rabbinge (1993). Base on it, the "water limited" yield (attainable yield) of a crop plant has been defined. Taking into account these factors, Europe is divided into several environmental zones, reflecting dominating weather patterns (Jongman et al., 2006). The Western part of the continent and British Islands are characterized by a humid climate, highly suitable for cultivating most of C_3 crops. Mild winters and sufficiently high precipitation in spring are prerequisites for high yields of cereals also supporting good yields of sugar beets, potatoes and maize in summer months (Supit et al., 2010). For example, in Belgium and Germany, harvested yields of wheat are about 75% of wheat yield potential, whereas French farmers reach about 85% of sugar beet yield potential. In Belgium and The Netherlands, farmers are able to harvest *ca* 90% of maize yield potential (FAOSTAT, 2011).

Most of the CE area is located in the Continental zone, which covers the northern part of the region and also northern part of the Balkan Peninsula. The Pannonion zone extends from the Black Sea up to Alps. The southern part of the region, including Albania, belongs to Mediterranean Mountains and Mediterranean North zones. Main attributes of dominating climatic patterns are irregular precipitations in summer months and frequently occurring droughts, negatively affecting plant growth and harvestable yield. For example, maize yield potential in Romania is calculated at the level of 13.0 t ha^{-1}, but grain yields harvested by farmers within the period 2005-2009 amounted to 3.2 t ha^{-1}. In Poland, yield potential of potatoes is fixed at the level of *ca* 40 t ha^{-1}, however tuber yields harvested by farmers are below 50% of this level (FAOSTAT, 2011; Supit et al., 2010). In spite of climatic disadvantages, a spatial analysis of the YG for grain crops undertaken by Neumann et al. (2010) indicates, that CE is a region with great opportunity for intensifying the production of wheat and maize. Therefore, current yield gaps in this region cannot be explained solely by seasonal weather variability.

2.2 Soil cover: Origin and distribution

Soil cover in CE is not uniform, taking into account prevailing soil types. The gradient of diversification extends from north to the south in this mega region. The natural borders of distinct soil types are Sudety and Carpathian Mountains. They were the first natural barriers seriously limiting the transgression of Fenno-Scandian and Alp's ice sheets during the Pleistocene epoch towards south of Europe. At the same time, edges of both transgressing glaciations were natural borders of climatic zones. The average annual temperature at the glacial edge was -6°C, but on the permafrost zone extending several kilometres to the south or east, close to 0°C. Extremely low temperatures during ice-sheet transgression created a high pressure gradient favourable to long-distance transport of air-born silt particles. Wind mineral deposits, termed as loess, covered large areas of CE, mainly south of Carpathian and Sudety. Loess became an excellent parent material for Chernozems and chernozem-like soils. Severe climatic conditions in the permafrost zone resulted in intensive weathering of surface rock layer. Consequently, huge layers of loam materials were formed both *in situ* or transported as alluvial deposits, being precursors in the development of Cambisols (Catt, 2001; Plant, et al., 2005).

The mineralogical composition of soil parent material is therefore, significantly different on both sides of Sudety and Carpathian Mountains. European countries lying over the northern part, i.e., Poland and also eastern part of Germany, are dominated by soils originating from sands, loamy sands and sandy loams. Consequently, the current soil cover is mostly consisted of associations of Luvisols and Podzoluvisols. In addition, in Poland *ca* 22% of the soil cover, partly used by farmers, is classified as associations of Arenosols and Podzols. Soil cover in other countries of the CE represents mainly Cambisols (Czech Republic, Serbia &

Montenegro, Slovakia, Slovenia) contributing to more than 20%. Another attribute of the region south of Carpathians is associations of Chernozems (Bulgaria, Hungary, Romania, Serbia & Montenegro), (Table 1).

Country/soil types	Cambi-sols[1]	Cherno-zems[2]	Fluvisols	Luvisols[3]	Total carbon stock[5] t ha[-1] soil depth, m	
					0.0 – 0.3	0.0 – 1.0
Albania (AL)	31.6	10.6	5.5	25.3	67.0	131.0
Bulgaria (BU)	20.0	22.0	9.0	31.0	69.6	146.2
Croatia (CR)	23.6	0.9	2.4	12.6	64.9	117.8
Czech Republic (CZ)	45.1	13.2	5.9	19.3	69.3	135.0
Hungary (HU)	10.8	21.8	17.4	6.5	81.1	184.4
Macedonia (MC)	40.6	1.3	5.1	9.7	68.0	124.8
Poland (PL)	14.6	3.5	4.7	27.3	62.0	119.9
Romania (RO)	23.2	28.5	10.7	26.3	72.5	155.0
Serbia & Mt[4] (S&M)	44.9	21.8	8.6	11.9	73.6	149.8
Slovak Rep. (SK)	46.8	11.6	6.5	29.0	67.1	129.3
Albania (AL)	45.3	0.0	5.4	6.4	68.8	124.5

[1] & vertisols;[2] & Chernozems-like; [3] & Rendzinas; [4]Montenegro; [5] based on Batjes (2002)

Table 1. Main soil associations (average coverage) of the Central European countries

Using soil potassium (K) supply as the key indicator of soil fertility level, it can be assumed, that soil type significantly reflects inherent parent material properties, i.e., soil quality. Nikolova, (1998) has classified soils in three main groups based on their K-sorption characteristics (Table 2). The highest K-fixation capacity is attributed to soils originated from loess or loams from Pliocene and Quaternary deposits. The lowest potential of K supply occurs in soils formed from sandy materials. Therefore, relatively to this rating, CE countries may be additionally divided into two main groups, i.e., Poland and parts of Bulgaria *versus* all other countries. In Poland, soils with low and even very low K sorption capacity are prevalent. About 45% of the agricultural area in Poland is covered by soils originated from sands and loamy-sands. This explains the serious limitation of crop productivity by insufficient soil K supply (Grzebisz & Fotyma, 2007).

K-sorption ability classes	Particles < 0.01 mm, %	CEC cmol$_{(+)}$ kg^{-1} soil	Soil types
High	> 50	> 40	Vertisols, Planasols, Kastanozems
Medium	30 - 50	20 - 40	Haplic & Luvic Chernozems, Molic Planasols
Low	< 30	< 20	Luvisols, Eutric Planasols

[1]Nikolova (1998)

Table 2. General potassium sorption characteristics of main soil types in Central Europe (source: Nikolova, 1998)

3. Agriculturally induced constraints - Fertilizer management

The recorded yield increase of most crops in the last 60 years is a result of two main factors. The first is the progress in plant breeding, which took place in the 60s and 70s during the *Green Revolution* (Sinclair et al., 2004). However, fulfilling nutritional requirements of modern, high yielding varieties requires a well balanced supply of nutrients. The second factor contributing to modern yield increases is fertilizer use. In intensive agriculture, it is related to the consumption of fertilizers, such as N, P and K. Therefore, a reliable estimation of agriculturally induced constraints for crop production in CE requires an insight into past and current nutrient management strategies.

3.1 Long-term patterns of fertilizers' consumption

Data on historical fertilizer consumption in the period 1986-2008 were obtained from the IFA databank (2011). Fertilizer use per hectare was calculated by dividing the total use of N, P and K fertilizers in a given year per actual area of arable land, as the target area of application. The quantities of fertilizers annually used by farmers over the studied period were both nutrient and country specific (Table 3). During the period 1986-2008, total consumption of N-P-K fertilizers in CE countries has undergone significant changes with respect to:

a. the quantity of fertilizers annually applied per hectare;
b. the structure of applied fertilizers, as related to P:N and K:N ratios.

Statistical parameters	Countries[1]											
	AL	B&H	BU	CR	CZ	HU	MC	PL	RO	S&M	SV	SL
Nitrogen												
Average	45.0	31.1	63.3	100.7	93.1	72.4	46.7	71.0	39.9	45.8	70.3	200.0
SD[2]	44.1	23.1	37.2	38.9	22.3	25.1	13.6	18.2	20.1	16.4	31.3	129.9
CV[3], %	97.9	74.2	58.8	38.6	23.9	34.6	29.0	25.7	50.5	35.7	44.5	64.9
RI$_N$	86.6	75.3	68.0	53.3	41.4	52.6	45.8	37.7	62.9	51.0	60.9	80.0
Phosphorus												
Average	14.9	14.4	15.2	44.0	32.2	21.3	22.8	30.5	16.5	14.4	30.4	69.8
SD[2]	14.7	12.1	23.2	15.4	32.3	21.4	8.10	16.3	13.1	12.1	33.3	33.7
CV[3], %	98.7	84.5	152.8	35.0	100.5	100.4	35.5	53.4	79.6	84.6	109.4	48.3
RI$_P$	91.4	81.2	97.6	42.5	83.9	85.4	53.1	65.7	80.0	83.8	86.6	64.1
Potassium												
Average	2.4	13.9	7.1	48.1	30.3	26.0	17.5	36.6	6.4	15.7	29.1	85.1
SD[2]	2.1	11.3	10.6	20.9	34.3	26.9	9.3	18.3	8.5	10.0	34.8	46.1
CV[3], %	87.4	81.4	149.3	43.3	113.3	103.6	53.4	49.9	132.4	64.1	119.7	54.2
RI$_K$	96.5	80.0	97.0	53.7	89.2	86.3	68.4	61.9	93.1	70.6	88.8	71.8

[1] IFA data bank, accessed 2011-05-24. [2]standard deviation, [3]coefficient of variation,

Table 3. Consumption of N, P and K fertilizers (in kg ha[-1]) in Central European countries during the period 1986-2008 and the reduction index (RI) for each nutrient – a statistical overview. [1]Country acronyms are provided in Table 1.

The long-term course of fertilizers' consumption showed some resemblances. Therefore, the whole investigated period has been divided into three well defined phases: (1) High Consumption Level (HCL); (2) Collapse and Transition (CT); (3) Post Transition (PT) (Fig. 1). Two distinct procedures have been applied to make a reliable estimate of the length of each phase. The HCL was separated from the CT when there was more than 25% change in the baseline – yearly values. The PT was established using a linear regression model, assuming statistically proven change.

The key attribute of the first phase, lasting from four to five years was characterized by high levels of consumed fertilizers, mainly N, irrespective of the country (Fig. 1). The consumption of other nutrients, i.e., P and K was country specific. The CT phase was characterized by a dramatic, sudden decrease in use of each fertilizer. This process occurred for one or two years in some countries (Czech Republic, Croatia, Hungary, Poland, Slovakia, Slovenia), but in others up to 10 years, showing prolonged depression. The reason of this drastic decline was well described in economy the *scissor phenomena* (Cochrane 2004). The PT phase has been appearing in three distinct forms, i) increase (PT-I, above the baseline HCL level); ii) restoration (PT-R, significant trend, but below the baseline, HCL level), iii) stagnation (PT-S, no significant changes over time).

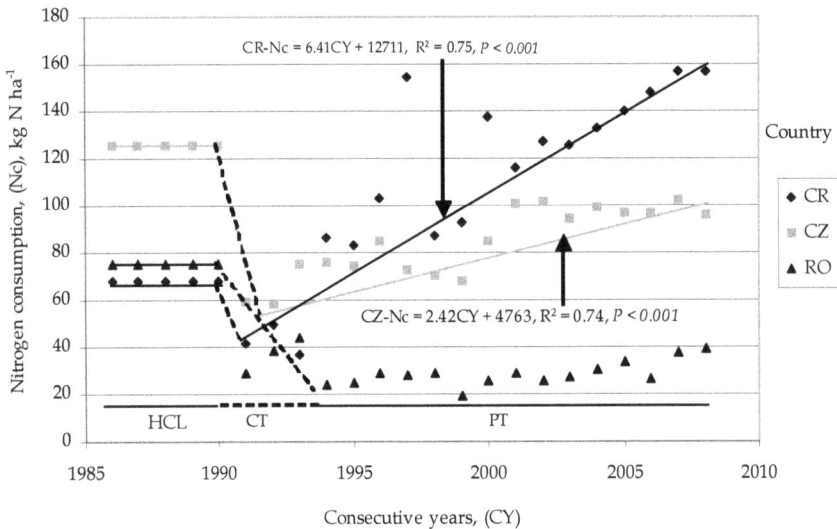

Fig. 1. Patterns of N fertilizers' consumption in selected Central European countries - Croatia (CR), Czech Republic (CZ) and Romania (RO) during the 1986-2008 period (based on: IFA database, accessed: 2011-05-24). Significant ($P<0.05$) trends are indicated with solid lines and linear regression equations given. The CT periods are marked with dashed lines.

Nitrogen, considered as the key nutrient in crop production, but at the same time potentially unfriendly to the environment, requires much deeper attention. Its consumption in the HCL period was high, varying around 100 kg N ha^{-1} for most countries in the region (Table 3). The lowest consumption (defined by 0.25 percentile) took place during the CT phase. The degree of collapse can be simply assessed by using a fertilizer use reduction index, (RI):

$$RI = [1- (0.25P/0.9P) \cdot 100\%]$$

where: 0.25P and 0.9P are the 25th and 90th percentiles of fertilizer consumption over the period 1986-2008.

Fertilizer N management in CE during the PT phase relies on consumption trends, evaluated for a period of 18 years, i.e., since 1991 onwards. On the basis of elaborated regression models, countries of the region have been divided into three defined groups:

1. Increase (PT-I): Croatia, Slovenia;
2. Restoration (PT-R): Czech Republic, Hungary, Poland, Slovakia, Slovenia;
3. Stagnation(PT-S): Albania, Bosnia & Herzegovina, Macedonia, Bulgaria, Romania, Serbia & Montenegro.

Phosphorus is a nutrient of basic importance for any crop plant, because all seed crops are highly sensitive to its deficiency during the onset of ripening. In the HCL phase, the consumption of P fertilizers was in the range of *ca* 35 to 95 kg P_2O_5 ha-1 (Table 3). During the period of study, its use underwent a great, mostly negative change. Its consumption trends evaluated for years 1991-2008 showed much greater variability compared to N fertilizers' use in CE countries:

1. Increase: Croatia, Slovenia;
2. Restoration: Poland, Hungary, Macedonia;
3. Stagnation: Albania, Czech, Slovakia, Serbia & Montenegro;
4. Recession: Bosnia & Herzegovina, Bulgaria, Romania.

The occurrence of the fourth group stresses the fact that P fertilizers' consumption was below the levels required to support current crop production, leading to *nutrient mining* in three countries of the region.

Potassium is generally considered as a nutrient significantly affecting water management in plants during their life cycle (Cakmak, 2005). Therefore, this nutrient requires special care on farm in regions like CE that experience frequent drought during the vegetative growth period (Falloon & Betts, 2010). Potassium fertilizers' consumption during the HCL period varied among countries of the region from *ca* 5 kg K_2O ha-1 (Albania) to more than 100 kg K_2O ha-1 (former Czechoslovakia) (Table 3). As in the case of P, the greatest decline in K consumption occurred in countries significantly increasing the use of nitrogen, like the Czech and Slovak Republics, and also Hungary (estimated from the potassium fertilizer use reduction index, RI_K).

The second indicator of fertilizer management is in the proportions of macronutrients consumed, indicated during the study by the P:N and K:N ratios. The P:N ratio did not show a clear trend for two countries, Albania and Bosnia & Herzegovina. In other countries, negative tendencies were detected, appearing as the increasing value of P:N ratio, meaning that proportionately more N fertilizer than P fertilizer was applied by farmers. This process took place mainly during CT and PT phases. Its course during the PT phase outlines three decisive patterns of P fertilizers' management:

a. negative: Czech Republic, Croatia, Serbia & Montenegro, Romania;
b. positive: Hungary, Macedonia, Poland;
c. stagnation: Slovakia.

The first trend covers the whole study period, differing only in the annual rate of decrease, as indicated by the coefficient of the linear regression equation. In the case of Romania, it was almost four times faster than for Croatia (Fig. 2). The second pattern occurred in countries like Hungary and Poland, showing a slightly positive trend, since 1991.

Trends of K:N ratios show similar patterns as described for P:N ratios (Fig. 3). Three patterns of ratios can be also distinguished, but comprising different sets of countries:

a. negative: Macedonia, Slovenia;
b. positive: Hungary;
c. stagnation: Croatia, Czech, Poland, Slovakia, Bulgaria and Romania (since 1993), Serbia & Montenegro (since 2000).

All above-described attributes of long-term fertilizers' consumption were evaluated in Table 4, to develop a ranking based on three types of nutrient management in CE countries, as follows:

1. *Extensive*; generally showing low consumption of all fertilizers, including N; the most negative attribute is the low P and K inputs, leading to crop plants reliance on inherent soil P and K supply and potentially leading to nutrient mining (Albania, Bulgaria, Macedonia, Serbia & Montenegro, Romania);
2. *Unbalanced*; high consumption of fertilizer N, but at the same time not balanced by adequate use of other nutrients, such as P and K, at least (Croatia, Czech, Slovakia, Slovenia);
3. *Balanced*; high consumption of N fertilizer, which is in part balanced by relatively high consumption of P and K fertilizers (Hungary, Poland).

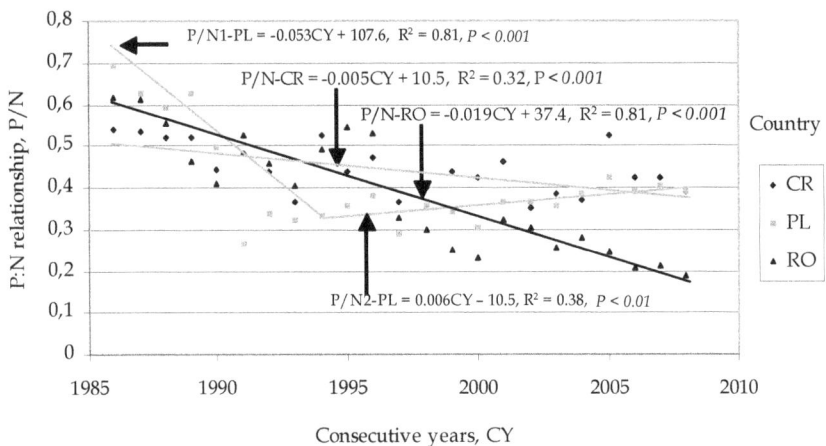

Fig. 2. Typical patterns of P:N ratios in selected Central European countries - Croatia (CR), Poland (PL) and Romania (RO) during the period 1986-2008. Significant ($P<0.05$) negative linear equations are shown for CR and RO, while the PL data fitted to a negative linear equation in the period 1986-1996 and thereafter showed a positive linear relationship.

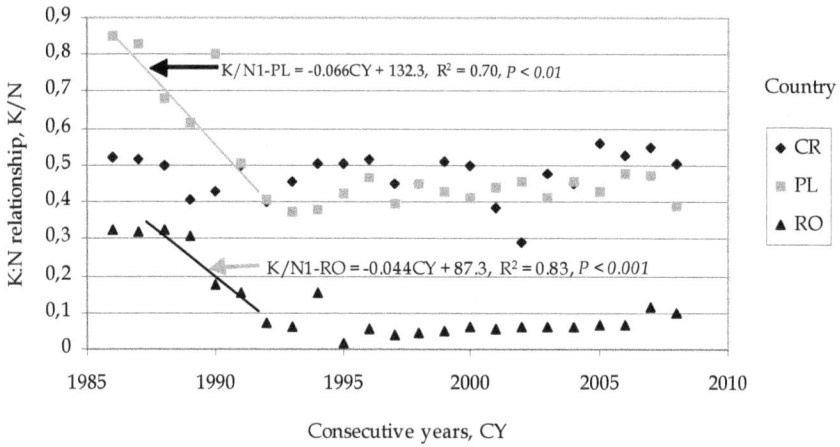

Fig. 3. Typical patterns of K:N ratios in selected Central European countries - Croatia (CR), Poland (PL) and Romania (RO) in the period 1986-2008. Significant ($P<0.05$) negative linear equations are shown for CR and RO in the 1986-1992 period, thereafter showed non-significant trends. For CR the K:N trend was non-significant for the whole period.

Estimated Parameters	Country[3]											
	AL	B&H	BU	CR	CZ	HU	MC	PL	RO	S&M	SK	SL
Nitrogen	0[1]	0	0	++	+	+	0	+	0	0	+	+/-
Phosphorus	0	-	-	++	0	+	+	+	-	0	0	+/-
Potassium	0	0	0	++	0	+	0	+	0	0	0	0
P:N	0	0	0	-	-	+	0	++	-	0	0	-
K:N	0	0	0	0	0	+	-	0	0	0	0	-
Ranking[2]	0	0	0	2	2	1	0	1	0	0	2	2

[1] evaluation of trends: ++ - increase; +- restoration; 0 - stagnation; - - decrease; +/- - time variable;
[2] fertilizer management ranking: 0 - extensive; 1 - balanced; 2 - unbalanced.
[3] Country acronyms are provided in Table 1.

Table 4. Fertilizer consumption trends –summary and country management ranking for countries in Central Europe during the period 1986 to 2008.

3.2 Soil fertility status

Soil is considered as a slowly renewable or non-renewable resource that requires the same level of protection as other limited resources, like phosphorus rock. There is strong agreement that productivity of arable land depends on both a soil's fertility potential (soil quality, inherent soil fertility) and present state of soil properties (soil health) to sustain primary production and other ecological functions. The minimum data set to describe current fertility status of arable land, i.e., soil health, should contain indicators reflecting both past and actual organic matter content, soil pH and management of nutrient soil resources (Atkinson et al., 2005; Karlen et al., 2003).

Total organic carbon (TOC) stock in the topsoil (0.0-0.3 m) or its content (%) in arable soils is a good descriptor of short-term trends in soil organic matter (SOM) management. However, the organic carbon stock measured to a depth of 1 m in the soil profile can be considered as an indicator of soil fertility potential. The actual content of TOC in an averaged arable soil unit was calculated by weighted mean content of organic carbon for main soil types (Batjes, 2002) and its percentage share in the soil cover for each CE country (see Table 1). Analysis of this data showed that TOC in the average topsoil unit of CE countries is insufficient, ranging from *ca* 60 (Poland) to 80 (Hungary) t TOC ha^{-1}, i.e., about 1.4 to 1.8% TOC. Differences between CE countries can be evaluated by TOC stock in the soil profile to a depth of 1 m. Based on this criterion, countries can be classified into three groups:

a.   low (≤ 120 t TOC ha^{-1}): Croatia, Poland;
b.   medium, (121-160 t TOC ha^{-1}): Albania, Bulgaria, Czech Republic, Macedonia, Romania, Serbia & Montenegro, Slovakia, Slovenia;
c.   high (> 161 t TOC ha^{-1}): Hungary.

Phosphorus and Potassium are key nutrients for achieving high productivity of arable soils. Available resources of both nutrients are the most important indicators of soil health, i.e., its current productivity. The country status of current soil nutrient management was evaluated by means of soil fertility index (SFI) expressed as the sum of the percentage share of arable land characterised by the very high, high and half of medium rating of available P or K (Grzebisz & Fotyma, 2007). It has been assumed, that SFI for P and K exceeding 70% of arable land share is a prerequisite for high yield of any crop. In the 1980s, this target was achieved in many countries of the region, like Czechoslovakia, Hungary and some countries of the former Yugoslavia (Serbia & Montenegro, Slovenia), (Fig. 4). In these countries, the rates of applied fertilizers were far in excess of crop removal capacity. Poland, in spite of currently high K consumption, differs from other CE countries. The sufficiency level for P and K of 70% was never reached, due to much lower inherent soil potential for nutrient supply. At the beginning of 2000, significant changes of P and K fertility were recorded. The increase of the P sufficiency level was recorded only in Slovenia and Poland. In Slovenia, the reason was high consumption of P fertilizers at the beginning of the 21st century. In Poland, the increase is related to sufficiently well balanced use of N and P fertilizers and slightly lower yields, than in the 80s. In all other CE countries, the percentage of P-fertile soils generally decreased, nevertheless, the rate of the decrease differed across the region. A slight decrease occurred in Czech and Slovak Republics, as well as in Hungary. The greatest drop was recorded in Bulgaria followed by Serbia and Montenegro. The potassium SFI presents much more alarming trend. As reported by Grzebisz & Fotyma (2007) for countries of the Northern Agricultural Zone (Czech, Hungary, Poland, Slovakia), the K sufficiency level decreased below the target value within two decades. Based on K fertilizers' consumption trends, it can be assumed that the same trend occurs also in the Southern Zone.

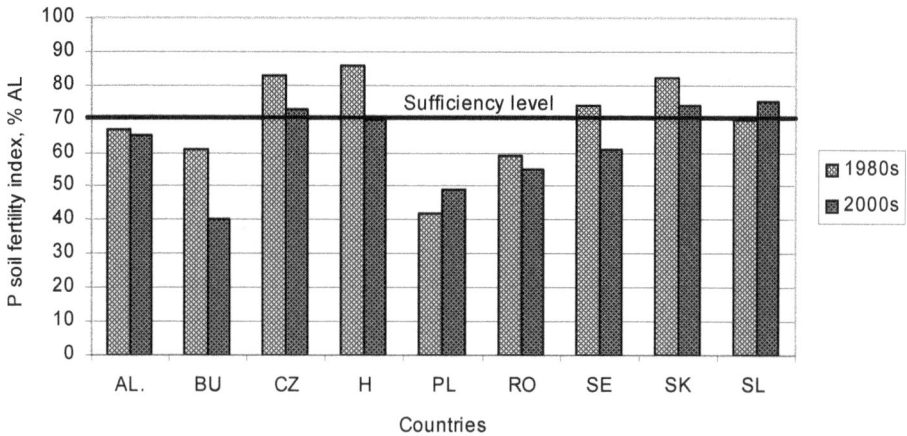

Fig. 4. Phosphorus soil fertility indices (expressed as the percentage of arable land, AL) for selected CE countries during the period 1986 - 2008 (based on: Csatho et al., 2007). The sufficiency level was assumed to be 70%, as described by Grzebisz & Fotyma (2007). Country acronyms are given in Table 1.

4. Soil productivity assessment

4.1 Long-term trends in crop yields

All CE countries grow the same crops and cereal species predominate. The area sown with cereals occupies 64% of arable land in Croatia, 52% - in the Czech Republic, 68% - in Poland and 59% – in Romania. In the Czech Republic, cereals' production is dominated by wheat and barley, but in Poland by wheat and rye. In Romania and Croatia, maize is the key grain crop. Sugar beets, due to EU regulation in the last 10 years, are of minor importance, but this crop is a good indicator of soil quality. Therefore, it has been assumed that these crops would reflect fairly well changes in current soil fertility status, as a consequence of applied nutrient management strategies. This assumption is well documented by observed variability of yields of wheat, maize and sugar beets in selected countries of CE (Table 5). The highest variability was found for Romania and the lowest for Poland. Romania is a country with an extensive pattern of fertilizers' management, typical in the Southern Agricultural Zone of Europe. Poland belongs to countries of the Northern Zone, but presents a balanced patterns of fertilizers' management (Poland, Hungary). Other selected countries are members of the Unbalanced fertilizing group of CE countries.

The long-term yield trend can be used as an indicator of key crops' sensitivity to past and current fertilizers' management. As in the case of fertilizers' consumption trends, three consecutive phases of yield development over the period 1986-2008 may be distinguished:

1. Actual-Standard, (AS);
2. Recession, (RC);
3. Restoration, (RS), or Stagnation, (SG).

The observed trends of harvested yields were most pronounced for wheat, other cereals and for seed crops like oil-seed rape (Grzebisz et al., 2010). In the first phase of the study, taking place at the end of the 80s, yields of wheat reached on average the highest recorded levels. Therefore, this particular period is considered as a standard level (AS) for evaluating subsequent phases (Fig. 5). The highest yield decline, i.e., the recession phase (RC) in response to the collapse in fertilizer consumption took place in the Northern Zone, but

Crops	Croatia		Czech Republic		Poland		Romania	
	x ± SD[5]	CV[6], %	x ± SD	CV, %	x ± SD	CV, %	x ± SD	CV, %
Wheat	4.0±0.5[2]	12.5	4.8±0.5	10.2	3.6±0.3	9.0	2.7±0.6	21.2
	3.9±0.3[3]	8.6	5.1±0.4	7.8	3.7±0.2	5.6	3.1±0.4	14.1
	4.4±0.9[4]	21.6	5.2±0.6	11.3	3.9±0.4	10.0	2.9±0.8	26.5
Maize	4.9±1.0	20.5	5.8±1.3	21.4	5.2±0.8	14.3	3.1±0.8	24.4
	4.0±0.9	21.7	5.1±0.5	9.3	4.9±0.2	4.1	2.9±0.5	18.0
	5.5±1.3	22.9	6.9±0.5	7.6	5.6±0.8	15.7	3.4±1.1	31.5
Sugar beet	40.4±7.7	19.2	42.4±7.6	18.0	38.1±5.5	14.5	22.8±5.1	22.2
	39.8±4.9	12.3	34.5±1.5	4.2	34.6±1.9	5.6	21.3±2.9	13.6
	48.7±5.2	10.8	53.1±4.9	4.9	45.1±3.9	8.8	30.4±3.3	10.8

[1]FAOSTAT, available online 2011-05-24; [2]1986-2008, [3]1986-1990, [4]2004-2008, [5]standard deviation, [6]coefficient of variation, %

Table 5. Statistical overview of key crops' yields (t ha^{-1}) in selected Central European countries, during the period 1986-2008 in comparison to the AS phase and the 2004-2008 periods.

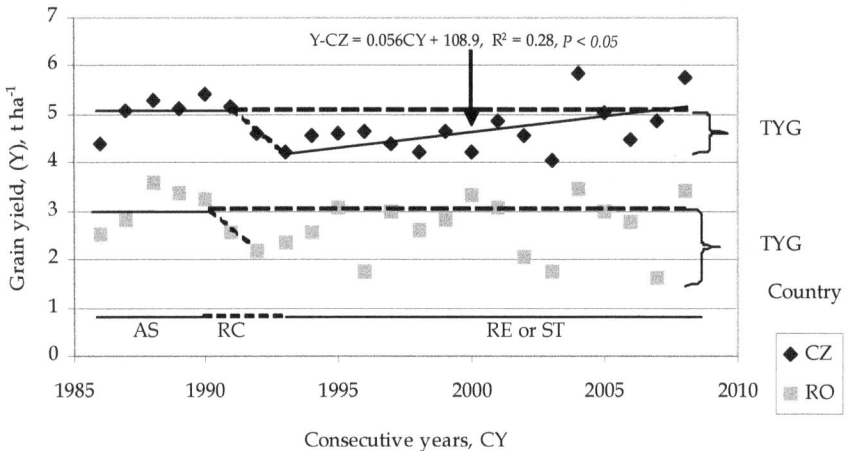

Legend: level of harvested yield: AS: Actual-Standard, RC: Recession, RE: Restoration, ST: Stagnation. Dashed line refers to AS level of yield, i.e., an average yield harvested in the period 1986-1990. TYG: Temporary Yield Gap.

Fig. 5. Long-term trends of wheat production in the Czech Republic (CZ) and Romania (RO) during the period 1986-2008. Significant ($P<0.05$) positive linear trend is shown for the Czech Republic, whereas non-significant for Romania, in the period 1992-2008.

lasted for only two or three years. In other countries, mainly representing the Southern Zone, it was also fast or showed sustained depression. The third phase of yield development was manifested in two distinct trends. The first "restoration" trend has shown a positive yield gain since yield recession. It was revealed in the Northern Zone countries and also for Croatia and Slovenia. The second trend has not shown any significant change in yield since the recession, resulting in yield stagnation. The most interesting attribute of this trend is its high year-to-year variability. The difference between the actually harvested and the AS yield has been termed as the *temporary yield gap* (TYG) (Fig. 5). The time required to recover to the AS level was long, but country specific. In the Northern Zone, it lasted for *ca* 10-12 years, exceeding the AS level in 2004 or 2005. In most other countries, representing the Southern Zone, it has not been reached up to 2008. For maize and sugar beets, the recession was also observed. In the recovery period, maize yields were generally positive, but the rate of the yield increase was much higher in the Balanced and Unbalanced groups than in the Extensive group. Sugar beets yield increase was very high at the beginning of the 90s, exceeding the level reported in the 80s. The main reason was a reduction in area sown with sugar beets, which forced farmers to cultivate this crop on more fertile soils.

4.2 The concept of partial factor fertilizer nitrogen productivity, (PFP$_N$)

This part of the study relies on two main conceptual pillars. The first one assumes, that each hectare of arable soil, regardless of the current crop is supplied with the same amount of nutrients, including nitrogen. The second assumption takes into account the fact, that the unit productivity of in-season applied fertilizer N, known as PFP$_N$, is a reliable scientific tool to make simple, but quick discrimination of factors limiting attainable crop yield (Dobermann & Casman, 2002). Therefore, the PFP$_N$ index was used to evaluate some, but selected CE countries that represent the three patterns of current fertilizer management (Table 4). Croatia and the Czech Republic have been selected as examples for the Unbalanced group, whereas Poland is the Balanced group, and Romania represents the Extensive group of fertilizers' use.

The PFP$_N$ index was obtained by dividing the actual yield of a respective crop harvested within a given year by the annual amount of N fertilizers consumed by the country, assuming it was applied at an equal rate for all currently cultivated crops. Operationally, the PFP$_N$ index consists of two components:

a. yield of crop (Y), expressed in kg, or t ha^{-1};
b. nitrogen fertilizer rate (F$_N$), expressed as kg N ha^{-1}.

$$PFP_N = Y/F_N$$

Based on the long-term trends of PFP$_N$ indices for each of the key crops, two different types of PFP$_N$ indices can be distinguished:

1. The *real* partial factor productivity of applied N – (*r*PFP$_N$); it reflects actual yields as limited by both water and N availability;
2. The *maximal* partial factor productivity of applied N (*m*PFP$_N$); it was calculated as the 4th quartile of *r*PFP$_N$ indices; it refers to yields obtained under ample water supply.

To make a reliable estimate of the yield gap for some crops cultivated in the CE, the *corrected* partial factor productivity of applied N (*c*PFP$_N$) instead of *m*PFP$_N$ has been applied.

Operationally, it has been calculated as the average of the 4th quartile, but excluding extremely elevated rPFP$_N$ values, which emerged in most of investigated countries at the beginning of the 90s. Indices of N productivity (PFP$_N$: rPFP$_N$ and cPFP$_N$) were then used to calculate the yield gap.

The calculated rPFP$_N$ indices showed, as expected, high year-to-year variability for all tested crops (Table 6). They were crop and country specific when averaged over the 1986-2008 period:

a. wheat: CR < CZ = PL < RO;
b. maize: CR < CZ < PL < RO;
c. sugar beets: CR = CZ < PL < RO.

These ranges are not in line with long-term average yields of all crops, which followed generally the order: CZ > CR > PL > RO.

Maximum productivity of tested crops as evaluated for the 4th quartile of rPFP$_N$ indices was much higher, rising on average by 43%. However, the increase was country specific, and represents 63% for Croatia, 28% for the Czech Republic, 30% for Poland and 48% for Romania. The concept of mPFP$_N$ assumes an increase of rPFP$_N$ indices pursuant to higher efficiency of applied fertilizer nitrogen, due to ample water supply. Its reliability has been corroborated by low coefficients of variation, implicitly indicating the potential productivity of applied fertilizer nitrogen on the background of soil quality. The sequences of countries for each crop are as follows:

a. wheat: CZ = PL < CR < RO;
b. maize: CR = CZ < PL < RO;,
c. sugar beets: CZ < PL < CR < RO.

As presented in Table 6, coefficients of variations of cPFP$_N$ indices were significantly lower in comparison to mPFP$_N$, stressing on the reliability of the conducted analysis.

Relationships between sets of developed rPFP$_N$ and applied fertilizer nitrogen rates can also be a useful tool for making long-term assessment of nitrogen efficiency. This relationship fits the best the power function:

$$r\text{PFP}_N = 911.2N^{-067} \text{ for } R^2 = 0.77 \text{ and } n = 92$$

where: N – annual rate of fertilizer nitrogen consumption, kg N ha^{-1}.

Analysis of the general curve shape reveals decreasing yield per unit of N fertilizer applied, with highest values of rPFP$_N$. It has been observed, that irrespective of the applied N rate, long-term productivity of fertilizer nitrogen increased in the order: Romania < Poland < Croatia < the Czech Republic. However, the obtained high rPFP$_N$ values are, as a rule, related to lower yields of harvested crops. This pattern of crop response refers mostly to Bulgaria, Romania and former Yugoslavia countries, excluding Croatia and Slovenia. As pointed out by Fixen (2004), the target of efficient N fertilizer management based on high fertilizer N use efficiency is counter-productive. In this N supply system plants are forced to take up an appreciable amount of N from its soil resources, which decreases their rate of growth in critical stages of yield performance. Therefore "the elevated" level of the PFP$_N$ could be used as an attribute of the extensive type of agriculture production.

PFP$_N$ index	Croatia		Czech Republic		Poland		Romania	
	x ± SD[1]	CV[2], %	x ± SD	CV, %	x ± SD	CV, %	x ± SD	CV, %
Wheat, kg yield kg N^{-1}								
rPFP$_N$	48±24	49.8	54±13	23.5	54±15	27.5	81±33	40.5
mPFP$_N$	80±23	28.8	70±11	16.0	72±15	21.2	121±14	11.9
cPFP$_N$	63±11	17.4	62±4	5.9	62±2	3.0	112±8	6.9
Maize, kg yield kg N^{-1}								
rPFP$_N$	56±26	46.4	66±18	28.1	78±21	26.3	95±45	47.8
mPFP$_N$	89±29	28.8	89±7	7.6	103±11	10.2	149±20	13.5
cPFP$_N$	69±10	15.1	89±7	7.6	96±2	2.4	137±9	6.6
Sugar beets, kg yield kg N^{-1}								
rPFP$_N$	471±224	47.7	478±120	25.0	566±136	24.0	695±266	38.3
mPFP$_N$	766±218	28.5	588±49	8.25	728±67	9.3	977±122	12.5
cPFP$_N$	613±86	14.1	588±49	8.25	675±35	5.2	905±36	4.0

[1]standard deviation, [2]coefficient of variation, %

Table 6. Partial factor fertilizer N productivity statistics for key crops in selected Central European countries with unbalanced (Croatia, Czech Republic), balanced (Poland) and extensive (Romania) fertilizers' management, during the period 1986-2008.

4.3 Yield gap (YG) and its attributes

Actual yields of crops harvested by farmers in all areas of the world are much lower than the yield potential of currently cultivated varieties under defined soil and climatic conditions. This "virtual" un-harvested portion of yield is generally termed the *yield gap* (YG), (Evans & Fisher, 1999, Dobermann & Cassman, 2002). However, the CE countries experience another phenomenon, known as *the temporary yield gap* (TYG), (Grzebisz et al., 2010). The main reason of its emergence is the collapse in fertilizer use, leading to elevation of the rPFP$_N$ index. Dibb (2000) relates extremely high N efficiency to low yield, and this trend was fully corroborated by the crop yield data from countries with extensive fertilizer management. Nevertheless, the countries of the Northern Agricultural Zone (Czech Republic, Hungary, Poland, Slovakia), in spite of drastic decrease of consumed fertilizers in the early 90s, were characterized by high yields of wheat and other cereal crops (FAOSTAT, 2011, Grzebisz et al., 2010). The observed rPFP$_N$ indices, extremely elevated at the beginning of the 90s, were a result of the residual effect of applied fertilizer N. This phenomenon could be manifested only under growth conditions created by ample water and sufficiently high supply of immobile nutrients such as P and K, in turn balancing N use by currently growing crops.

In the first step of yield gap calculation, the cPFP$_N$ index was applied to assess the maximal attainable yield, (Y$_{AM}$). The yield gap was then calculated as a difference between really harvestable yield, (Y$_A$) and Y$_{AM}$:

$$Y_{AM} = cPFP_N \cdot D_N$$

$$YG = Y_A - Y_{AM}$$

where,

D_N is the rate of annually applied fertilizer N, (kg N ha^{-1}), Y_A is the actual yield in t ha^{-1} for a given year and Y_{AM} is the maximal attainable yield, (t ha^{-1}).

The observed similarities of the general yield gap pattern over time for key crops allow to distinguish three consecutive phases of its development in the period 1986-2008 (Fig. 6):

1. First – primary deep yield gap; to the left of the transition point (TP);
2. Second – point of extra yield gain, TP;
3. Third- secondary yield gap; to the right of the TP, alternatively: yield gap stagnation as indicated by the lack of significant ($P < 0.05$) linear trend response.

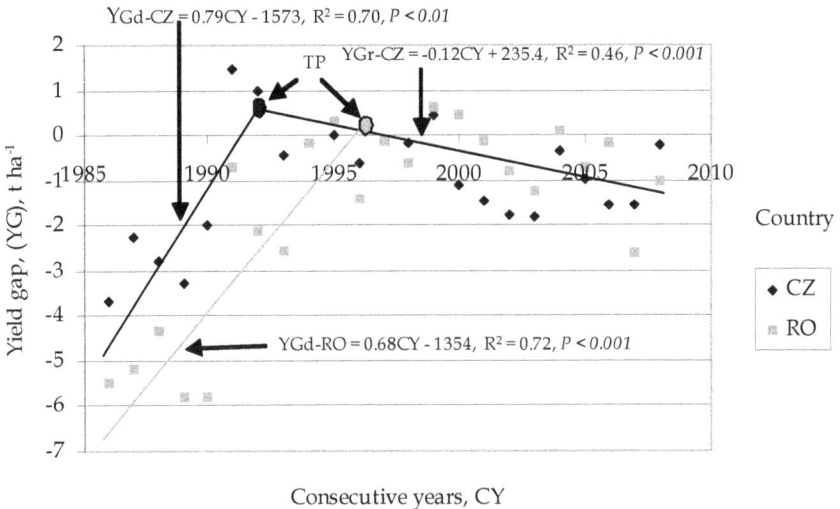

Consecutive years, CY

Legend: TP - Transition Point; YGd - yield gap decline; YGr - yield gap restoration.

Fig. 6. Evolution of the yield gap in wheat in the Czech Republic (CZ) and Romania (RO) during the period 1986 -2008. Significant ($P<0.05$) positive linear equations are shown for the Czech Republic in the period 1986-92 and for Romania in the period 1986-1995, thereafter showed a negative linear relationship for the Czech Republic and non-significant trend for Romania.

Wheat is the best example of the yield gap course in the CE countries, stressing on the difference in fertilizer management. The main dissimilarities among countries do not refer to the first phase. The YG was deep in spite of relatively high yield, indicating thus a low capacity of contemporary cultivated varieties to achieve further gains in yield. Much more important were YG length and number of years to reach the TP. The time to reach the TP was shorter (6(7) years) in countries such as the Czech and Slovak Republics, Hungary and Poland and also in Croatia and Slovenia, corresponding to rapid rebuilding of fertilizer markets in these countries. In countries with extensive fertilizers' management, the length of the first phase extended up to mid-90s.

The YG evaluation shows that the current nutrient management in CE countries can overcome this phenomenon, especially in the case of wheat and maize. A TYG can be most easily negated in countries belonging to Unbalanced or Balanced nutrient management groups such as the Czech and Slovak Republics, Hungary and Poland and also Croatia, and Slovenia. In countries with extensive fertilizers' management, the TYG has not been overcome, and enters the third phase of yield development, stagnation, due to low use of fertilizers, in turn resulting in the very low and unstable level of annually harvested yields.

4.4 Yield prognosis – The unassimilated nitrogen concept

The medium or long-term food and environmental policy strategies for any country or particular region are supported by yield prognosis considered as an operational tool. However, the *ex-ante* prognosis depends on the reliability of the data used. Two approaches, both relying on the unit productivity of fertilizer nitrogen, (PFP_N) have been considered. The first assumes extrapolation of real and improved yields, based on the linear model (Reilly & Fuglie, 1998). The second one is based on long term trends of the YG, but transformed into unit of temporarily lost fertilizer nitrogen, termed for the purposes of this study as *unassimilated* nitrogen. This approach as related to nitrogen management has been used to make a prognosis of yields of key crops in the CE countries.

Operationally, the concept presented in this chapter assumes, that the total amount of applied fertilizer nitrogen (F_N) is simply divided into three main pools: i) assimilated (Fa_N) – nitrogen taken up by meanwhile grown crop, ii) unassimilated nitrogen (F_{uaN}) – temporary out of use by plants and/or iii) lost from the field (F_{LN}). The quantitative assessment of the F_{uaN} pool relies on the defined YG procedure, individually related to each investigated crop:

$$F_N = F_{aN} + F_{uaN} + F_{LN}$$

$$F_{uaN} = YG/cPFP_N$$

where,

F_{uaN} – unassimilated nitrogen, kg N ha[-1]; YG - yield gap; cPFPN – the corrected partial factor productivity of applied N, kg yield kg N-1

In the first step of yield prognosis, the *Unassimilated Nitrogen Indices* ($I_{ua}N$) should be computed. Therefore, the original set of YG data for each of the studied crops was transformed into a quantitative amount of nitrogen lost temporarily from the system, marked as minus (-$I_{ua}N$). In the second step of the analysis, both sets of actual (Y_A) and maximum attainable yields (Y_M) of each crop were regressed against corresponding set of $_{ua}N$ indices. The obtained regression models clearly show that actual yields over the period of 20 years did not respond significantly to the $I_{ua}N$, as reported in Fig. 7 for wheat in Romania. The positive and simultaneously significant trends were, however, achieved, when the maximum attainable yield (Y_M) instead of the actual yield (Y_A) was introduced into the regression model. It can be therefore hypothesized that any virtual loss of nitrogen, i.e., the negative $_{ua}$N indices reflect the potential status for a yield increase and *vice versa*. Unfortunately, the positive indices of the $_{ua}$N simply describe a state of N soil mining, which in turn causes direct yield decline. This is consistent with the opinion of Dibb (2000) and typical for many parts of the world, including most CE countries.

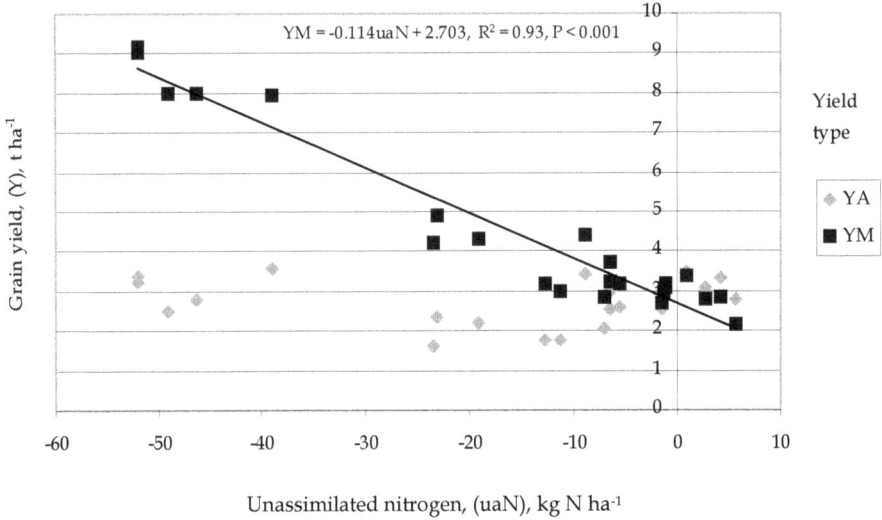

Fig. 7. Yield prognosis of winter wheat based on residual nitrogen - a case of Romania (RO). Significant ($P<0.05$) linear equation represents the trend for the attainable yields in the 1986-2008 period. Legend: Y_A – actual yield; Y_M – maximum yield.

Reliable yield prognosis can be performed, assuming a certain level of expected yield or by focusing any improvement of production technology on higher N use efficiency. The first approach can be best presented for wheat. The maximum attainable yield for this crop in Poland and the Czech Republic is fixed at the level of 6.5 t ha^{-1}, but for Romania at 4.5 t ha^{-1} (Rabbinge & van Diepen, 2000). The amount of $_{ua}$N required to be incorporated into grain yield was calculated by the models and amounted to 48, 27 and 33 kg ha^{-1}, respectively. It is equal to 58%, 27% and 100% of annual N rate applied in these countries, and does not seem a realistic approach. The same rules refer to other key crops in the study. Therefore, the second concept relying on the ability of currently cultivated crops to fix an extra amount of $_{ua}$N seems to be a more realistic approach for making any reliable yield prognosis (Table 7). This assumption is highly promising, especially for countries presenting an extensive type of agriculture production, for example, Romania. The expected wheat increase in response to the use of extra 20 kg N ha^{-1} is possible, considering the water limited yield, fixed at 4.5 t ha^{-1}. The prognosis for the Czech Republic is still below the water limited yield, being at 6.5 t ha^{-1}. In Poland, the calculated yields are at the same level as Romania, but soils are less suitable for wheat production. Therefore, any grain yield increase seems to be a great challenge for farmers in all countries of the CE. In the light of yields prognosis for sugar beets, the fixed level of $_{ua}$N at 20 kg N ha^{-1} is generally too low. The level of 60 t ha^{-1} of beets seems to be the most reliable short-term target. Thereby, the amount of $_{ua}$N successfully managed should be established at the level of 30 kg ha^{-1}.

Crop/Country	Linear regression model	Yield at 20 kg $_{ua}$N t ha^{-1}
Wheat		
Croatia	$Y_M = -0.064\,_{ua}N + 3.989$ for $R^2 = 0.96$	5.27
Czech Rep.	$Y_M = -0.063\,_{ua}N + 4.777$ for $R^2 = 0.88$	6.04
Poland	$Y_M = -0.060\,_{ua}N + 3.623$ for $R^2 = 0.92$	4.82
Romania	$Y_M = -0.114\,_{ua}N + 2.703$ for $R^2 = 0.93$	4.98
Maize		
Croatia	$Y_M = -0.069\,_{ua}N + 4.852$ for $R^2 = 0.86$	6.23
Czech Rep.	$Y_M = -0.064\,_{ua}N + 6.470$ for $R^2 = 0.69$	7.76
Poland	$Y_M = -0.085\,_{ua}N + 5.431$ for $R^2 = 0.83$	7.13
Romania	$Y_M = -0.121\,_{ua}N + 3.348$ for $R^2 = 0.94$	5.78
Sugar beets		
Croatia	$Y_M = -0.605\,_{ua}N + 40.677$ for $R^2 = 0.89$	52.78
Czech Rep.	$Y_M = -0.458\,_{ua}N + 45.872$ for $R^2 = 0.76$	55.03
Poland	$Y_M = -0.609\,_{ua}N + 39.801$ for $R^2 = 0.81$	51.98
Romania	$Y_M = -0.822\,_{ua}N + 24.046$ for $R^2 = 0.93$	40,49

Table 7. Yield prognosis based on the unassimilated quantity of nitrogen ($_{ua}$N)

5. Soil productivity improvements: The concept of N:P:K fertilizers balance

The simple comparison of CE countries with respect to nutrient management clearly shows, that annual inputs of nitrogen are too high with respect to phosphorus and potassium. Therefore, the main reasons of general yield stagnation in most of the CE countries are long-term negative balances of P and K, caused by low input of both mineral and organic fertilizers. As a consequence of soil P and K long-term mining, the percentage share of soils rich in both nutrients declined in the last two decades below the sufficiency level, which was achieved in many countries in 80s (Csatho at al., 2007; Grzebisz & Fotyma, 2007).

It has been clearly stressed that in *the transition phase of yield development*, the PFP$_N$ indices reached the highest values, resulting in yield gain. This phenomenon emerged in years 1990, 1991 for Poland and Hungary and in 1991, 1992 for the Czech and Slovak Republics. In all these years, the supply of P and K fertilizers was low. In the Czech Republic, ratios between amounts of applied N-P-K fertilizers, expressed as N:P$_2$O$_5$:K$_2$O was 1:0.25:0.25 on average. This structure, in spite of decreased fertilizers' consumption, allowed farmers to achieve the same yields of all crops as in the period 1986-1990. The present structure of nutrient use is as 1:0.15:0.15 for the Czech Republic, and 1:0.4:0.4 for Poland. These disparate nutrient ratios stress the importance of N:P:K balance for high crop production. The proposed N:P:K ratio for the Czech Republic, based on the cPFP$_N$ index, can be established at the level of 1:0.25:0.25 in *good years*, i.e., under ample water supply or 1:0.25:0.50 in *unfavorable years*, i.e., under expected water stress. Hence, keeping the current level of fertilizers' consumption on arable land, ca 130 kg ha^{-1}, one can predict the right amount of nutrients to apply. In Poland, the total amount of currently consumed fertilizers is almost the same as in the Czech

Republic. Unfortunately, soils in Poland are much poorer in potassium and also in phosphorus, being therefore, highly sensitive to external supply of nutrients. This conclusion is corroborated by positive trends of yields in response to P and K fertilizers application, as found for all crops since 1991 onwards. Consequently, under favorable weather conditions, or on soils naturally rich in potassium, the suggested formulation of N-P-K fertilizers is as follows: 1:0.33:0.66, but in years with high probability of drought or on areas sensitive to drought, a slightly different formulation is recommended, namely 1.0:0.5:1.0.

It is not easy to formulate any efficient fertilizing strategy for Romania. Since mid- 90s, maize and wheat have shown yield stagnation at a very low level. The main factors responsible for this situation are i) low N fertilization rates, ii) imbalanced use of P and K fertilizers. Therefore, yields of these two crops are, in fact, dependent only on the seasonal weather course, fluctuating for *ca* 100% on a yearly scale basis. The main reason is unbalanced consumption of basic fertilizers, which leads to permanent soil mining, especially potassium. The $N:P_2O_5:K_2O$ relationships, averaged for the period 2004-2008, as 1:0.27:0.09 is not sufficient to get any yield increase or even its stabilization. Potassium accumulates in plant tissues in much higher amounts and is considered as useful nutrient in water management (Cakmak, 2005). Thereby, this nutrient seems to be crucial for crop production in Romania, even taking into account, that almost 50% of land area is covered by soil naturally rich in minerals containing K. In spite of low N consumption, but unbalanced nutrient management, there is still a space for an extra N loss for both key crops, i.e., wheat and maize (see Fig. 7). Therefore, it could be concluded, that any yield increase depends on both significant N rate rise and simultaneous change of fertilizer formulations. The operational N-P-K fertilizers ratios suggested for Romania are as follows: 1:0.33:0.17 in *favorable* and 1:0.4:0.4 in *unfavorable* years. The second scenario should aim at increasing plant survival under drought conditions.

6. Thoughts and prospects

The new era of agriculture development in the Central Europe revealed how actual agriculture outcomes in this region depend on three key factors, presented in descending order: i) weather conditions during the growth season, ii) inherent soil fertility, iii) nitrogen fertilizer consumption. At present, the first factor is decisive for crop production. High year-to-year yield variability reflects the imbalanced use of soil and external resources such as fertilizers. Therefore, maximum attainable yields, i.e., limited by water supply, for example for wheat, are not attainable in all countries of the region. In Germany, also part of the Central European region, actual yields are above the "water limited" level (Rabbinge & Diepen, 2000; Supit et al., 2010).

Inherent differences in the nature of soil quality between the CE countries, as a background of nutrient management, are strictly related to the soil origin (Plant et al., 2005). Except for Poland, all other countries are relatively rich in soils originated from loams (Cambisols) and loess (Chernozems). These soils exhibit a high inherent potential for supplying K and other cations, such as calcium and magnesium (Nikolova, 1998).

However, at present nutritional requirements of high yielding crops, most of the CE countries are almost entirely oriented on the third yield forming factor, i.e., nitrogen fertilizer use. It has been well recognized, that a balanced supply of external sources of slowly mobile elements, such as phosphorus and potassium is broadly expected (Atkinson et al., 2005; Struik & Bonciarelli, 1997). Consequently, the average actual and attainable yields of main crops in CE countries are much more related to the soil-adjusted potential, than to amounts of currently applied fertilizers. As a result, CE countries show at present much greater differences in agricultural production than at the end of the 80s.

The observed unfavorable patterns of N fertilizer long-term consumption creates problems not only for agriculture production, but are also potentially dangerous for the environment as a potential source of *reactive nitrogen* (Roberts, 2006). There are three main questions to be urgently answered by specialists in all countries of the CE region: i) do farmers really recognize an increasing yield gap?, ii) what is the reason for farmers and their advisory services to tolerate low efficiency of nitrogen? iii) do farmers and advisory services recognize environmental aspects of the increasing amount of residual nitrogen?! In order to get answers to these questions one has to be familiar with factors contributing to the improvement in N fertilizer use efficiency:

1. Farm management of N in terms of i) N rate quantification, ii) time of N fertilizers application and iii) method of N fertilizers application.
2. Management of nutrients responsible for the efficiency of N uptake and it in-plant transformation;
3. Management of production factors other than N.

The first problem is not only of technical nature, because it affects in-season nitrogen use efficiency, and in turn actual yields' variability. However, CE countries representing the restoration type of yield development, but at the same time unbalanced pattern of consumed fertilizers do not indicate needs for increasing actually applied N rates. This strategy is, to some extent, the core of yield improvement in countries presenting stagnated type of fertilizers' management, provided that any increase in N use will be compensated by adequate amounts of applied phosphorus and potassium.

It could finally be concluded, that the realization of both goals of agricultural production in the CE countries, i.e., i) long-term yield stabilization and even its increase and ii) unit productivity of the applied fertilizer N increase, in turn decreasing amounts of the residual N, requires in the coming future changes in the structure of fertilizers' consumption, keeping in the Unbalanced group of countries the same level of applied fertilizers. The problem of nutrients supply to a growing plant does not however refer only to the balanced amounts of the applied fertilizers, but also to the crop accessibility to soil natural nutrient pools. This problem is inherently related to conditions of root system growth in the soil body. There are a lot of factors limiting roots accessibility to water and nutrients, even in soils of high natural fertility, which have been forgotten in the last 50 years, but the most important is soil acidy and related toxicity of aluminum (Atkinson et al., 2005; Diatta et al., 2010; Marschner, 1991; Struik and Bonciarelli, 1997). The key strategic target for farmers is to increase soil volume directly occupied by plant roots in order to improve their access to unavailable pools of nutrients. Three main areas of soil productivity improvement *via* increasing crop plants accessibility to nutrients in the subsoil, at least may be considered: i) regulation of soil reaction ii) increase of organic matter content iii) removing of any kind of hardpans. It seems highly probable, that in the case of Poland all three groups of measures are important. In other countries of the CE soil acidity should be considered as an agronomical problem for farmers conducting production on Luvisols.

7. References

Atkinson, A.; Black, K. & Dawson, L. (2005). Prospects, advantages and limitations of future crop production systems dependent upon the management processes. *Annals of Applied Biology* Vol. 146, pp. 203-215, ISSN 1744-7348

Batjes, H. (2002). Carbon and nitrogen stocks in the soils of Central and Eastern Europe. *Soil Use Management*, Vol. 18, pp. 324-329, ISSN 1475-2743

Cakmak, I. (2005). The role of potassium in alleviating detrimental effects of abiotic stresses in plants. *Journal of Plant Nutrition Soil Science*, Vol. 168, pp. 521-530, ISSN 1436-8730

Catt, J. (2001). The agricultural importance of loess. *Earth-Science Reviews*, Vol. 54, pp. 213-229, ISSN 0012-8252

Cochrane, N. (2004). EU Enlargement: Implications for U.S._EU agricultural relations. *U.S.-EU Food and Agriculture Comaprison*/WRS-04-04, 78-86. Available online at: http://www.ers.usda.gov/publications/WRS0404/WRS0404.pdf   (Access 2011.05.24).

Csatho, P.; Sisak, I. & Radimszky, L. (2007). Agriculture as a source of phosphorus causing eutrophication in Central and Eastern Europe. *Soil Use management*, Vol. 23 (suppl. 1), pp. 36-56, ISSN 1475-2743

Csatho, P. & Radimszky, L. (2009). Two Worlds within EU27: Sharp Contrasts in Organic and Mineral Nitrogen-Phosphorus Use, Nitrogen-Phosphorus Balances, and Soil Phosphorus Status: Widening and Deepening Gap between Western and Central Europe. *Communications in Soil Science and Plant Analysis*, Vol. 40, pp. 999-1019, ISSN 0010-3624

Diatta, J.; Bocianowski, J. & Skubiszewska, A. (2010). Sulphate-based aluminum phytotoxicity mitigation under strong soil acidification. *Fresenius Environmental Bulletin*, Vol. 19, No. 12b, pp. 3185-3192, ISSN 1018-4619

Dibb, D. (2000). The mysteries (myths) of nutrient use efficiency. *Better Crops*, Vol. 84, No. 3, pp. 3-5, ISSN 0006-0089

Dobermann, A. & Cassman, K. (2002). Plant nutrient management for enhanced productivity in intensive grain production systems of the United States and Asia. *Plant and Soil*, Vol. 247, pp. 153-175, ISSN 1573-5036

Evans, L. & Fisher, R. (1999). Yield potential: its definition, measurement and significance. *Crop Science*, Vol. 39, pp. 1544-1551, ISSN 1896-0782

Falloon, P. & Betts, R. (2010). Climate impacts on European agriculture and water management in the context of adaptation and mitigation – the importance of integrated approach. *Science of Total Environment*, Vol. 408, pp. 5667-5687, ISSN 0048-9697

FAOSTAT. Available online at: http://faostat.fao.org/default.aspx. (Access 2011.05.24).

Fixen, P. (2004). Phases in narrowing the yield gap. *Better Crops*, Vol. 88, No. 4., p. 3, ISSN 0006-0089

Fronzek, S. & Carter, T. (2007). Assessing uncertainties in climate change impacts on resource potential for Europe based on projections from RCMs and GCMs. *Climate Change*, Vol. 81, pp. 357-371, ISSN 1758-6798

Grzebisz, W. & Fotyma, M. (2007). Recommendations and use of potassium fertilizers in Central-Eastern Europe (CEE). In: *Proceedings International Fertiliser Society*, No. 621, York, UK, pp 24, ISBN 978-0-85310-258-8

Grzebisz, W.; Diatta, J. & Haerdter, R. (2010). Fertilizer consumption patterns in Central European countries – effect on actual yield development trends in 1986-2005 years - a comparative study of the Czech Republic and Poland. *Journal of Central European Agriculture*, Vol. 11, No. 1, pp. 73-82, ISSN 1332-9049

IFADATA Statistics online. (2011). www.fertilizer.org/ifa/statistics/ifadata /dataline.asp, (Access, 2011.05.24)

Jongman, R.; Bunce, R. & Metzger, R. (2006). Objectives and applications of a statistical environmental stratification of Europe. *Landscape Ecology*, 21, 409-419, ISSN 1572-9761

Karlen, D.; Ditzler, C. & Andrews, S. (2003). Soil quality: why and how?! *Geoderma*, Vol. 114, pp. 145-156, ISSN 0016-7061

Kelch, D. & Osborne, S. (2001). Crop production capacity in Europe" Economic Research Service/USDA, *Agricultural outlook/*March 2001, 19-23, ISSN 0099-1066

Marschner, H. (1991). Mechanisms of adaptation of plants to acid soils. *Plant and Soil*, 134, 1-20, ISSN 1573-5036

Neumann, K.; Verburg, P. & Stehfest, E. (2010). The yield gap of global production: a spatial analysis. *Agricultural Systems*, Vol. 103, pp. 316-326, ISSN 0308-521X

Nikolova, M. (1998). The soil potassium resources and the efficiency of potassium fertilizers in Bulgaria. Nikola Poushkarow Inst. of Soil Science and Agroecology, Sofia, Bulgaria. *IPI, Country Report*, 12, 39 pp.

Olesen, J. & Bindi, M. (2002). Consequences of climate change for European agricultural productivity, land use and policy. *European Journal of Agronomy*, Vol. 16, pp. 239-262, ISSN 1161-0301

Olesen, J.; Trnka, M. & Kersebaum, K. (2011). Impacts and adaptation of European crop production systems to climate change. *European Journal of Agronomy*, Vol. 34, pp. 96-112, ISSN 1161-0301

Plant, J.; Whittaker, A. & Demitriades, B. (2005). The geological and tectonic framework of Europe. *Geochemical Atlas of Europe*. Part 1. *Background Information, Methodology* Ed. Salminen R. Maps. GTK, FOREGS, 20 pp., ISBN 951-690-956-6

Rabbinge, R. (1993). The ecological background of food production. In: *Crop protection and sustainable agriculture*. Wiley, Chichester (Ciba Foundation Symposium 177, 2-29, ISBN 10 0471939447

Rabbinge, R. & Diepen, C. (2000). Changes in agriculture and land use in Europe. *European Journal of Agronomy*, Vol. 13, pp. 85-100, ISSN 1161-0301

Reilly, J., Fuglie, K., (1998). Future yield growth in field crops: what evidence exists? *Soil & Tillage Research* Vol. 47, pp. 275-290. ISSN 0168-1987

Roberts, T. (2006). Improving Nutrient Use Efficiency. *Proceedings of the IFA Agriculture Conference: Optimizing Resource Use Efficiency for Sustainable Intensification of Agriculture*, 27 february-2 March, 2006, Kunming, China, pp 8.

Rosegrant, M.; Paisner, M. & Meijer, S. (2001). Global Food Projections to 2020, Emerging trends and alternative futures. *IFPRI*, pp 206, ISBN 0-89629-640-7

Sinclair, T.; Purcell, L. & Sneller, C. (2004). Crop transformation and the challenge to increase yield potential. *Trends in Plant Science*, Vol. 9, No. 2, pp. 70-75, ISSN 1360/85

Struik, P. & Bonciarelli, F. (1997). Resource use at the cropping system level. *European Journal of Agronomy*, Vol. 7, pp. 133-143, ISSN 1161-0301

Supit, I.; van Diepen, C. & de Wit, A. (2010). Recent changes in the climatic yield potential of various crops in Europe. *Agricultural Systems*, Vol. 103, pp. 683-694, ISSN 0308-521X

11

Nutrient Management in Silvopastoral Systems for Economically and Environmentally Sustainable Cattle Production: A Case Study from Colombia

Liliana Mahecha and Joaquin Angulo
GRICA Research Group, University of Antioquia,
Colombia

1. Introduction

In recent years, livestock production has received negative publicity due to environmental degradation. Critics charge that the expansion of cattle production around the world has destroyed the forest, increased soil erosion, and has contaminated the environment. These negative effects have been caused by poor decisions in the production system. Nevertheless, there are possible solutions. Given ongoing climate change, fears of environmental contamination, and global market competition, silvopastoral systems emerge as a valuable alternative to develop an economic, productive, and environmentally-friendly system for livestock raising in the world. The purpose of this chapter is to give a general description of the silvopastoral systems and to provide a synopsis of the main productive and environmental benefits obtained by using them for dairy and beef cattle production. Most of information in this chapter is inspired by the Colombian experience during the past 20 years.

2. What are silvopastoral systems?

In Colombia and much of Latin America, the traditional cattle production systems have been based on grass monoculture (treeless pastures) as the main food source. According to **Steinfeld et al. (2006),** the development of this kind of cattle production system has led to extensive deforestation, soil degradation, and contamination of water bodies and the environment. In Colombia, the annual deforestation rate of 300.000 ha for expansion of pasturelands has seen this land use more than double from 14.6 to 35.5 million ha between 1960 and 1995, while natural forests and agriculture have declined in area from 94.6 to 72.4 million ha (**Lavh et al., 1998**). This land use transformation has homogenized and simplified the ecosystems and has negatively impacted the quality of the environment and its ecological diversity (**Mahecha, 2002; Giraldo et al. 2011**). The problem is exacerbated by the increasing degradation of soils and grasslands in areas used by cattle, which has reached levels of 73.4, 68.5 and 94.1% in the Córdoba, Sucre and Atlantic departments, respectively (**CORPOICA, 2010**). Ironically, this transformation has not made cattle production systems more effective, economically and productively. The traditional cattle production system in

Picture 1. Lucerna cows grazing in a high density silvopastoral system. Picture: Liliana Mahecha, in Hatico Natural Reserve

Colombia exhibits low productivity and competitiveness (**Colombian Ministry of Agriculture and Rural Development, 2005**) expressed as low number of animals per ha (average 1.2), low birth rate (less than 69%) and low weight daily gain (between 350-450 g/d) (**Mahecha et al. 2002**). These parameters have shown only modest productivity increments during the last 10 years, with a few exceptions. It is predicted that the situation will worsen in coming years due to the global climate change. According to the **IPCC (2007)**, the air temperature will increase from 2.4 to 6.4° C with an average of 4.0°C increasing 0.2°C per decade, which is expected to cause a decline in animal performance (**Nardone et al., 2010**).

Unfortunately, this kind of production system and its negative environmental impacts have raised criticism about livestock raising around the world. However, appropriate production strategies with correct environmental management would help in restoring degraded ecosystems, provide environmental services (**Calle et al., 2002**), and improve productivity (**Mahecha, 2003**). One of the strategies that can be used to reduce the impact of this activity is the implementation of silvopastoral systems (SPS).

SPS are animal production systems that combine fodder plants, such as grasses and leguminous herbs, with shrubs and trees for animal nutrition and complementary uses (**Mahecha, 2002; Murgueitio et al., 2011**). The strategic management of nutrients and the interactions between components of the system has an important impact on the environment, productivity, and society.

3. What types of silvopastoral systems exit?

Trees used in SPS can be from natural vegetation or planted for timber, industrial products, fodder and fruit or specifically for animal production (providing shadow, fodder, seeds, wood) (**Mahecha, 2002**). Therefore, there are several types of SPS such as protein and energy banks (cut and carry systems), live fences, wind barriers, and low and high density silvopastures (**Picture 2, Mahecha, 2002**). The advantages of the high density silvopastures are the best known. The selected type depends on the topography, type of soil, and the presence of strategic areas for water, soil or biodiversity conservation. For that reason, it is important to have in mind the use of the different areas of the farm according to needs: protecting from livestock trampling and grazing the fragile areas or areas that are important to conserve the biodiversity or water, grazing in appropriate areas of pastures with low and high density of trees, and using areas to produce fodder (cutting and carrying) where direct access of cattle is not recommended because they would increase erosion, thus direct grazing should be avoided (**Murgueitio and Ibrahim 2001**).

Picture 2. Silvopastoral systems: protein bank (*Cratylia argentea* bank, Candelaria farm, University of Antioquia, high density, low density (*Prosopis juliflora* trees, Hatico Natural Reserve), and wind barriers (Colombia coffee area). Picture: Liliana Mahecha.

4. Main shrubs and trees used in silvopastoral systems

One of the main ongoing research activities of research on going about in SPS is the identification and characterization of different shrubs and trees that can be used as forage for animals, especially incorporating them into the SPS for direct intake by the animals. In Colombia, several species have been evaluated such as: *Leucaena leucocephala, Tithonia diversifolia, Crescentia cujete, Erythrina fusca, Gliricidia sepium, Guazuma ulmifolia, Moringa Oleifera, Cratylia argentea, Acacia decurrens, Sambucus peruviana.* **Table 1** summarizes some nutritional characteristics of these plants for use as animal feed. Besides forage plants, other species have been identified in Colombia to be for wood, shade, and/or nitrogen fixation (**Table 2**). The farmers' decision making about what kind of tree should be incorporated in a SPS is influenced by a combination of factors: soil characteristics, purpose of production, type of grass involved, weather conditions, and other factors (**Mahecha, 2003**). However, one of the main aspects to consider is the relationship between the different components of the system because complex interactions occur between livestock, trees, and pasture in SPS. For example, the key factor of the success of SPS that involves timber trees is to achieve a compromise between the two sources of economic benefit. Grazing is possible when the tree canopy allows light to reach the understory layer, with recommended canopy covers less than 50% (**Pasalodos-Tato, 2009**). Light infiltration to the understory was affected in a silvo-pastoral system with *Eucalyptus tereticornis* and *Panicum maximum* in the San Sebastian reforestation program in Magdalena Department of Colombia when tree height was greater than 10 m using tree density of 3 x 1.5 m (**Mahecha et al. 2007**). In all cases, it is very important to promote nutrient recycling to support plant communities and livestock sustainably.

Species	Crude protein	Ether extract	Neutral detergent fiber	Acid detergent fiber	Calcium	Phosphorous	Source
Acacia decurrens	17.8	3.5	39.2	30.6	0.7	0.3	Chamorro and Rey, 2009
Sambucus peruviana	23.8	5.2	19.4	17.3	9.2	0.9	Chamorro and Rey, 2009
Montanoa cuadrangularis	25.9		48.8	34.6	-	-	Chamorro and Rey, 2009
Tithonia diversifolia	22.6	2.3	35.3	30.4	2.1	0.4	Mahecha and Rosales, 2005
Leucaena leucocephala	26.3			14.1	1.2	0.2	Mahecha et al. (2000)
Guazuma ulmifolia	13-17	-	46.1[1]	29.4[1]	0.9	0.3	Calle and Murgueitio, (2011) [1]Manriquez et al. (2011)
Crescentia cujete (fruit)	12.5	14.8	-	-	0.4	0.4	Calle et al. 2011
Erythrina fusca							
Gliricidia sepium	10.4	-	55.0	41.8	-	-	Lucero (2009)

Table 1. Chemical composition (%) of the forage of some trees used in silvopasture (SPS) in Colombia

5. Impact of the silvopastoral systems on the soil

The incorporation of trees (shrubs and/or trees) in SPS increases soil fertility, improves soil structure, and reduces erosion processes. **Ramirez (1998)** found that the presence of legumes trees in pastures led to an increase in the content of soil nutrients such as nitrogen (N), phosphorus (P), and carbon (C) at depth of 10-30 cm, compared to grass monoculture (**Table 2**). **Rodriguez (1985)** also found higher amounts of organic matter, N and Ca in soil of SPS containing *E.poeppigiana* trees and *P. purpureum* grass compared to monoculture grass. These results have been explained by the increased recycling of nutrients, N_2 fixation, the extensive rooting of trees and greater activity of soil macro and micro fauna given the greater mass of litter and organic residues from diverse plant species and livestock (**Mahecha, 2002**).

Treatment	Soil depth								
	0-10 cm			10-20 cm			20-30 cm		
	N gkg-1	P ppm	C gkg-1	N gkg-1	P ppm	C gkg-1	N gkg-1	P ppm	C gkg-1
SPS	1.4	29	16.8	1.1	25	14.0	2.2	15	9.2
Grass monoculture	0.8	16	10.0	0.6	16	7.0	1.2	15	4.8

Source: Ramirez (1998)

Table 2. Concentration of total N, P, and C in different soil depths of a silvopastoral system (SPS of *C.plectostachyus* grass + *Leucaena leucocephala* and *P. juliflora* compared to grass monoculture

Nutrient cycling and fixation: the management of grass with trees and/or shrubs recycles nutrients extracted from the soil when vegetation (roots, leaves, fruits) dies and decomposes, from manure of grazing animals and residues from tree pruning (**Sadhegian et al. 1998**). A positive balance was found after one year for N (+16 kg/ha) and P (+1 kg/ha) in silvopastoral systems comprised by native grass and *Leucaena leucocephala* compared to grass monoculture where the balance was negative for N (-15 kg/ha) and P (-6 kg/ha) (**Crespo et al. 1998**). Additionally, most of trees used in SPS are legumes that have the capacity to fix nitrogen from the atmosphere through the association with bacteria living in root nodules. These bacteria can change inert N_2 to biologically useful NH_3, which is then converted to protein in the plant (**Lidemann and Gloves, 2008**). In the SPS of Colombia, legumes provide the main input of nitrogen for pastures. Such systems can substantially reduce inputs of chemical fertilizers and have the added benefit of improving feed quality for grazing animals, especially in the high density SPS. **Ramírez (1997)** found high productivity of forage without using urea fertilizer was achieved by the introduction of shrubs of *Leucaena leucocephala* and trees of *Prosopis juliflora* in the plots (**Table 3**). Other non-legume plants may also be beneficial for soil fertility, such as *Tithonia diversifolia* (**Mahecha and Rosales, 2005**). It is unclear whether the ability to restore degraded soils by *T. diversifolia* is because of the association with mycorrhizal fungi, which are efficient at capturing soil phosphorus or because of the exudation of organic acids by roots that allows for efficient assimilation of phosphorus and other nutrients (**Calle and Murgueitio, 2010**).

Treatments	T1	T2	T3
Fresh forage (t/ha/year)	108.4	121.4	81.7
Dry matter (%)	30.9	31.6	28.4
C. plectostachyus (t DM/ha/year)	33.4	38.3	23.2

T1 - *C. plectostachyus* + *L. leucocephala* (10000 plantas/ha), *P. juliflora* (10 trees/ha)
T2 - *C. plectostachyus* + *P. juliflora* (18 trees/ha) + 400 kg urea/ha/year
T3 - *C. plectostachyus* + 800 kg urea/ha/year.

Table 3. Productivity of *C. plectostachyus* associated with *L. leucocephala* and *P. juliflora*

Rooting depth: the wide, deep root systems of trees in SPS increases the available area for nutrient capture and helps maintain nutrient stocks by reducing leaching losses or by taking up nutrients from deeper soil layers (**Beer et al. 2003**).

Higher diversity and activity of micro and macro fauna: the higher content of organic matter in soil and the improvement of the microclimate (moisture and temperature) due to the presence of trees in SPS promotes the biological activity of the macro and micro fauna, resulting in a greater mineralization and availability of soil nutrients. In addition, organic matter is incorporated gradually into the soil by the action of soil fauna. This helps to improve soil stability, due to the production of stable soil aggregates, and water infiltration capacity through pores constructed by the macrofauna, earthworms in particular (**Belsky et al. 1993**). A study carried out in Caqueta, Colombia compared soil fauna in two production systems: native grass and grass plus leguminous trees (SPS); after 3 years, SPS had 59 taxa of macro-invertebrates at family level and a total of 913 individuals per sampling unit while native grass (monoculture) had values of 30 and 305 individuals, respectively (**Gómez and Velasquez, 1992**). In another study carried out in Cuba, it was found 300 individuals / m^2 in soil from SPS compared to 170 individuals / m^2 in soil from treeless improved grass (**Sanchez, 1998**). Similarly, **Velasco et al. (1999)** found higher numbers of endomycorrhizal fungi and earthworms in soil from SPS of *A.mangium* and *Brachiaria humidicola* compared to grass monoculture. In the same way, **Pardo-Locarno (2009)** found higher earthworm populations in soil from high density SPS compared to other land uses at the Hatico Natural Reserve in Colombia (**Figure 1**). Likewise, **Giraldo et al. (2011)** found that the adoption of SPS promotes the recovery of ecological processes regulated by the increase of dung beetles in the Colombian Andes compared to treeless improved pastures. These authors report that changes in the number of dung beetles is considered an important indicator of land-use change and pasture health. The activities of these beetles are linked to a wide variety of ecological processes, including the incorporation of organic matter into the soil and the control of haematophagous flies and gastrointestinal parasites that breed in manure and affect domestic animals and humans. In the same way, **Vallejo et al. (2010)** assessed the effect of a silvopastoral chronosequence in a tropical region of Colombia on soil microbiological and physico/chemical properties, considering three production systems: monoculture grass conventional pasture (CP), native forest (F), and a silvopastoral system (SPS) chronosequence with ages of 3 to 6 (SPS$_3$), 8 to 10 (SPS$_8$), or 12 to 15 (SPS$_{12}$) years. SPS$_{12}$ showed the highest microbial biomass and enzyme activities on a per unit C basis and was consistently and significantly different from CP. Additionally, microbiological to C ratios were significantly affected by SPS establishment age ($P < 0.05$). The low microbiological responses were consistent with high penetration resistance and bulk density of CP, which indicates that the SPS are improving soil quality. This study presented quantitative data that SPS stimulated soil microbial biomass and enzyme activities, which indicates greater potential to carry out biogeochemical process, and that SPS provides a more favorable microbial habitat than CP.

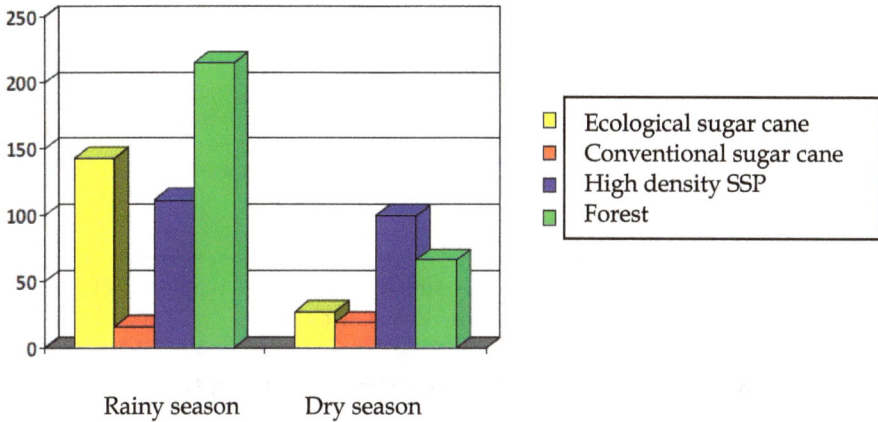

Fig. 1. Density of earthworm per 0.75 m^3 in four land uses in the Hatico Natural Reserve, Colombia. Source: **Pardo-Locarno (2009)**

Erosion control: trees in SPS fulfill protect the soil from the direct effects of sun, wind and water (**Fassbender, 1993**). **Gomez and Velasquez (1999)** showed that the loss of soil in areas without protection of trees is higher than in forest (**Table 4**). The control of erosion by trees is due to several factors: reducing the impact of rain, increasing the infiltration and increasing the stability of the organic matter.

The effect of SPS on the physical characteristic of the soil was evaluated in the "Mainstreaming sustainable cattle ranching project" carried out in Costa Rica. Results showed lower runoff and erosion in SPS compared to treeless improved pastures. Likewise, soil in SPS had higher infiltration rates which improves its ability to retain water, reduce runoff, and contributes to the regulation of water cycle (**Figure 2**) (**Rios, 2006**).

	Slope	Stocking rate	Loss of soil (ton/ha/year)
Forest	32	---	0.61
Grass monoculture	22	1.5	8.23
Bare soil	24	---	20.4

Source: Gómez and Velasquez (1999)

Table 4. Loss of soil per year in Caquetá, Colombia with and without tree coverage

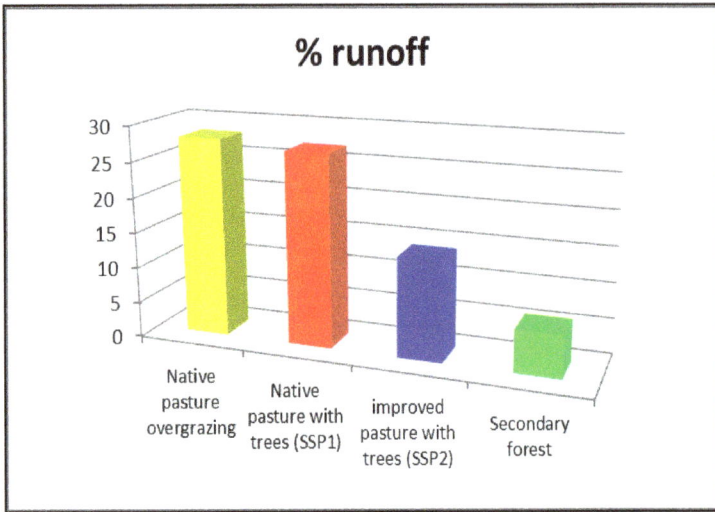

Fig. 2. Average superficial runoff rate for 50 rainfall events in different livestock production systems around the Jabonal river, Costa Rica. Source: **Rios (2006).**

6. Impact of the silvopastoral systems on the environment

The removal of forest for implementation of treeless improved pastures for cattle production impacts negatively the carbon content in the soil due to increase in surface temperature that accelerates the oxidation of organic compounds. The most profound difference is seen when comparing forest to degraded pastures, where carbon stocks are 42 t C/ha in forest and decline to 34.5 t C/ha in pastures (**Ibrahim, 2001**) This decrease in organic carbon content leads to loss soil fertility and increases the emission of greenhouse gases. Reforestation, SPS, and protection of forest in the tropics (for instance, increasing forest cover from 300 to 600 thousand ha) would retain and store between 36 to 71 Pg of C over 50 years (**Ibrahim, 2001**). In a study carried out in Colombia´s Amazon region, the carbon sequestration in pasture and silvopastoral systems under conservation management was evaluated. Results of 5 years of research (2002-2007) show that improved and well-managed pasture and SPS can contribute to the recovery of degraded areas as C-improved systems (**Table 5**).

Land use system	Total C in soil (t/ha per 1m-eq)	%	Total soil in pasture (t/ha)	%	Total C in fine roots (t/ha)	%	Total C in thick roots, trunks, and leaves (t/ha)	table	Total C in system (t/ha)
Native forest	181[a]	61.7	-	-	-	-	112.4	38.3	293.4
B.decumbens + legume	172[b]	98.1	0.9	0.5	2.4	1.4	-	-	175.3
B.humidicola	159[c]	96.6	1.1	0.7	4.5	2.7	-	-	164.6
Degraded pasture	129[d]	97.4	0.9	0.7	2.6	1.9	-	-	132.5

Means with different letters differ statistically ($P<0.05$), Source: Amezquita et al. (2008)

Table 5. Carbon in soil and biomass of tropical rainforests in Colombia's Amazon region

7. Impact of the silvopastoral systems on the forage availability and quality

The introduction of legumes trees in pastures (SPS) improves the quality of forage of the associated grass compared to forage of treeless improved pastures. It is also possible to increase the total amount of forage for animals, but this depends on factors such as the management of trees, environmental conditions, tree density and species. In a study carried out by **Mahecha et al. (2000)** in the Hatico Natural Reserve located in Valle del Cauca, Colombia, a SPS of *Cynodon plectostachyus*, *Leucaena leucocephala* and *Prosopis juliflora* showed a forage production of *C.plectostachyus* of 25 t DM/ha/year without fertilization and under the influence of the climate patterns known as El Niño-Southern Oscillation (ENSO). This production was higher than that found by **Ramirez (1998)** in treeless improved pasture of *C.Plectostachyus* (21 t DM / ha / year) on the same farm under no influence of ENSO when fertilizer was applied (400 kg urea / ha / year). In addition, the protein content and in vitro digestibility of grass in the SPS (12% and 64.7%, respectively), with rotational grazing every 42 days, was better than those reported in the literature for treeless improved pasture. It is also important to keep in mind that in the SPS, besides forage production from the grass (25 t/ha/year), cattle also graze on forage from the leaves and thin branches of *Leucaena leucocephala* (4.3 t DM / ha / year) and leguminous pods of *Prosopis* trees (0.4 t DM /ha/ year) that falls to the ground twice a year. Thus, this kind of high density SPS offered a total oc 29.9 t DM /ha / year compared with 21 t MS / ha / year from grass alone in the treeless improved pasture, based on the values reported by **Ramirez (1998).**

8. Impact of the silvopastoral systems on the beef cattle production

Mahecha et al. (2011) evaluated the animal performance and carcass features of two breeds of dual purpose cattle grazing on intensive SPS. The animals were divided in two groups: G1: mostly crossbred genotype F1 * Zebu, G2: animals with F3 * Brahman. Animals were evaluated for 263 d in SPS of *L.leucocephala*, *C. plectostachyus* and timber trees. The average daily weight gain was 863 and 796 g/animal/d for G1 and G2 respectively. G1 had 60.66 cm^2 of *longissimus dorsi* eye area (AOL) and 6.77 mm of backfat thickness (BT) while G2 had 64.55 cm^2 AOL and 6.44 mm BT, respectively. These results represent the high quality of the forage in the SPS since animals did not receive dietary supplements. In contrast, **Vasquez et al. (2005)** report an average daily weight gain of 578 g/animal/day, AOL between 43-54 cm^2, and BT between 3.0-4.3 mm in different improved systems in treeless pastures.

Although animal performance is positive in SPS, it is necessary to consider that the success depends of the positive relationship between the components of the system. **Mahecha et al. (2007)** evaluated the effect of tree height on the grass availability and quality and on the daily weight gain of young Zebu steers in a SPS of *Eucalyptus tereticornis* and *Panicum maximum* during the dry season in the reforestation program "San Sebastian" in Madalena, Colombia. In addition, the impact of the animals on the tree growth and on soil fertility was estimated. The treatments were: T1: *Eucalyptus tereticornis* + *Panicum maximum*, density 3 x 1.50 m, height of trees 5 m. T2: *Eucalyptus tereticornis* + *Panicum maximum*, density 3 x 1.50 m., height of trees 10 m. Each treatment was accompanied by a plot with similar conditions, but without the presence of animals (T4) to be able to compare tree growth with and without cattle present. These authors demonstrated that animal performance, soil fertility, and tree growth can be positive in a short-term, but could be altered in a long-term because SPS with very high tree density can affect the light availability to grass and the cycling of nutrients. After 140 days of evaluation, T1 gained 0.491 g/animal/d and T2 0.245 g/animal/d and there was no effect of animals on the tree growth, but the fertility of soils was lower in the SPS than in the area without trees (**Table 6**). These results indicate that the change in soil fertility was not too drastic to affect the tree growth in the short term, but long-term management strategies should consider reducing the tree density to promote nutrient cycling in the system.

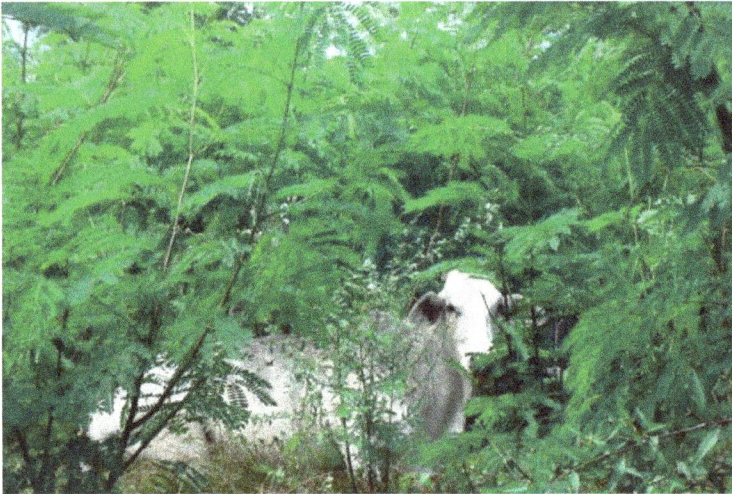

Picture 3. Beef cattle in a SPS of *Leucaena leuocephala* and grass. Picture from John Jairo Lopera, El Porvenir farm, Cesar, Colombia

	Texture and pH	N, gkg^{-1}	Organic matter, gkg^{-1}	P, ppm	Ca, cmol(+)kg^{-1}	Mg, cmol(+)kg^{-1}	K, cmol(+)kg^{-1}
T1	F-A 5.16	2.0	42.0	35.1	2.83	0.56	0.32
T2	F-A 5.31	1.9	41.0	33.	6.52	1.77	0.43
T3	F-A 5.74	4.2	103.0	35.	1.08	0.26	0.48
T4	F-A 5.68	4.1	102.0	49.0	4.35	1.43	0.64

T1= Plot with trees of 5 m without animals; T2= Plot with trees of 5 m with animals; T3= Plot with trees of 10 m without animals; T4= Plot with trees of 10 m with animals

Table 6. Effect of tree height and presence of animals in silvopastoral systems on the soil fertility

Nutrient Management in Silvopastoral Systems for Economically and Environmentally Sustainable Cattle Production:
A Case Study from Colombia

211

9. Impact of the silvopastoral systems on the dairy cattle production

The area to be covered by SPS on farms varies between 15-100% and in all cases, the stocking rate and milk productivity is increased by the transformation of the pasturelands without trees. However, farms vary a lot according to the intensity of trees and shrubs that are included in the SPS, which determines the different relationships between components and the cycling of nutrients, with increases that fluctuate between 87.5 and 166.6 % for stocking rate, and 20-35% for milk production (**Murgueitio et al. 2006**). The increase of the land use with high density SPS of *C.pleactostachyus*, *L.leucocephala* and *P.juliflora* in the Hatico Natural Reserve, Colombia, has led to an increase the stocking rate 31% and the milk production 95% during the last 10 years (**Table 7**) (**Molina et al. 2009**). This SPS also offers higher quality of milk fat related to a higher composition of conjugated linoleic acid (CLA), which has been reported as beneficial for human health (**Mahecha et al. 2008**). Improvement in milk production have also found with *Tithonia diversifolia* used as protein bank for cutting and carrying its forage for animals. **Mahecha et al. (2007b)** evaluated the quantity and quality of milk of F1 cows Holstein x Zebu supplemented with forage of *T.diversifolia* as a partial replacement of concentrate food. The authors did not find significant differences in milk yield during the rainy and dry season; cows fed 100% concentrate food produced 12.5 and 11.71 L /day (rainy and dry season, respectively) compared to 12.4 and 12.16 L /day (rainy and dry season, respectively) of cows fed a concentrate consisting of 35% *T.diversifolia* forage. Likewise, the milk protein level was increased from 3.51% to 3.82% (100% concentrate and 35% replacement, respectively). New experiments are being carrying out with *T.diversifolia* in high density SPS, which allows grazing animals direct access to this potential forage on acid soils of Caqueta, Colombia with low phosphorus and high aluminum saturation.

Picture 4. Lucerna dairy cows in a high density SSP. Hatico Natural Reserve, Colombia.
Picture from Liliana Mahecha

Variables	Year											
	1996	1997	1998	1999	2000	2001	2002	2003	2004	2005	2006	2007
Total area ha	89	89	89	73	51	63	49.6	49.6	49.6	52.8	56	61
Area without Leucaena ha	75.4	68.2	61.8	29.8	0	0	0	0	0	3.2	0	0
Area with Leucaena ha	13.6	20.8	27.2	43.2	51.0	62.2	48.6	49.6	49.6	49.6	56	61
Number of milking cows	299	286	259	266	230	304	259	244	252	260	246	248
Stocking rate cows/ha/year	3.35	3.21	2.91	3.74	4.5	4.82	5.22	5.04	5.08	4.92	4.4	4.1
Milk production L/ha/year	7 436	8 298	9 770	11 684	17 025	16 798	18 290	18 486	17 857	16 501	14 306	14 473

Source: Molina et al (2009)

Table 7. Effect of the increase of SPS on the land use on the stocking rate and milk production in the Hatico Natural Reserve, Colombia

In summary, this chapter showed that livestock production system can be environmentally friendly, efficient and productive when diversified plant communities are supported through the use of SPS. This system relies on optimizing relationships between components (soils-plants-animals)   to obtain appropriate nutrient balances/recycling, high vegetal biomass production and the multiple benefits for grazing cattle mentioned above.

10. References

Amezquita M.C, Amezquita E, Casasola F, Ramirez B.L, Giraldo H, Gomez M.E, Llanderal T, Velazquez J, Ibrahim M.A. 2008. C stocks and sequestration,. En: Carbon sequestration in tropical grassland ecosystems. Edited by L. 't Mannetje, M.C. Amézquita, P. Buurman and M.A. Ibrahim.

Belsky AJ, Mwonga SM, Duxbury JM. 1993. Effects of widely spaced trees and livestock grazing on understory enviroments in tropical savannas. Agroforestry Systems, 24:1-20.

Beer C.A, Harvey M, Ibrahim M, Harmand E, Somarriba E and Jiménez F. 2003. XII world forestry congress. Quebec, Canada.

Chamorro D and Obando A.M. 2009. El component arbóreo como dinamizador del sistema de producción de leche en el trópico alto colombiano. Experiencias de Corpoica. Tibaitatá. In: Murgueitio, E., Cuartas, C., Naranjo, J.F. (Eds.), Ganadería del futuro: Investigación para el desarrollo. Fundación CIPAV. Cali, Colombia, pp. 351-397.

Calle Z and Murgueitio E. 2011. El guasimo: uno de los árboles más adaptables a los sistemas silvopastoriles del trópico americano. Carta Fedegan 121.

Calle Z, Murgueitio E, Botero L.M. 2011. El Totumo, árbol de la Américas para la ganadería moderna. Carta Fedegán 122:64-73.

Calle Z and Murgueitio E. 2010. Botón de oro. Arbusto de gran utilidad para sistemas ganaderos de tierra caliente y de montaña. Carta Fedegán 108:54-59.

Crespo G, Castillo E, Rodriguez I. 1998. Estudio del reciclaje de N, P y K en dos sistemas de producción de vacunos de carne en pastoreo. En Memorias III taller Internacional silvopastoril realizado del 23-27 de noviembre 1998. Cuba. 234-236.

Fassbender H. 1993. Modelos edafológicos de sistemas agroforestales; 20. Edición. CATIE. Turrialba. Costa Rica. 490 p.

Giraldo C, Escobar F, Chara J, Calle Z. 2011. The adoption of silvopastoral systems promotes the recovery of ecological processes regulated by dung beetles in the Colombian Andes. Insect Conservation and Diversity 4: 115–122.

Gómez JE, Velásquez JE. 1999. Proceso integral de recuperación y manejo de praderas, condición fundamental para el desarrollo ganadero en Caquetá.

Boletín Técnico Corpoica-Pronatta. 42 p.

Ibrahim M. 2001. Potencialidades de los sistemas silvopastoriles para la generación de servicios ambientales. Conferencia electronica en potencialidades de los sistemas Silvopastoriles para la generación de servicios ambientales. FAO.

IPCC, 2007. Climate Change 2007: The Physical Science Basis. Contribution of Working Group I to the Fourth Assessment Report of the Intergovernmental Panel on Climate Change Solomon, S., D. Qin, M. Manning, Z. Chen, M. Marquis, K.B. Averyt, M. Tignor and H.L. Miller (eds.). Cambridge University Press, Cambridge, United Kingdom and New York, NY, USA, 996 pp.

Lindemann W.C and Glover C.R. 2008. Nitrogen Fixation by Legumes. Guide A-129. College of Agriculture and Home economics.

Lucero C. 2009. Evaluación agronómica de G. ulmifolia a dos densidades de siembra en sistemas silvopastoriles con B. arrecta. Consulta: 30 de mayo, 2010 de http://www.agronet.gov.co

Mahecha L, L.; Durán C., C.V.; Rosales M. 2000. Análisis de la relación planta-animal desde el punto de vista nutrición en un sistema silvopastoril de Cynodon plectostachyus, Leucaena leucocephala y Prosopis juliflora en el Valle del Cauca. Acta Agron , 50(1 y 2): 59-70.

Mahecha, L. 2002. El silvopastoreo: una alternativa de producción que disminuye el impacto ambiental de la ganadería bovina. Rev Col Cienc Pec Vol. 15(2):226-231.

Mahecha L, Gallego L, Pelaez F. 2002. Situación actual de la ganadería de carne en Colombia y alternativas para impulsar su competitividad y sostenibilidad.. Rev Col Cienc Pec Vol. 15: 2, 2002.

Mahecha, L. 2003. Importancia de los sistemas silvopastoriles y principales limitantes para su implementación en la ganadería colombiana. Rev Col Cienc Pec Vol. 16(1): 11-18.

Mahecha L and Rosales M 2005 Valor nutricional del follaje de Botón de Oro Tithonia diversifolia (Hemsl.) Gray, en la producción animal en el trópico. Livestock Research for Rural Development. Volume 17, Artículo 100. http://www.cipav.org.co/lrrd/lrrd17/9/mahe17100.htm

Mahecha L, Monsalve M A y Arroyave J F 2007a. Evaluation of grass availability and growth of Zebu steers in a silvo-pastoral system based on Eucalyptus tereticornis and Panicum maximum in the San Sebastian reforestation program in Magdalena Department of Colombia. Livestock Research for Rural Development. Volume 19, Article #94. Retrieved March 20, 2011, from http://www.lrrd.org/lrrd19/7/mahe19094.htm

Mahecha L, Escobar J.P, Suárez J.F y Restrepo L.F. 2007b. Tithonia diversifolia (hemsl.) Gray (Botón de oro) as forage supplement of Holstein x Zebu cows. Livestock Research for Rural Development. Volume 19, Article #16. Retrieved October 17, 2009, from http://www.lrrd.org/lrrd19/2/mahe19016.htm

Mahecha L, Angulo J, Salazar B, Cerón M, Gallo J, Molina C.H, Molina E.J, Suárez J.F, Lopera J.J and Olivera M. 2008. Supplementation with bypass fat in silvopastoral systems diminishes the ratio of milk saturated/unsaturated fatty acids. Tropical Animal Health and Production 40(3): 209-216.

Mahecha L, Murgueitio MM, Angulo J, Olivera M, Zapata A, Cuartas C.A, Naranjo J.F, Murgueitio E. 2011. Desempeño animal y características de la canal de dos grupos raciales de bovinos doble propósito pastoreando en Sistemas Silvopastoriles Intensivos. Rev Colomb Cienc Pecu 24 (4), dic 2011.

Manriquez-Mendoza L, Lopez-Ortiz S, Perez-Hernandez P, Ortega-Jimenez E, Lopez-Tecpoyotl Z, Villaruel-Fuentes M. 2011. Agronomic and forage characterization of Guazuma ulmifolia Lam. Tropical and subtropical ecosystems 14: 453-463.

Molina C.H, Molina C.H, Molina E.J. 2005. La Reserva Natural El Haico: Ganaderia competitiva y sostenible basada en el silvopastoreo intensivo . Carta Fedegán 95: 74-76.

Molina C.H, Molina-Durán C.H, Molina E.J, Molina, J.P. 2009. Carne, leche y mejor ambiente en el sistema silvopastoril con Leucaena leucocephala. In: Murgueitio, E., Cuartas, C., Naranjo, J.F. (Eds.), Ganadería del futuro: Investigación para el desarrollo. Fundación CIPAV. Cali, Colombia, pp. 41–65.

Murgueitio E, Calle Z, Uribe F, Calle A, Solorio B. 2011. Native trees and shrubs for the productive rehabilitation of tropical cattle ranching lands. Forest Ecology and Management 261:1654–1663.

Murgueitio E, Cuellar P, Ibrahim M, Gobbi J, Cuartas C.A, Naranjo J.F, Zapata A, Mejía C.E, Zuluaga A.F, Cassasola F. 2006. Adopción de sistemas agroforestales pecuarios. Adoption of agroforestry systems for animal production.Pastos y Forrajes 29 (4): 365-379.

Murgueitio, E., Ibrahim, M., 2001. Agroforestería pecuaria para la reconversión de la ganadería en Latinoamérica. Livest. Res. Rural Dev. 13 (3), http://www.lrrd.org/lrrd13/3/murg133.htm.

Nardone A.B, Ronchi, et al. 2010. Effects of climate changes on animal production and sustainability of livestock systems. Livestock Science 130 (1-3): 57-69.

Pardo-Locarno, L. C. 2009. Macroinvertebrados edáficolas en agroecosistemas del municipio de El Cerrito (Valle), con énfasis en la comunidad de escarabajos Melolonthidae (Coleoptera: Scarabaeoidea). Doctorado, Universidad del Valle.

Pasalodos-Tato M, Pukkala T, Rigueiro-Rodríguez A, Fernández-Núñez E and Mosquera-Losada M.R. 2009. Optimal Management of Pinus radiata Silvopastoral Systems Established on Abandoned Agricultural Land in Galicia (North-Western Spain). Silva Fennica 43(5): 831-845.

Ramirez H. Evaluación agronómica de dos sistemas silvopastoriles integrados por pasto estrella, Leucaena y Algarrobo forrajero. Tesis de Grado Universidad Nacional Bogotá. 1998.

Rios N, Jimenez F, Ibrahim M, Andrade H, Sancho F. 2006. Parámetros hidrológicos y de cobertura vegetal en sistemas de producción ganadera en la zona de recarga de la cuenca del río jabonal , Costa Rica.

Rodríguez FRA. 1985. Producción de biomasa de Poró gigante (*E. Poeppigiana*) y king grass (*P. purpureum * P. thypoides*) intercalados en función de la densidad de siembra y la biomasa de poda del Poró. CATIE. Tesis Mg. Sc.

Sadeghian S, Rivera JM, Gómez ME. 1998. Impacto de la ganadería sobre las características físicas, químicas y biológicas de suelos en los andes de Colombia. En: Memorias de la conferencia electrónica sobre agroforestería para la producción animal en América Latina, realizada de abril a septiembre de 1998. CIPAVFAO. p 123-141.

Sánchez M. 1998. Sistemas agroforestales para intensificar de manera sostenible la producción animal en América Latina Tropical. En: Memorias de la conferencia electrónica sobre agroforestería para la producción animal en América Latina, realizada de abril a septiembre de 1998. CIPAV-FAO. 1-13.

Steinfeld H, Gerber P, Wassenaar T, Castel V, Rosales M, De Haan C. 2006. Livestock's Long Shadow, Environmental Issues and Options. LEAD-FAO. Rome.

Vallejo V, Roldan F, Richard D. 2010. Soil enzymatic activities and microbial biomass in an integrated agroforestry chronosequence compared to monoculture and a native forest of Colombia. In: Colombia Biology And Fertility Of Soils ISSN: 0178-2762 ed: Springer v.1 fasc.1 p.1 - 1 ,2010.

Velasco J, Camargo J, Andrade H, Ibrahim M.1999. Mejoramiento del suelo por *Acacia mangium* en un sistema silvopastoril con *B. humidicola*. En: Memorias

VI Seminario Internacional sobre sistemas agropecuarios sostenibles. 28-30 de Octubre 1999. Realizado por la Fundación CIPAV y LA FAO. Cali, Colombia.

Vásquez R, Pulido J, Abuabara Y, Martínez R, Abadía B, Arreaza L.C, Silva J, Sánchez L.E, Ballesteros H, Muñoz C, Rivero T, Nivia A, Barrera G. 2005. Patrones tecnológicos y calidad de la carne bovina en el Caribe colombiano. Corpoica. Plan de modernización de la Ganadería. Ed. Camilo Baqu

12

Sustainable Management of Fruits Orchards Using Organic Matter and Cover Crops: A Case Study from Brazil

Sarita Leonel[1], Erval Rafael Damatto Junior[2],
Dayana Portes Ramos Bueno[1] and Carlos Renato Alves Ragoso[1]
[1]UNESP. FCA – Depto. de Produção Vegetal, Botucatu, SP,
[2]APTA – Pólo Regional Vale do Ribeira, Registro, SP,
Brasil

1. Introduction

Few rural productive activities are as diversified and versatile as fruit farming and few countries are deeply engaged as Brazil in the cultivation of all existing species. This is reason enough for the world to keep an eye towards what is happening in the Brazilian plantations, which range from huge modern commercial orchards covering 2.266.791 ha ha. Fruit farming is scattered throughout all the Brazilian states and as an economic activity, it involves about 5 million people, either directly or indirectly. It comprises 30 organized belts and every year some belts are created. This makes Brazil rank as third global producer, with an annual harvest of 42 million tons, but represents only 2% of the global fruit trade, attesting to heavy domestic consumption (BRAZILIAN FRUIT YEARBOOK, 2010).

The world gets regular supplies of more than a dozen different types of fruit produced in various regions across Brazil. This gave origin to more than 30 fruit belts, each of them benefiting from regional geography, soils and climatic conditions to achieve high yields in orchards that generate thousands of jobs, drive the economy and position Brazil's states to participate in the global economy.

Sustainable agriculture is a keystone for development in Brazil and other parts of the world, as it seeks to preserve environmental quality whilst allowing for economic growth Integrated fruit production fulfils such purpose as it aims for crop sustainability, balanced nutrient cycles, preservation and improvement of soil fertility and biological diversity as essential components. At the same time, Brazilian fruit producers must consider , environmental protection, economical profits, and social and health demands of both farm workers and fruit consumers.

Fertilization using organic materials is complementary to chemical fertilization and especially necessary for integrated fruit production. Reliance on organic fertilizers for fruit production is already a reality in practically all countries in Europe. Brazil has tried to adapt its crops to such system because it has a great potential to supply fruit to countries in the northern hemisphere, mainly during the period between their harvests. However, its production processes still need to be fitted to the demands of foreigner consumers. There is increasing demand for organic products in the international market and such products obtain better prices than conventional products. Another positive aspect of organic products is the quality control by non-governmental organizations, which guarantees the producer will receive a premium for their organic product. From a technical and scientific point of view, the challenges imposed by organic agriculture are huge. Extensive research is needed to develop appropriate operational technologies to enhance the productivity of organic crops.

As there are few papers describing research with organic fertilization in fruit orchards, the present study aimed to evaluate the effects of organic matter in the form of cattle, coat and green manure, as well as cover crops, on the soil fertility, plant nutrition, growth and yield of some fruit orchards in Brazil. The goal of this work was to obtain nutritionally equilibrated and productive plants. In addition, some quality characteristics of the fruit were also evaluated to determine how organic fertilization would influence such characteristics.

2. Fig tree orchard management using organic fertilization

The fig tree originated from Asia Minor and Syria, in the Mediterranean region, and was first cultured and selected by Arabs and Jews in Southwest Asia. It is one of the oldest plants cultivated in the world – since prehistoric times – and is considered by ancient people as a symbol of honor and fertility. Fig trees are cultivated in approximately 40 countries in the world, predominantly in the Mediterranean region, but is adapted to several climates and can be cultivated in both subtropical and temperate regions. Fig tree cultivation is interesting for Brazil, as it may lead to a market for exported Brazilian figs during the period between fig harvests in Turkey, which is the world's main producer of figs. Brazil produces a large volume of figs (25.500 tonnes per year), and 20 to 30% of the total volume produced in the country is destined for the export market.

Fig trees thrive in several soil types, provided it is deep, well drained and rich in organic matter. According to the experience of producers, the higher the organic fertilizer input, the better the results for both yield and fruit quality. Another reason for the interest in organic fertilization is the growth of the certified organic market, since certified products can be sold for higher prices.

However, scientific research into the fig tree requirements in organic fertilization is scarce and not conclusive. Balanced mineral nutrition during the plant formation phase guarantees good harvests in the plant productive phase. However, Fernandes & Buzetti (2000), emphasized that research recommending the best sources, doses, periods and methods of application of organic fertilizers are scarce.

The use of organic matter in fruit orchards favors the growth and development of plants as well as the soil properties. Organic matter is responsible for maintaining soil life, with benefits observed in the soil physical, chemical and biological properties. According to Kiehl (1985), several published works on the utilization of organic matter from different sources demonstrated the importance of such material as a soil fertilizer and conditioner. Animal manure is an excellent organic fertilizer, due to its content of all essential plant macro- and micro-nutrients, its chemical composition is highly variable depending on the animal species, age and feeding, the bedding material and the methods employed to prepare the fertilizer.

Campo Dall'Orto et al. (1996) noted that organic fertilizers have low and unbalanced nutrient levels, relative to plant requirements, but do possess the necessary nutrients for plant growth, which must be considered when selecting fertilization rates. However, those authors emphasized that special attention should be given to the time lapse for the release of nutrients from organic fertilizers to the plants.

The Minas Gerais Satate Soil Fertility Commitee (1989) and Mielniczuk (1999) reported that the time required for nutrient conversion organic to mineral forms varies with the nutrient. For nitrogen (N), approximately 50%, 20% and 30% of the total N in an organic fertilizer are released in the first, second and third year after application, respectively. The nutrient release from organic fertilizers is slower than that from mineral fertilizers, since the former depends on the mineralization of organic material. Such data must be carefully considered in the calculation of application rates for organic fertilizers, since plant-available N concentration can be different from that obtained with conventional mineral fertilizers, which are generally water soluble and readily release N for absorption.

In this section, we report the results of organic fertilization on the soil fertility, plants nutrition and yield of fig trees (*Ficus carica*, L.) cultivar 'Roxo de Valinhos' during four crop cycles.

The experiment was carried out in Botucatu, São Paulo State, Brazil, 22° 52' 47" S and 48° 25' 12" W, 810 m altitude. The local climate is classified as Cwa after Koppen (Curi, 1972), with rainy summers and dry winters, and the highest average temperature is above 22°C (Cunha et al., 1999). The soil is Rhodolic Haplo Udalf, according to the criteria established by United States Soil Conservation Service (1988). The initial chemical characteristics of the 20 cm topsoil, determined based on the methodology of Raij and Quaggio (1983), were: pH ($CaCl_2$) = 4.2; organic matter (O.M.) = 24 g dm^{-3}; P (resin) = 3 mg dm^{-3}; H + Al = 84 $mmol_c$ dm^{-3}; K = 1.4 $mmol_c$ dm^{-3}; Ca = 12 $mmol_c$ dm^{-3}; Mg = 5 $mmol_c$ dm^{-3}; Sum of bases = 18 $mmol_c$ dm^{-3}; cation exchange capacity (CEC) = 102 $mmol_c$ dm^{-3}; Percent Base Saturation = 18%.

Treatments consisted of cattle manure rates according to the recommended N level for the fig tree: control (no fertilizer), 25%, 50%, 75%, 100%, 125% and 150% of the recommended N level (Table 1). Since these levels depend on the tree size and as the N content in cattle manure varied (Table 2), the actual quantity of applied manure varied every year (Table 1). Half of the manure level was applied in August and the remainder in September, every year from 2002 to 2005.

Treatment (% crop N requirement)	N levels				Manure levels- dry weight			
	1st year	2nd year	3rd year	4th year	1st year	2nd year	3rd year	4th year
	---------------(g plant-1)--------------				------------ (kg plant-1) ------------			
0 %	0	0	0	0	0	0	0	0
25 %	10	20	30	30	0.8	0.82	1.9	3.6
50 %	20	40	60	60	1.6	1.64	3.8	7.3
75 %	30	60	90	90	2.4	2.46	5.7	11.0
100 %	40	80	120	120	3.2	3.28	7.6	14.3
125 %	50	100	150	150	4.0	4.10	9.5	18.0
150 %	60	120	180	180	4.8	4.92	11.4	22.0

Table 1. Nitrogen levels applied with cattle manure for fig tree fertilizations in four crop cycles (August 2002 to August 2005). The experiment was conducted in Botucatu, São Paulo State, Brazil.

Sample	N	P$_2$O$_5$	K$_2$O	humidity	Organic matter	C	Ca	Mg	S
				-------(g kg^{-1}dry matter) -------					
1	0.13	0.15	0.17	4.1	4.3	2.38	0.15	0.05	0.02
2	0.24	0.21	0.15	3.7	4.1	2.30	0.11	0.04	0.03
3	0.16	0.09	0.09	4.2	3.9	2.17	0.13	0.08	0.02
4	0.11	0.12	0.08	3.0	3.4	1.89	0.84	0.04	0.03

Sample	Fe	Cu	Mn	Na	Zn	pH*	C/N
		------(mg kg^{-1} dry matter)------					
1	11300	120	146	N/D	190	7.60	19
2	7950	302	308	1860	300	7.20	9
3	14300	6	5250	580	286	14.00	14
4	24950	156	278	1460	208	7.97	17

* (in H$_2$O). Laboratory analyses of chemical fertilizers.

Table 2. Chemical analysis of cattle manure used for fig tree fertilization (Samples were taken as follows - 1: August 2002; 2: June 2003; 3: July 2004; 4: July 2005.). The experiment was conducted in Botucatu, São Paulo State, Brazil.

Values	pH	Organic matter	P	K	Ca	Mg	Sum bases	Effective CEC	Percent Base saturation
	CaCl$_2$	g dm^{-3}	mgdm^{-3}	-------------- mmol$_c$ dm^{-3} -------------					
Mean	5.5	25	34	4.5	43	20	67	96	66
CV (%)	7.22	17.07	55.55	21.47	31.27	29.18	29.17	15.00	18.11
Values	B	Cu	Fe	Mn	Zn				
	----------------- mg dm^{-3} ------------------								
Mean	0.39	6.2	25	7.7	4.2				
[1]CV (%)	36.67	11.96	15.45	26.72	41.19				

[1]CV = coefficient of variation

Table 3. Macro- and micro-nutrients, pH, organic matter, sum of bases, effective CEC, and percent base saturation (%) of the soil where fig trees were cultivated. Soils were collected after four years of fertilization with cattle manure (December 2005), from an experimental site in Botucatu, São Paulo State, Brazil.

Soil analyses after four applications of cattle manure, compared with the initial soil analysis described in the text, indicated that cattle manure treatments did not influence macro- and micronutrient levels at 0–20cm depth (Table 3); the obtained values were considered appropriate for the fig tree (Raij & Quaggio, 1983). Although N concentration   in cattle manure was low (Table 2), the calculation of levels based on the recommended N levels gave a satisfactory input of manure, with no need of readjustment. Besides N, cattle manure contains other macro- and micro-nutrients that are essential to the development of plants, but is necessary to consider that the amount of manure needed to supply the N level was reasonable in this study case, because there was a great quantity of cattle manure for consumption in the experimental farm.

Significant responses to the application of cattle manure were obtained starting in the first crop cycle. On the other hand, plant yield during the orchard formation phase was below the mean values obtained by producers who use conventional fertilizers (about 9 to 20 kg plant[-1]). One interesting point is that the addition of organic material to the soil aims not only to supply nutrients, but also (and mainly) to improve the soil physical and biological properties which, although not measured herein, may have contributed to the yield increase over three crop cycles (Figure 1). However, the obtained data does not allow the assessment of how such effects occurred and which mechanisms were involved (LEONEL & DAMATTO JÚNIOR, 2008).

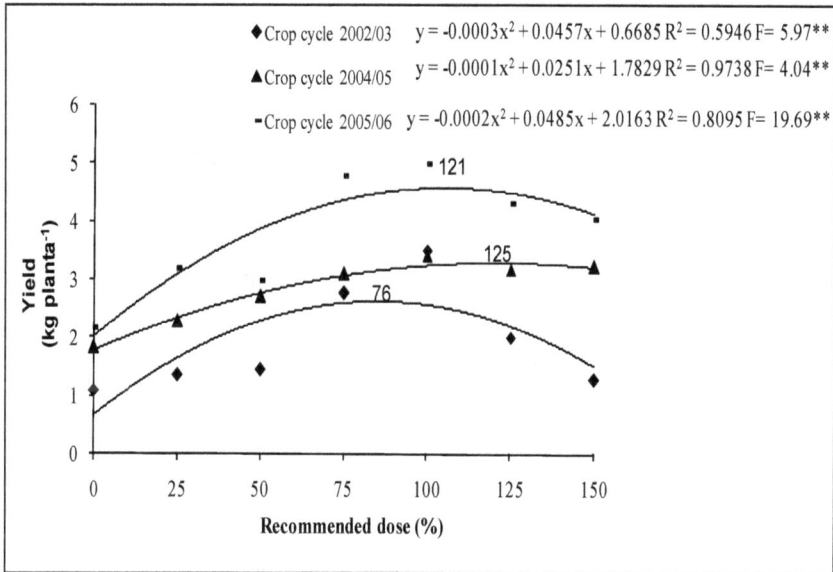

Fig. 1. Yield (kg plant[-1]) of fig tree fertilized with cattle manure at different nitrogen levels corresponding to 0 to 150% of the recommended N dose (%). Data were from three crop cycles and the experiment was conducted at Botucatu, São Paulo State, Brazil.

In Small Meander Valley, Aegean Region of Turkey, where dried fig production is concentrated, Eryuce et al. (1995) selected an orchard in a lowland area where the trees were ten-year-old, cultivar Calyrmina receiving conventional chemical fertilization, and did not observe any effect of fertilization on fruit length, width and weight. In Jiangsu Province, situated in eastern China, where the annual average temperature ranges from 13°C to 16°C, the climate is semi-humid and subtropical, and the annual average rainfall is 800 to 1200 mm, Jinseng et al. (1997) studied the effects of supplementary fertilization with nitrogen, phosphorus and potassium on fig trees in one crop cycle. These authors did not report differences among treatments or positive effects on the yield due to fertilizer application. In the present experiment, the effects on fruit diameter and weight were slighter and the main increases were due to more fruits per plant, rather than a change in fruit size. The fig tree has continuous growth; thus, increased yield is due to fruit production along an elongated stem rather than an increase in fruit size.

Caetano et al. (2006) studied the effects of organic fertilization on fig yield using treatments without (0 kg plant[-1]) and with (10 kg plant[-1]) cattle manure. There were significant differences among the evaluated levels and the yield was 6597 g plant[-1] for the treatment without fertilizer and 7517 g plant[-1] for the treatment receiving 10 kg manure. Such results indicate a 13.9% increase in the yield due to the addition of organic matter, besides the benefits to the soil, which were not assessed. However, similar to the present work, the economical viability of this application was not investigated. In the current experiment, the yield increased in each evaluated crop cycle when comparing the 0% recommended dose (control) to the 100% recommended dose, and the average increase was 45.1% considering all four cycles.

3. Use of organic fertilizer for bananas

The banana is one of the most consumed fruits in the world, exploited in most tropical countries, reaching a world production of 70.7 million tons in 2009; Brazil is responsible for approximately 10% of this total. Brazil has around 500 thousand hectares planted with bananas and an annual production of around seven million tons, nearly all of which is destined for the internal market (Brazilian Institute for Geographical Statistics – IBGE 2009). Producing bananas under an organic system has appeared as an alternative so the final product can reach a whole segment of specific consumers with greater purchasing powers (BRAZILIAN FRUIT YEARBOOK, 2010), as well as being a means to reduce costs, as chemical fertilizers have incurred high cost increases over the last few years. In addition, organic fertilization is a more sustainable method of cultivation, creating less impact on the environment.

Even though banana cultivation of great importance at present to generate income and employment for the country, there is still a shortage of studies related to the use of organic fertilizer for banana nutrition. Since the banana tree demands high nutrient inputs and this factor has not always been given due attention, many of the banana producing locations have historically been under fertilized, which has led to low production and susceptibility to disease.

Although it is cultivated in various types of soil, the banana plant prefers soils rich in organic matter, well drained, and with a loamy soil that has a good water retention capacity and favorable topography (Rangel, 1997). In reality, many of the producing regions are located on soil poor in nutrients and also with a low organic matter index; the inadequate supply of nutrients is one of the main causes for low yields in banana cultivation. According to Lahav & Turner (1983), only a part of the necessary nutrients for banana plants can be supplied through natural soil reserves, the rest has to be incorporated into the soil-plant system, which should be made through crop waste matter or fertilizers.

Therefore it is necessary to use fertilizers to meet the demands of cultivating this crop and, according to Teixeira (2000) the recommendations for using N and K based fertilizers vary greatly, depending on the location of the banana plantation. This is because recommendations for using fertilizers should take into consideration the soil/climatic conditions, plant variety, production ceilings, local banana cultivation management practices, available resources and the plant's response to the application of nutrients.

Moreira (1987) points out that fertilization should be carried out during periods of greatest demand for nutrients, for example; during the vegetative growth phase and when the fruit first appears, which is when the greatest demand for N occurs, whilst the period of fattening and fruit ripening is when the greatest demand for K occurs.

Organic fertilizers are excellent source of nutrients for the banana plant, and common types available are animal manure, agro-industrial waste, straw and organic composts. These fertilizers contain all the necessary nutrients for the plants, such as nitrogen, phosphorus, potassium, calcium, manganese, sulfur, apart from micronutrients (Kiehl, 1985). Another benefit of adding organic material to the soil is in improving its physical attributes, helping to maintain humidity, as well as being responsible for assisting in some beneficial chemical reactions, such as in the concatenation interaction of toxic elements and micronutrients, increasing the cation exchange capacity and buffering pH levels, apart from increasing biological diversity. All these benefits created by the application of organic fertilizers come about through trying to meet the real needs of our soil, because tropical soils have their limitations in respect to chemical properties, with low levels of nutrients and little organic matter, which hinders the effective development of plants.

Various studies on the use of organic fertilizer demonstrated improvements in the physical, chemical and biological characteristics of the soil, as well as encouraging the growth of the plant, in addition to increasing the yield. Lahav & Turner (1983) reported that applying up to 80 t.ha^{-1}.year^{-1} of stable waste on the soil encouraged growth and advanced flowering and fruiting of the banana plant. In a study using different top soils, in a partnership with a banana farm, Espindola et al. (2006) found that peanut waste fodder (*Arachis pintoi*) showed an increased speed in its decomposition, while spontaneous vegetation such as predominant guinea grass (*Panicum maximum*), demonstrated a slower rate, indicating that legumes liberated N, Ca and Mg rapidly, while spontaneous vegetation liberated P.

According to Borges (2008) the application of nitrogen can be accomplished through incorporating green fertilizers, especially legumes, which permits the production of organic matter and the supply of nitrogen, so that during the formation phase of the banana plant, it is recommended to plant legumes between the lines of plants, leaving it as green manure mulch. In the semi-arid region of North East Brazil, Silva et al. (2008) found that a mixture of legumes, grasses and oilseed in the understory of a banana plantation, to be chopped and left on the soil surface as a mulch and green manure, was a beneficial practice that contributed to the diversity of species, both above and below the soil, giving greater protection for the soil and a greater diversity of nutrients within the system.

Damatto Junior et al. (2006a) found that applying organic compost (sawdust and bovine manure) to banana plants (Prata anã) led to increases in pH, organic matter, phosphorous, calcium, CEC, and the soil saturation base level. However, it did not affect K and Mg levels (Table 4). The nutritional assessment of the Prata anã banana leaves, fertilized with different doses of organic compost indicated average levels of 30 mg N kg^{-1}; 2.0 mg kg^{-1} of P; 31 mg kg^{-1} of K; 9 mg kg^{-1} of Ca; 3.2 mg kg^{-1} of Mg and 2.5 mg kg^{-1} of S, during the flowering season (Damatto Junior et al., 2006 b).

Treatments	pH	O.M.	P	K	Ca	Mg	SB	CEC	Percent Base Saturation
	CaCl₂	g/dm³	mg/dm³	---------------- mmol₊/dm³ -------------					
T1: 0 kg compost/plant	5.4 c	32 b	54 b	1.2 a	59 b	17 a	77 d	108 d	69 c
T2: 43 kg compost/plant	5.6 bc	33 b	85 ab	1.3 a	74 b	19 a	93 cd	124 cd	73 bc
T3: 86 kg compost/plant	5.7 abc	35 ab	93 ab	1.3 a	90 ab	22 a	113 bc	141 bc	79 ab
T4: 129 kg compost/plant	6.0 a	43 a	153 a	1.4 a	114 a	22 a	137 ab	161 ab	85 a
T5: 172 kg compost/plant	5.9 ab	40 ab	135 ab	1.3 a	119 a	19 a	151 a	178 a	84 a
Averages	5,7	37	104	1,3	91	20	114	142	78
Coefficient of variation (%)	3	12	41	19	20	18	15	9	5

Averages followed by different letters in the columns differ by Tukey test at 5% of probability.

Table 4. Macronutrient concentrations, pH, organic matter (O.M.), sum of bases (SB), effective CEC, and percent base saturation values of an Alfisol cultivated with banana plants fertilized with different compost rates in Botucatu, Sao Paulo, Brazil from November 2002 to May 2004.

Other studies on the use of organic fertilizers with banana cultivation proved to be efficient. Borges et al. (2002) found that during the first cycle of Banana de Terra, when 267 kg of N/ha/year had been applied in the form of farmyard manure, an increase in the number of fruit per bunch was achieved, as well as increasing the average length of the fruit, in relation to an equivalent input of chemical N fertilizer.

A farmer growing the Prata anã banana, fertilized with organic compost, in the town of Botucatu, Sao Paulo obtained an average yield of 26.2 t/ha during the first harvest, producing bunches with an average weight of 16.4 kg, and 9 stalks on each bunch (DAMATTO JÚNIOR, 2005), this yield is higher than the national average, which was 19.8 t ha[-1] in the year 2008 (FAO, 2011). While studying the effect of the residual levels of the organic compost during the Prata anã plant's second cycle, Damatto Junior et al. (2006 c) found that there was no residual effect with increasing levels of organic compost, although the average productivity was satisfactory at 27.3 t/ha.

In the State of Mato Grosso, Marílio et al. (2006) applied bovine manure, natural phosphate and leaf bio-fertilizer and using a covering of bean chaff mulch on five genotypes of banana plants, confirmed the viability of the banana production within an organic system.

Another important aspect to be taken into consideration when using organic fertilizer is the time taken for the organic residue to decompose and consequently liberate the nutrients contained within it. According to Campo Dall'Orto et al., (1996) the liberation of nutrients from organic fertilizer is much slower than from mineral fertilizers, as this is dependent on the mineralization of the organic matter. Bartz et al. (1995) reported that 50% of the applied N was mineralized in the first cultivation cycle and 20% in the second cycle, while P mineralization was 60% in the first cultivation cycle and 20% in the second.

When calculating the amount of fertilizer to use for banana plants, producers should consider the cultivated waste from the previous harvest, as the cultivated organic waste is considerable. This includes pseudo steam, leaves, husks and raques, which are an important source of nutrients. Various writers stress the importance of this sort of maintenance. In Vale do Ribeira in the state of São Paulo, Moreira (1987) estimated that a banana plantation can produce up to 200 t/ha/year by using the cultivated waste from the banana plants as a source of organic material for plants under development. Cintra (1998) pointed out that by using the cultivated waste from the banana plants as a form of mulch coverage; this represented a substantial source of organic material. Rodriguez et al. (2006), in Mexico, found that working the waste from the harvest between the banana plant rows was an excellent practice for increasing levels of organic matter in the soil by up to 1.0 to 1.5%. In addition, this practice reduced production costs and preserved productive top soil in that region.

Besides its nutritional aspect and in improving soil conditions, positive results were noted by using organic fertilizer for the banana plants in pest control and diseases, as it was observed that plants under the organic system presented a greater tolerance against attack from pests and mainly from diseases, which favors a longer productive period for the banana plant in the field. According to observations made by Penteado (2006), the cultivated plants within the organic system tend to be healthier and more resistant to diseases and insect pests than the plants grown using chemical fertilizers.

Banana in the Ribeira Valley region is susceptible to Sigatoka-negra (a disease caused by the *Mycosphaerella* fungus), but disease levels were no different in the organic system than conventionally managed plantations. The best performing species were IAC 2001 (19.9 kg per bunch) and Tropical (10.6 kg per bunch).

According to Pereira et al. (1996), organic composts can act on soil phytopathogens because they can produce inhibitory chemicals such as antibiotics or indirectly by encouraging the antagonist population, which will favor an increase in the useful life of the plant. This biological control is important, as banana plantations suffer big losses from the "Mal-do-Panama", *(Panama disease)* , a fungal disease caused by *Fusarium oxysporum*, which reduces the useful productive life of the banana plant, and can only be combated by the antagonistic micro-organisms present within organic materials.

4. The influence of organic fertilizer on production and quality of citrus fruits

According to Melarato (1989) most soil cultivated with citrus plants in South East Brazil demonstrated high soil acidity and toxic aluminum concentrations, low cation retention levels and low levels of saturated bases. These constraints are, due to a high degree of soil instability, because the parent materials are poor in base cations and need to be understood and overcome to optimize yields of citrus fruit.

Alva & Paramasivam (1998) mentioned that the production of citrus fruit is largely influenced by the supply of N, due to the fact that this nutrient regulates the rate of photosynthesis and the synthesis of carbohydrates (Kato, 1986), the specific weight of the leaves (Syvertsen & Smith, 1984); the total bio-mass production and the allocation of carbon in plant organs (Lea-Cox et al., 2001). Due to these reasons, N is considered as the most important nutrient in citrus plants and is instrumental in growth, yield and quality of the fruit (Quaggio et al., 2005). Within this context, the use of organic N sources seems to be useful in the pursuit of sustainable and productive systems that will incorporate locally-available resources, satisfying human needs while maintaining or improving the environmental quality and conserving natural resources.

According to Pereira Neto (2007) the amounts of organic fertilizers to be applied vary according to the characteristics of the final product, the soil, the climate, type of cultivation, agricultural activity, method of fertilization, among other aspects. Considering these factors, researchers can perform fertilization experiments within representative orchards that test the levels of nutrients in soils and plants, the fertilizer source and the timing of application in orchards in a precise manner for assessing the nutrient requirements and responses of the citrus plants (Koller, 2009).

As a result of these needs, Ramos et al. (2010) assessed the effects of organic compost treatment on soil fertility, relative to the yield and quality of the 'Poncã' (*Citrus reticulata* Blanco) tangerine, cultivated in São Manuel-SP, where N treatments of 0, 50, 100, 150 and 200% the recommended dose were tested, corresponding to 0, 41.5, 83, 124.5 and 166 kg compost /plant respectively. Their findings confirmed that in the case of applying 200% nitrogen treatment, the was an increase in soil pH and greater concentrations of phosphorus, calcium, total basic cations, effective CEC, percent base saturation, manganese and zinc were observed in the soil. The (0%N) treatment reduced pH (and consequently a smaller H+A1 value), organic matter, phosphorus, potassium, calcium, total basic cations, CEC, Percent Base Saturation, sulphur, copper and zinc levels, indicating that applying organic compost alters the soil fertility (Table 5 and 6). The same authors observed that after fertilization there was an increase in pH, phosphorus, potassium, manganese, V%, sulphur, copper and zinc levels and a decrease in H+A1, boron, iron and manganese for practically all compost-amended treatments.

Treatments	pH	MO	P	S	Al^{3+}	H+Al	K	Ca	Mg	SB	CEC	Percent Base Saturation
	$Cacl_2$	g/dm^3	mg/dm^3			---mmol$_c$/dm^3---						
Before fertilization (September 2009)												
T1	5,7	9,7	43,7	3,7	0	14,0	1,4	21,3	8,3	31,3	45,0	69,3
T2	6,1	7,5	20,5	3,0	0	11,5	1,7	17,5	5,7	25,0	36,7	68,2
T3	6,1	8,0	34,5	3,2	0	12,2	1,8	29,0	7,5	38,2	50,2	75,2
T4	6,4	9,5	45,5	4,2	0	11,5	1,7	40,2	8,5	50,7	62,7	80,0
T5	6,1	10,7	44,0	4,0	0	13,0	1,4	35,0	7,7	44,2	57,0	75,7
After fertilization (February 2010)												
T1	5,8	6,7	17,5	5,7	0	10,5	1,3	16,5	7,5	25,0	35,5	70,2
T2	6,3	10,5	22,7	6,5	0	9,7	1,8	24,5	7,0	33,2	42,5	77,0
T3	6,1	11,7	34,2	6,0	0	9,5	2,4	26,5	9,0	38,2	47,7	78,0
T4	6,5	11,7	48,2	7,7	0	9,0	2,7	37,0	9,7	49,5	58,5	82,5
T5	6,5	9,7	50,5	7,2	0	9,0	2,5	49,7	9,5	61,7	70,7	83,7

Table 5. Average concentrations of macronutrient, pH, organic matter (O.M.), sum of bases (SB), effective CEC, and base saturation values (V) in the soil cultivated with the 'Poncã' tangerine before and after fertilization using organic compost (moisture of the horse, cattle and bird manure) treatment (São Manuel/SP, 2009).

Treatments	B	Cu	Fe	Mn	Zn
			--mg/dm3--		
Before fertilization (September 2009)					
T1	0,24	0,75	23,00	10,07	1,52
T2	0,20	0,62	15,75	10,42	2,50
T3	0,17	0,95	19,75	8,75	3,27
T4	0,24	1,30	17,50	10,27	4,70
T5	0,20	1,20	20,75	10,12	4,30
After fertilization (February 2010)					
T1	0,17	0,90	9,75	6,52	2,30
T2	0,12	1,00	9,25	5,85	3,27
T3	0,16	1,42	10,5	6,17	4,92
T4	0,16	1,97	12,0	6,00	8,12
T5	0,15	1,87	11,0	7,25	8,97

T1: 0%N; T2: 50%N; T3: 100%N; T4: 150%N; T5: 200%N

Table 6. Average concentrations of micronutrients in the soil cultivated with the 'Poncã' before and after fertilization using organic compost (moisture of calcareous sugarcane bagace and birds manure) treatments (São Manuel/SP, 2009).

Citrus yields in treatments 1, 2, 3 and 4 (Table 7) were hardly different, ranging from 24.2 to 26.2 t/ha. However, using the recommended nitrogen treatment of 200% there was a yield of 43 t/ha (104 kg/plant), a significant increase of 16 t/ha. As a result of this high productivity in the 200% N treatment, a system of classification of the fruit was performed to see if this greater productivity affected the size of the fruit. Therefore, the fruit was classified: small (≤69mm), medium (70-81mm) and large (≥82 mm) and according to the results presented in Table 7, it was noted that treatments 3 and 5 presented the greater percentage of the large and medium sized fruit and a smaller percentage of the small sized fruit. This demonstrates that the greater yields achieved with treatment 5 did not affect the fruit size.

Treatments	Yield (kg/plant)	Productivity (t/ha)	Big (≥82 mm)	Medium (70-81mm)	Small (≤69mm)
T1	59	24,6	7%	25%	68%
T2	62	25,8	5%	26%	69%
T3	58	24,2	14%	30%	56%
T4	63	26,2	7%	23%	70%
T5	104	43	11%	27%	62%

T1: 0%N; T2: 50%N; T3: 100%N; T4: 150%N; T5: 200%N.

Table 7. Yield, productivity and size classification (big, medium and small) of the 'Poncã' tangerine fruit (Citrus reticulata, Blanco), with organic compost (moisture of calcareous, sugarcane bagace and birds manure) treatments (São Manuel/SP, 2009).

However, Panzenhagen et al. (1999) observed that the treatments achieved a greater fruit yield in the third category (diameters less than 5,7 cm), which would indicate that the greater total fruit yield was attributed to the sharp increase in the yield of the smaller sized fruit, of poor quality and commercially undesirable. This leads to the conclusion that higher levels of N are positively related to an increase in fruit yield and with a decrease in their average weight.

With respect to the quality of the fruit, little difference was observed on the average weight, juice yield, technological index, pH, acidity, soluble solids and Vitamin C ratio. However, some authors claim that the quality of citric fruit is also influenced by fertilizing. Pozzan and Tribone (2005) claimed that studies such as those by Chapman (1968), Embleton and Jones (1973), Reese and Koo (1975) and Koo and Reese (1976) concluded that the greatest effect from nutrition in the quality of the fruit, is related to the supply of nitrogen and potassium; the former (N), tends to reduce the size of the fruit and increase the quantity of juice, soluble solids and acidity. On the other hand, high levels of potassium (K) proportionally decrease the concentration of soluble solids and increase the acidity of the juice by significantly increasing the size of the fruit and the thickness of the skin.

Through these various studies we can see that there still exists some controversy regarding the influence of fertilization on soil fertility, yield and quality of fruit. Although Ramos et al. (2010) compiled data on fertilization practices for a productive cycle of citrus orchards, it is possible to achieve good results with organic fertilizer, which influences the soil fertility and the yield, although does not interfere with the size or quality of the fruit.

5. Alternative management of intercropping in citrus

The greatest challenge for citriculturists is to obtain a product that fulfills the needs of consumers (which can be either the fresh fruit market or the processing industry) at a price that allows a profit, relative to the investment and the annual expenses demanded by the crop.

Citriculture in São Paulo State, Brazil is concentrated in regions with low fertility soils. Demattê & Vitti (1992) summarized soil characteristics in this area and found that 65% of the evaluated areas presented medium-texture soils (clay content of 15%–35% at the soil surface); 30% had sandy-texture soils (clay content of up to 15%); and about 5% were clayish soils. This forces the producer to use external inputs to achieve good yields, increasing thus the costs of citrus production.

Considering the above-mentioned factors – reducing production costs, maintaining soil chemical and physical properties with fewer inputs, and increasing competitiveness in the global market, assuring the citriculturist's profit – suggests that the use of green manure, catch crop or cover crop could be beneficial for citrus production in Sao Paulo, Brazil.

Green manuring consists in incorporating into the soil the non-decomposed biomass of plants locally cultivated or imported to preserve and/or improve the productivity of cultivable lands. According to scientific studies and practical evidence, green manure acts on diverse soil fertility aspects by: protecting against the rain impact and the direct incidence of sun rays; rupturing soil layers that were compacted over the years; increasing organic matter content; increasing water infiltration and retention capacity; decreasing Al and Mn toxicity due to complexation of these compounds with organic matter and pH buffering capacity of organic residues; promoting nutrient recycling; extracting and mobilizing nutrients from the deepest layers of the soil and subsoil including Ca, Mg, K, P and micronutrients; extracting fixed phosphorus; fixing atmospheric N in a symbiotic manner with plants of the Leguminosae; and inhibiting germination and growth of invasive plants either by allelopathic effects or by competition for light (Von Osterroht, 2002).

There is little information on the response of soil chemistry and fertility following the introduction of green manure in *Citrus* orchards in Brazil. The following research aimed to evaluate soil chemistry following the introduction of green manure in an orchard of orange trees 'Pêra' (*Citrus sinensis* Osbeck).

The soil at the studied farm had a texture as Oxic Quartzipsamments. Orange trees were budded on 'Rangpur' lime trees and planted in a space of 7 x 4 m in 1996. Four different treatments corresponding to the evaluated green manure types were employed: jack bean (JB) (*Canavalia ensiformis* DC), lablab (LL) (*Dolichos lablab* L.), pigeon pea (PP) (*Cajanus cajan* L. Millsp), and brachiaria (BQ) (*Brachiaria brizantha* Hochst ex A. Rich. Stapf) as a control. They were sown in December of both 2003 and 2004 and were mowed and transferred to the plant rows during full flowering. The tests included chemical analysis of the soil in and between plant rows (Tables 8, 9, 10 and 11), macro and micronutrient levels in green manure and control (Table 12), dry matter accumulation in orange trees (Table 13) and assessment of the productivity (Table 14).

Depth	pH	O M	P	H+Al	K	Ca	Mg	SB	CEC	V%
(cm)	CaCl$_2$	g dm^{-3}	mg dm^{-3}	------------------mmol$_c$ dm^{-3}------------------						
0-20	7.4	12	18	14	0.4	40	32	72	86	84
20-40	5.0	11	6	21	0.3	16	8	25	46	54

V = Percent base saturation. OM = organic matter. SB = sum of bases. CEC = effective cation exchange capacity (CEC)

Table 8. Results of soil chemical analysis in plant rows before treatment with green manure (collected in 0-20 and 20-40 cm depth). The experiment was conducted in a citrus orchard at the FCA/UNESP site in Botucatu, Sao Paulo, Brazil.

Depth	B	Cu	Fe	Mn	Zn
(cm)	-----------------------------------mg dm^{-3}-----------------------------------				
0-20	0.20	0.7	20	1.3	0.8
20-40	0.13	0.6	31	1.3	0.8

Table 9. Soil micronutrient concentrations in plant rows before treatment with green manure. The experiment was conducted in a citrus orchard at the FCA/UNESP site in Botucatu, São Paulo, Brazil.

The values of pH slightly varied among treatments, ranging between 5.0 and 5.8 in the first experimental year and between 5.0 and 5.4 in the second year (Table 11), which lower than the ideal values for such plants (about 6.0-6.5). The decline in soil pH, relative to the initial soil analysis (Table 8), can be attributed to the soil acidification caused nitrification, considering that two H^+ ions are released for every molecule of ammonium that is nitrified (Silva et al. 1999). Another possibility is that decaying plant residues release organic acids, which also promote soil acidification. Another possibility is that it was stimulated uptake of cations by the citrus trees and cation uptake is accompanied by H^+ release from the root surface to maintain electrical neutrality in the plant.

Year	Treatment	Depth (cm)	B	Cu	Fe	Mn	Zn
			------------------------------mg dm⁻³------------------------				
1st	JB	0 - 20	0.16	1.2	34	3.4	1.5
		20-40	0.12	1.0	33	2.5	1.2
1st	PP	0 - 20	0.15	2.1	35	3.5	1.5
		20-40	0.10	1.6	32	2.5	1.6
1st	LL	0 - 20	0.17	1.4	38	4.2	1.9
		20-40	0.16	1.1	40	3.1	1.5
1st	BQ	0 - 20	0.19	2.0	49	3.7	1.8
		20-40	0.19	1.8	42	3.5	2.1
2nd	JB	0 - 20	0.18	1.3	59	2.8	1.3
		20-40	0.14	0.8	44	1.3	0.7
2nd	PP	0 - 20	0.21	1.6	48	3.3	1.4
		20-40	0.16	0.6	39	0.8	0.4
2nd	LL	0 - 20	0.21	0.9	49	2.4	0.7
		20-40	0.18	0.6	35	0.9	0.5
2nd	BQ	0 - 20	0.20	1.3	49	2.5	1.1
		20-40	0.15	0.6	35	1.0	0.6

*jack bean (JB) (*Canavalia ensiformis* DC), lablab (LL) (*Dolichos lablab* L.), pigeon pea (PP) (*Cajanus cajan* L. Millsp), and brachiaria (BQ) (*Brachiaria brizantha* Hochst ex A. Rich. Stapf).

Table 10. Soil micronutrient concentrations at the layers 0-20 and 20-40 cm in plant rows after treatment with green manure* in the first (2003) and the second (2004) experimental years. The experiment was conducted in a citrus orchard at the FCA/UNESP site in Botucatu, Sao Paulo, Brazil.

Phosphorus levels at the layer 0–20 cm for JB, PP and LL treatments were considered high and those for BQ treatment were considered medium, according to GPACC (1994), compared with the initial soil chemical analysis; therefore, there was a considerable increase, 2-3 times greater, in P levels in the first experimental year (Table 11). In the following year, P concentrations declined to a similar level as the initial soil chemical analysis, i.e. medium P levels for JB, PP and BQ and low P levels for LL treatment. This indicates that maybe it had something to do with the time needed for P uptake or P fixation reactions to reduce the P concentration in soil solution in the first experimental year, altering thus the analysis results.

Potassium concentrations in the 0-20 cm layer were very low to low for all treatments during the evaluated years, which was consistent with the low fertility of the Oxic Quartzipsamments studied. It appeared that green manure did not provide sufficient K to increase the soil K concentration and highlights the need for additional fertilization to obtain high productivity and consequently high-quality citrus fruit.

According to Vitti et al. (1996), normal values for Ca/Mg ratio in the soil for citrus plants cultivation are about 4/1, which was not achieved with the treatments; the mean value obtained was around 1.4/1, mainly due to the utilization of dolomitic limestone for liming. Ca levels at the layer 20–40 cm should not be lower than 5 mmol$_c$ dm⁻³, which did not occur with any treatment. Mg levels were within the suitable limits for such plants which correspond to 4 and 8 mmol$_c$ dm⁻³ considering the layers 0–20 cm and 20–40 cm, respectively. Data in Table 8 show that the applied limestone had a noticeable mobility with considerably higher levels than the initial ones at the layer 20–40 cm. As the soil is extremely sandy and consequently excessively drained, limestoneprobably moved considerably, surpassing the 40 cm sampling depth. Other research works (Quaggio et al., 1982; Anjos & Rowell, 1983 and Wong, 1999) have demonstrated significant loss of Ca and Mg by leaching, even in soils that received limestone at suitable doses. Under laboratory conditions, Büll et al. (1991) observed that Ca and Mg loss by leaching increased proportionally to the limestone dose in soils with X, Y and Z textures.

Year	Treatment	Depth	pH CaCl$_2$	P mg dm^{-3}	K	Ca	Mg
		(cm)		----mmolc dm^{-3}-----			
1st	JB	0 - 20	5.8	48	1.0	29	23
		20-40	5.5	11	0.5	13	14
1st	PP	0 - 20	5.7	34	0.7	19	18
		20-40	5.6	18	0.6	14	12
1st	LL	0 - 20	5.4	36	0.5	16	12
		20-40	5.2	19	0.1	7	5
1st	BQ	0 - 20	5.0	28	0.7	11	9
		20-40	5.3	34	0.6	14	12
2nd	JB	0 - 20	5.3	16	0.9	16	11
		20-40	5.1	4	0.4	9	6
2nd	PP	0 - 20	5.4	16	0.8	16	11
		20-40	5.2	3	0.7	9	7
2nd	LL	0 - 20	4.9	11	1.0	11	6
		20-40	5.2	3	1.7	9	8
2nd	BQ	0 - 20	5.0	17	0.8	13	8
		20-40	4.9	4	0.3	8	6

*jack bean (JB) (*Canavalia ensiformis* DC), lablab (LL) (*Dolichos lablab* L.), pigeon pea (PP) (*Cajanus cajan* L. Millsp), and brachiaria (BQ) (*Brachiaria brizantha* Hochst ex A. Rich. Stapf).

Table 11. Soil macronutrient concentrations and pH at the layers 0-20 and 20-40 cm in plant rows after treatment with green manure* in the first (2003) and the second (2004) experimental years. The experiment was conducted in a citrus orchard at the FCA/UNESP site in Botucatu, Sao Paulo, Brazil.

B levels were considered low for all treatments in the first experimental year, and in the second year, JB kept showing low levels although increased of 0.02 mmol$_c$ dm $^{-3}$. The remaining treatments then presented medium levels although B was not applied via soil. Such situation can be explained by the relationship between pH and B availability, once B levels tend to be stable when pH is close to 5.0.

Cu levels were considered high in the first year and remained high in the second year for all treatments, except for LL which returned to medium levels. Mn levels were medium throughout the two experimental years, and increased compared to the initial soil chemical analysis. Similarly, the Zn concentration was initially at a medium level and remained medium or increased for the LL and BQ treatments (first year only). The Fe concentrations tended to increase throughout the experiment. The stability and, in some situations, the increase in the levels of micronutrients (Cu, Mn, Zn and Fe) were attributed to the soil acidification, as this reduced the pH and consequently increased their availability in the soil solution.

Green manure treatments presented higher levels of nutrients N, Ca, B, Fe and Zn, compared with BQ treatment (Table 12), corroborating the results of Weber & Passos (1991) and San Martin Mateis (2008), who reported that brachiaria, among other natural Gramineae of citrus orchard, have lower nutrient concentrations than Leguminosae.

Table 13. shows the dry matter yield of citrus trees, expressed as percentage and macro and micronutrient levels for green manure and control treatments over two years. Dry matter content at 20 cm deep had a slight increase of 3 g dm^{-3} in the first experimental year, compared with the chemical analysis of the soil before treatments, and a little decrease of 6 g dm^{-3} in the subsequent year. At the layer 20–40 cm, it also had an increase of up to 5 g dm^{-3} followed by a decrease of up to 10 g dm^{-3}.

There was a significant difference among treatments (Table 13); PP showed higher dry matter content (34.3%). In the second year, it was statistically different from JB and LL, but equal to BQ. In the first year, PP treatment showed the highest dry matter content, which significantly differed from the other treatments; in the second year, it was different from JB and LL (Ragoso et al., 2006).

	N	P	K	Ca	Mg	S	B	Cu	Fe	Mn	Zn
	g/kg							mg/kg			
JB	27	1.9	8	13	6	2	26	7	289	33	25
PP	21	1.7	7	6	4	2	24	9	364	30	26
LL	22	2	8	12	6	2	33	8	625	53	40
BQ	13	2	12	5	7	2	12	6	225	34	21

*jack bean (JB) (*Canavalia ensiformis* DC), lablab (LL) (*Dolichos lablab* L.), pigeon pea (PP) (*Cajanus cajan* L. Millsp), and brachiaria (BQ) (*Brachiaria brizantha* Hochst ex A. Rich. Stapf).

Table 12. Macro and micronutrient concentrations in four green manures*, averaged over two experimental years (2003 and 2004). The experimental treatments were applied to a citrus orchard at the FCA/UNESP site in Botucatu, Sao Paulo, Brazil

Treatments				
Year	JB	PP	LL	BQ
2004	25.7Aa	34.3Ba	21.7Aa	18.6Aa
2005	19.8Ab	32.0Ba	22.9Aa	27.2ABb

Coefficient of Variation = 22.2%	Minimum Significant Difference Treat.= 87	Minimum Significant Difference Year= 6.57

Table 13. Dry matter means (%) of orange trees subjected to four different treatments over two experimental years.

There was no difference in citrus productivity with green manure treatments in 2003 or 2004 (Table 14), although yields were up to 20% greater from trees receiving the legume-based green manures, compared to the grass-based control (BQ), by the second year of the study. This was attributed to the gradual decomposition and nutrient uptake from the nutrient-rich legume residues, compared to the control..

Green manure can provide the crop with a better nutrition, even though the soil macro and micronutrient content was slightly reduced in the second year. It would be beneficial to continue evaluating the experiment over a longer period of time, and analyze additional soil characteristics, such as density, water retention, texture, structure, porosity and thermal conductivity, in response to green manure application..

	Treatments			
Year	JB	PP	LL	BQ
2003	79.8Aa	83.9Aa	82.6Aa	81.7Aa
2004	102.2Ab	103.2Ab	103.0Ab	86.5Aa

CV = 21%   DMS Treat.= 29.60   DMS Year = 22.30

Means followed by the same letter in the same line and column are not statistically different by the Tukey at 5% significance. Uppercase letters = treatments; Lowercase letters = years.
Adapted from RAGOSO et al. (2006).

Table 14. Productivity (kg plant^{-1}) of orange trees receiving four green manure treatments in two years. The experimental treatments were applied to a citrus orchard at the FCA/UNESP site in Botucatu, Sao Paulo, Brazil

6. Final considerations

Based on the premise that organic fertilizers, including green manureare fundamental to rural sustainability, i.e. that locally produced compost and cultivation waste should be to their maximum potential, studies were conducted to evaluate their use in orchard production. Considering that the composition of organic materials available is variable, it is hard to provide general recommendations for organic fertilization and the use for diverse orchard crops.

One approach is to devise recommendations based on the N requirements of the crop, calculating the application rates based on the percentage of N present in the organic fertilizer. However, this method frequently suggests such high levels of organic fertilizer that it becomes too costly for producers to consider this option. One can consider organic fertilizer to be more than a simple N source, in that it possesses an entire suite of macro and micro nutrients required for nutrition of orchard crops; in addition, the high organic matter content of most organic fertilizers is beneficial for improving soil physical, chemical and biological properties, which can indirectly or directly improve tree growth i.e., improved rooting, warmer soil temperature, buffer pH, increase reaction surface for cation exchange, promote the growth of soil microorganisms including biocontrol organisms.

There still exist very few studies on managing orchards in Brazil with organic fertilizers only, without employing any chemical fertilizers. The experiments done until nowshow promising results, and these studies must be continued.

7. References

Agrianual 2001. Anuário Estatístico da Agricultura Brasileira. São Paulo: FNP Consultoria & Comércio, 2001. 492p.

Alva, A. K.; Paramasivam, S. Nitrogen management for high yield and quality of citrus in sandy soils. Soil Science Society America Journal, Madison, v. 62, n. 5, p. 1335-1342, 1998.

Anjos, J. T.; Rowell, D. L. Perdas de calcário em solos: efeitos de doses de óxido de cálcio, temperatura e períodos de secagem em colunas de solo. Revista. Brasileira de Ciência do Solo, Campinas, n. 7, p. 75-81, 1983.

Bartz, H.R. (Cood.) et al. Recomendações de adubação e calagem para os estados de Rio Grande do Sul e Santa Catarina. 3ed. Passo Fundo: SBCS-núcleo Regional Sul, 1995. 224p.

Borges, A.L.; Caldas, R.C.; Coelho, E.F. Fertirrigação com nitrogênio e potássio para bananeira cv. Grand Naine no Norte de Minas Gerais. In: CONGRESSO NACIONAL DE IRRIGAÇÃO E DRENAGEM, 12., 2002, Uberlândia. Anais... Uberlândia: ABID, 2002. 1CD.

Borges, A.L. Novas tendências para nutrição da bananeira: adubação orgânica e fertirrigação. In: WORKSHOP SOBRE INOVAÇÕES TECNOLÓGICAS EM BANANICULTURA, 1., 2008, Botucatu. Anais... Botucatu, 2008. 1CD.

BRAZILIAN FRUIT YEARBOOK 2010. Cleonice de Carvalho... [et al]. – Santa Cruz do Sul: Editora Gazeta Santa Cruz, 2010. 128 p. :il. ISSN 1808-4931.

Büll, L. T.; Fernandes, A. L.; Nakagawa, J. Influência da calagem na lixiviação de bases trocáveis em solos da região de Botucatu (SP), avaliada em condições de laboratório. Científica, São Paulo, v. 19, p. 37-46, 1991.

Caetano, L.C.S.; Carvalho, A.J.C.; Jasmim, J.M. Preliminary report on yield productivity and mineral composition of the fig trees as a function of boron and cattle manure fertilization in Brazil. Fruits, v.61, p.341-349, 2006.

Campo Dall'orto, F.A. et al. Frutas de clima temperado II: Figo, maçã, marmelo, pêra e pêssego em pomar compacto. In: RAIJ, B. van. et al. (Ed.). Recomendações de adubação e calagem para o Estado de São Paulo. 2 ed. Campinas: Fundação Instituto Agronômico (IAC), 1996, p.139-140.

Cintra, F.L.D. Manejo e conservação do solo em bananais. Revista Brasileira de Fruticultura, v.10, n.1, p. 65-73, 1988.

COMISSÃO DE FERTILIDADE DO SOLO DO ESTADO DE MINAS GERAIS. Recomendações para o uso de corretivos e fertilizantes em Minas Gerais, 4. Aproximação. Lavras, 1989. 176p.

Cunha, A.R.; Klosowski, E.S.; Galvani, E.; Scobedo, J.F.; Martins, D. Classificação climática para o município de Botucatu-SP, segundo Koppen. In: SIMPÓSIO EM ENERGIA NA AGRICULTURA, 1., 1999, Botucatu. Anais... Botucatu: UNESP-FCA, 1999. p.487-491.

Curi, P.R. Relações entre evaporação média pelo tanque IA-58 e evapotranspiração calculada pelas equações de Thornthwaite e Camargo, para o município de Botucatu. 1972. 88 f. Tese (Doutorado) – Faculdade de Ciências Médicas e Biológicas de Botucatu, Universidade Estadual Paulista, Botucatu, 1972.

Damatto Junior, E.R. Efeitos da adubação com composto orgânico na fertilidade do solo, desenvolvimento, produção e qualidade de frutos de bananeira `Prata-Anã´ (Musa AAB). 2005. 70f. Dissertação (Mestrado em Agronomia / Energia na Agricultura). Faculdade de Ciências Agronômicas, Universidade Estadual Paulista, Botucatu, 2005.

Damatto Junior, E.R.; Villas Bôas, R.L.; Leonel, S. Efeito residual de composto orgânico na segunda safra de bananeira 'Prata-anã'. In: REUNIÃO INTERNACIONAL ACORBAT, 17., 2006, Joinville. Anais... Joinville: ACORBAT/ACAFRUTA, 2006c. p.629-633.

Damatto Junior, E.R.; Villas Bôas, R.L.; Leonel, S.; Fernandes, D.M. Alterações em propriedades de solo adubado com doses de composto orgânico sob cultivo de bananeira. Revista Brasileira de Fruticultura, Jaboticabal, v.28, n.3, p.546-549, 2006a.

Damatto Junior, E.R.; Villas Bôas, R.L.; Leonel, S.; Fernandes, D.M. Avaliação nutricional em folhas de bananeira 'Prata-anã' adubadas com composto orgânico. Revista Brasileira de Fruticultura, Jaboticabal, v.28, n.1, p.109-112, 2006b.

Demattê, J. L.; Vitti, G. C. Alguns aspectos relacionados ao manejo de solos para os citros. In: SEMINÁRIO INTERNACIONAL DE CITROS, 2., 1992, Bebedouro. Anais... Campinas: FUNDAÇÃO CARGILL, 1992. p. 69-99.

Eryuce, N.; Caloloclu, H.; Audin, S.; Cokuysal, B.; Gerasco Poulos, D.; Olympios, C.H.; Passam, H. The effects of K and Mg fertilization on some quality characteristics and mineral nutrition of fig. Acta Horticulturae, n.379, p.199-204, 1995.

Espindola, J.A.A.; Guerra, J.G.M.; Almeida, D.L. De; Teixeira, M.G.; Urquiaga, S. Decomposição e liberação de nutrientes acumulados em leguminosas herbáceas perenes consorciadas com bananeira. Revista Brasileira de Ciência do Solo, Viçosa, v.30, n.2, p.321-328, 2006.

FAO – FOOD AND AGRICULTURE ORGANIZATION. Disponível em: <http://faostat.fao.org/faostat>. Acesso em: 30 junho 2011.

Fernandes, F.M., Buzetti, S. Fertilidade do solo e nutrição da figueira. In: CORRÊA, L. de S., BOLIANI, A.C. Cultura da figueira – do plantio à comercialização. Ilha Solteira: FUNEP/FAPESP, p. 69-85, 1999.

Gpacc – Grupo Paulista De Adubação E Calagem Para Citros. Recomendações de adubação e calagem para citros no Estado de São Paulo. Laranja, Cordeirópolis, p. 1-27, 1994. Edição especial.

IBGE - INSTITUTO BRASILEIRO DE GEOGRAFIA E ESTATÍSTICA, SIDRA. www. ibge.org.br. Acesso em 04 mar. 2011.

Jinsheng, Y.; Mi-Lin, Z.; Yang, J.S.; Mi, L.; Zhou, Y.H. Preliminary report on application ratis of single fertilizer with nitrogen, phosphorus and potasium in fig tree. Jiangsu Agricultural Sciences, v.4, p.47-49, 1997.

Kato, T. Nitrogen metabolism and utilization in citrus. Horticultural Reviews, New York, v.8, p.181-216, 1986.

Kiehl, E. J. Fertilizantes orgânicos. Piracicaba: Agronômica Ceres, 1985. 492 p.

Koller, O. C. Nutrição mineral e adubação. In: KOLLER, O. C. et al. Citricultura: cultura da tangerineiras: tecnologia de produção, pós-colheita e industrialização. Porto Alegre: Editora Rígel, 2009. p. 121-152.

Lahav, E; Turner, D. Banana nutrition. Berna: International Potash Institute, 1983. 62p. (IPI-Bulletin 7).

Lea-Cox, J. D.; Syvertsen, J. P.; Graetz, D. A. Springtime [15]nitrogen uptake, partitioning, and leaching losses from young bearing *Citrus* trees of differing nitrogen status. Journal American Society Horticultural Science, Alexandria, v. 126, n. 2, p. 242-251, 2001.

Leonel, S.; Damatto Junior, E.R. Efeitos do esterco de curral na fertilidade do solo, no estado nutricional e na produção da figueira. Revista Brasileira de Fruticultura, Jaboticabal, v. 30, n. 2, p. 534-539, 2008.

Marcílio, H. De C.; Andrade, A.L. De; Pereira, G. A.; Abreu, J.G. De; Santos, C.C. dos. Avaliação de genótipos de bananeira em sistema orgânico de produção. In: REUNIÃO INTERNACIONAL ACORBAT, 17., 2006, Joinville. Anais... Joinville: ACORBAT/ACAFRUTA, 2006. p.553-556.

Mattos Junior, D. De; Bataglia, O. C.; Quaggio, J. A. Nutrição dos citros. In: MATTOS JUNIOR, D. de et al. Citros. Campinas: Instituto Agronômico e Fundag, 2005. P. 199-215.

Melarato, M. S. A cobertura vegetal do solo na citricultura. In: SEMINÁRIO INTERNACIONAL DE CITROS – TRATOS CULTURAIS, 5.,1998, Bebedouro. Anais... Bebedouro: Fundação Cargill, 1998. p. 203-221.

Mielniczuk, J. Matéria orgânica e a sustentabilidade de sistemas agrícolas. In : SANTOS, G. de A. ; CAMARGO, F.A. de O. (Eds.). Fundamentos da matéria orgânica do solo: ecossistemas tropicais e subtropicais. Porto Alegre : Genesis, 1999. p.1-8.

Moreira, R.S. Banana: teoria e prática de cultivo. Campinas: Fundação Cargil, 1987. 335p.

Panzenhagen, N. V. et al. Respostas de tangerineiras 'Montegrina' à calagem e adubação orgânica e mineral. Pesquisa Agropecuária Brasileira, Brasília, v. 34, n. 4, p. 527-533, 1999.

Penteado, S.R. Adubação orgânica - Compostos orgânicos e Biofertilizantes. Campinas-SP. 2º Ed. Agrorganica, Fev/2006, 132 p.

Pereira, J.C.R; Zambolim, L.; Vale, F.X.R.; Chaves, G.M. Compostos orgânicos no controle de doenças de plantas. In: LUZ, W.C.; FERNANDES, J.M.; PRESTES; A.M.; PICININI, E.C (Eds.) Revisão Anual de Patologia de Plantas. Passo Fundo: RAPP, 1996, v.4, p. 353-379.

Pereira Neto, J. T. Manual de compostagem: processo de baixo custo. Viçosa: UFV, 2007. 81 p.

Pozzan, M.; Tribone, H. R. Colheita e qualidade do fruto. In: MATTOS JUNIOR, D. de. et al. Citros. Campinas: Instituto Agronômico e Fundag, 2005. p. 801-822.

Quaggio, J. A.; Dechen, A. R.; Raij, B. Van. Efeitos da aplicação de calcário e gesso sobre a produção de amendoim e lixiviação de bases no solo. Revista Brasileira Ciência do Solo, Campinas, n. 6, p. 189-94, 1982.

Quaggio, J. A.; Mattos Junior, D. De; Cantarella, H. Manejo da fertilidade do solo na citricultura. In: MATTOS JUNIOR, D. de et al. Citros. Campinas: Instituto Agronômico e Fundag, 2005. p. 485-507.

Ragoso, Carlos Renato Alves ; LEONEL, S. ; CROCCI, Adalberto José . Adubação verde em pomar cítrico. Revista Brasileira de Fruticultura JCR, Jaboticabal/SP, v. 28, n. 1, p. 69-72, 2006.

Raij, B. Van,; Quaggio, J. A. Métodos de análise de solo para fins de fertilidade. Campinas: Instituto Agronômico de Campinas, 1983. (Boletim Técnico, 81).

Ramos, D.P.; Dalastra, I.M.; Evangelista, R.M.; Leonel, S.; Mischan, M.M. Influência da posição do fruto na qualidade de laranja 'folha murcha' e tangerina 'poncã'. In: CONGRESSO BRASILEIRO DE FRUTICULTURA, 21., 2010, Natal. Anais... Natal, 2010. 1CD.

Rangel, A. Situação da bananicultura no Planalto Paulista. In: RUGGIERO, C. (Ed.) Bananicultura. Jaboticabal: Funep, 2001. p.18-28.

Rodríguez, V.L.; Molina, J.N.L.; Baez, O.L.; Hernandez, J.M.G. Manejo de la nutricion y fuente de matéria orgânica en el cultivo de platano (Musa AAA), município de Huehuetan, Chiapas. In: REUNIÃO INTERNACIONAL ACORBAT, 17., 2006, Joinville. Anais... Joinville: ACORBAT/ACAFRUTA, 2006. p.638-645.

San Martin Matheis, H.A. Uso contínuo de coberturas vegetais em citros: influência no banco de sementes, na comunidade infestante e nas características químicas do solo. 2008. 96 f. Tese/Doutorado em Fitotecnia- Programa de Pós-graduação em Fitotecnia, Escola Superior de Agricultura Luiz de Queiroz.

Silva, J. A. A.; Donadio, L. C.; Carlos, J. A. D. Adubação verde em citros. Jaboticabal: Funep, 1999. 37 p.

Silva, M.S.L. Da; Chaves, V.C.; Ribeiro, F.N.; Ferreira, G.B.; Mendonça, C.E.S.; Cunha, T.J.F. Espécies vegetais para adubação verde e/ou cobertura do solo em cultivo orgânico de fruteiras na região semi-árida do Nordeste Brasileiro. In: REUNIÃO BRASILEIRA DE MANEJO E CONSERVAÇÃO DO SOLO E DA ÁGUA, 17., 2008, Rio de Janeiro. Anais...Rio de Janeiro: SBCS, 2008. 1CD.

Syvertsen, J. P.; Smith, M. L. Light acclimation in Citrus leaves. I. changes in physical characteristicas, chlorophyll, and nitrogen content. Journal American Society Horticultural Science, Alexandria, v. 109, n. 6, p. 807-812, 1984.

Teixeira, L.A.J. Adubação nitrogenada e potássica em bananeira 'Nanicão' (Musa AAA subgrupo Cavendish) sob duas condições de irrigação. 2000. 132f. Tese (Doutorado em agronomia). Faculdade de Ciências Agrária e Veterinárias, Universidade Estadual Paulista, Jaboticabal, 2000.

UNITED STATES SOIL CONSERVATION. Soil Taxonomy: A Basic System of Soil Classification for Making and Interpreting Soil Surveys. 1988. 754 p.

Vitti, G. C. et al. Técnicas de utilização de calcário e gesso na cultura dos citros. In: Seminário Internacional de Citros - Nutrição e Adubação, 4., 1996, Bebedouro. Anais... Campinas: FUNDAÇÃO CARGILL, 1996. p. 131-60.

Von Osterroht, M. O que é uma adubação verde: princípios e ações. Agroecologia Hoje, n. 14, p. 9-11, mai/jun 2002.

Weber, O. B.; Passos, O. S. Adubação verde: aspectos relacionados a citricultura. Revista Brasileira de Fruticultura, Cruz das Almas, v. 13, n. 4, p. 295-303, out. 1991.

Wong. J. Disponibilidade de nutrientes e produção de soja nos solos de Mato Grosso do Sul em função da calagem. 1999. 81 f. Dissertação (Mestrado em Agronomia/Agricultura) - Faculdade de Ciências Agronômicas, Universidade Estadual Paulista, Botucatu, 1999.

13

Promoting Sustainable Crop-Livestock Integration Through Farmer's Participation and Integrated Soil Fertility Management in the Sahel of West Africa

Bidjokazo Fofana[1], Zacharie Zida[2] and Guillaume Ezui[3]
[1]IFDC - North and West Africa Division;
IFDC Mali - B.P. E 103, Badalabougou-Est Fleuve; Bamako,
[2]IFDC – North and West Africa Division,
IFDC-Ouagadougou – CMS 11 B.P. 82, Ouagadougou,
[3]IFDC-North and West Africa Division; IFDC-Lomé, BP 4483, Lomé,
[1]Mali
[2]Burkina Faso
[3]Togo

1. Introduction

The sahelian climatic zone extends over several countries of West Africa. The "Sahel" region is charac terized by low and variable rainfall, ranging from 300 to 600 mm annually. In the Sahel of West Africa (SWA), increasing human population combined with long-term weather changes are transforming pastoralist systems based on transhumance and communal grazing of rangelands, and cropping systems based on extensive shifting cultivation, to more sedentary, intensively managed enterprises (Winrock International, 1992). As livestock husbandry becomes more settled, rangeland grazing is substituted slowly by crop residues feeding and cropland forage production, and feed shortage is becoming a major constraint for livestock production in the dry season (6-7 months) (Savadogo, 2000). As a result, most livestock in the southern SWA are integrated into mixed farming systems, involving the integration of crops and livestock.

Livestock make a major contribution to the transfer of nutrient from rangeland to cropland. However, when grazing crop residues, livestock remove more nutrients from croplands than they return in manure (Powell and Williams 1993). Farmers owing many livestock have the potential capacity to maintain soil fertility by recycling nutrients in fecal materials. But the majority of farmers practicing mixed farming in the Sahel has relatively few livestock and therefore is less capable of counteracting the soil degradation caused by livestock in their cropland. Thus, common constraints to mixed farming systems in SWA include: declining soil fertility, soil erosion and lack of access to agricultural inputs, and farmers make little use of external nutrient inputs because of scarcity of resources and inefficient fertilizer use (Powell et al. 1996). Indigenous soil fertility management strategies in SWA have therefore been driven largely by an extensive use of resources. Although farmers use many locally derived soil amendments (crop residues, manure, or termitaria), they do so more in desperation than by choice. The reasons for low fertilizer use include among others the poor natural resource base for agriculture, erratic rainfall, sub-optimal crop management practices (e.g., poor weed control), unattractive cost:benefit ratios for using external inputs (Breman and Debrah 2003).

Integrated soil fertility management (ISFM) is an integrated crop - livestock system that provides opportunities for improving soil fertility and promoting sustainable intensification of cropland and staple food production. In this respect, efficient nutrient management, including the use of manure from ruminant livestock, combined with an appropriate use of mineral fertilizers, possibly incorporating a rotation with legumes (dual-purpose cowpea), is a crucial prerequisite for sustainable mixed farming systems in the SWA. To this end, research should start with farmer's participation and include their knowledge of traditional cropping systems. Authors observed that most African farmers have a thorough knowledge of their cropping systems, and there are several examples of research using farmers' knowledge of traditional cropping systems to guide research and develop solutions suitable for their needs (Chambers et. al., 1989, Quansah et al., 2001).

The objectives of the study were to *(i)* assess farmers' knowledge of soil fertility management in the desert margin zones, *(ii)* understand the reasons behind the traditional practices in order to *(iii)* identify and adapt potential solutions for sustainable mixed crop-livestock integration in SWA.

2. Material and methods

2.1 Experimental site

Field experiments were conducted near Djibo, the main district of the Soum province of Burkina Faso located between 13°44' and 14°50' N (latitude), and 0°32' and 2°07 W (longitude) (Figure 1). The experimental plots were chosen by selected sedentary pilot-farmers with livestock assets averaging 15 caws and 20 small ruminants (sheep and goat) per farm family. As such, there is a pressing need for cropland forage production. Farmers expressly choose, especially for fertilization trials, degraded land for experimental purpose, often due to expectations of increased soil fertility status on their infertile land through experimentation. Most of the experiments have been therefore conducted on infertile land.

Soil was acidic, contained 870 g kg^{-1} sand, 60 g kg^{-1} clay and 70 g kg^{-1}, and is classified as tropical ferruginous type (FAO-Unesco, 1973). The textural class is loamy-sand with low soil organic matter content and pronounced physical degradation. It's extremely coarse texture favors water infiltration, but the relatively high air temperature and low relative humidity most of the year, as well as the soil texture, do not favor soil organic matter accumulation. The main soil degradation is a result of nutrient depletion and water erosion. The topsoil (0-20 cm) at the trial sites before planting in 2003 had, on average, a pH of 5.7, an organic carbon content of 1.5 g kg-1 an available P (Bray) of 2.2 mg kg^{-1} and a CEC of 2.1 cmol kg^{-1}.

Djibo is located in the southern Sahel of Burkina Faso, characterized by a 3-month rainy season lasting from June to October, and a 8-month dry season from October to June. In 2003 and 2004, cumulative rainfall amounted to 577 and 290 mm, respectively. The region is traditionally dominated by a mixed crop-livestock farming system, including pearl millet, sorghum, cowpea and groundnut. Most of the experimental plots were located far (> 500m) from the homestead and had not received any organic inputs, except for animal droppings of grazing cattle. Prior to the experiments, fields were traditionally cropped with millet without application of manure or mineral fertilizers.

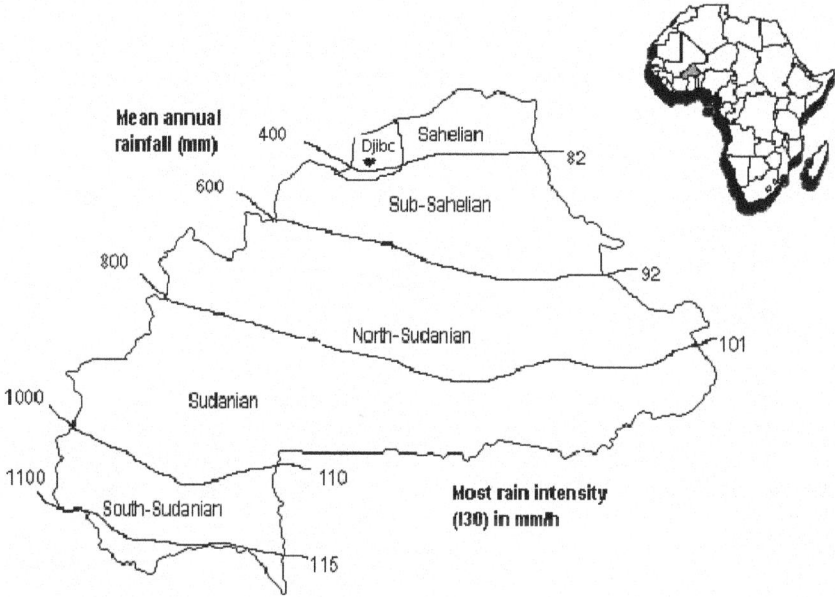

Fig. 1. Experimental site located in the Soum province (Djibo), in the north of Burkina Faso.

2.2 Farm household surveys

Farm household surveys were conducted to collect information on farmers' households including farm resources, current practices, constraints and opportunities with regard to the mixed crop-livestock system. Results were supposed to be used to set research priorities for soil fertility and livestock management options, and to identify pilot villages and farmers to collaborate with. To this end, a multi-visit participatory rural appraisal survey (PRAS) was conducted from May to December 2002 in the Soum province, located in the Sahel of Burkina Faso. This method characterized the agro-ecological zone on the basis of selected sites and extrapolated the results to a regional level (Ledoupil and Lidon, 1993). The PRAS was conducted in six (6) out of the nine (9) departments of Soum. It included 360 households living in 36 villages located in the departments of Aribinda, Baraboulé Kelbo, Nasoumbo, Pobé Mengao and Tongomayel. Farmers were selected in collaboration with national extension agents, a partner that has been involved from the start of the survey. They helped select farmers, focusing on different households that were representative of the area.

Selection criteria of departments, villages and pilot-farmers were well-defined to capture the dynamics that are so critical to the technology adoption process. Thus, the departments were chosen on the basis of herd size, population density and soil degradation. Farmers in SWA usually do not produce cropland forage and rely mainly on grazing pasture and crop residues for livestock nutrition. Pilot-villages were selected based on the soil fertility management practices including production and use of organic amendments (compost, crop residues, cover crops, manure), cropland forage and use of external inputs (fertilizer, improved variety, industrial byproducts for livestock). The PRAS concentrated on farmers owing at least ten cattle and/or small ruminants (goat or sheep), and inquired about livestock composition and dynamics, growing profitable cash crops and crops suitable for typical mixed cropping systems. Farm household information was collected with focus on human and natural resource needs and use for their activities, draught power, inputs and outputs of annual crop production on individual plots. Results of the PRAS were analyzed to distinguish production unit in different typologies defined as a household or group of households in which members work on the same fields and eat together (Samaké, 2003).

2.3 Modeling with SIMFIS (SImulating Mixed Farming In the Sahel)

The model is largely based on Struif Bontkes (1999) and composed of a number of subsystems including cropping systems, soil and livestock management. SIMFIS consists of a set of equations that are written in the Vensim language (Vensim DSS 5.3) and an EXCEL file (Struif Bontkes, 2006). The file contains general parameters pertaining to e.g. climate data, labor requirements, crops and prices. The parameters pertaining to a specific farm, e.g. number and size of fields and their soil fertility, number of animals (cattle, goat and sheep), crop and fertilizer use. The model therefore offers a way of identifying and quantifying significant interaction that occur between various components in mixed farming systems. Results of the participatory household survey were used as input data. By entering the basic data of a particular farm in an EXCEL sheet, SIMFIS can be used to quickly evaluate management alternatives that could be translated into operational support/guide for seasonal soil fertility management for an integrated crop-livestock production system. The model was used to evaluate farm management practices including crop rotation, soil organic amendments and crop residues recycling, and predict their impact on soil fertility evolution under actual soil fertility management practices.

2.4 Field experiments

Farmers were, under the supervision of extension agents, in charge of compost and manure collection/production, plot selection and for all farming operations. Participatory action research was conducted to promote ISFM options for cropland forage and cereal production. To this end, a field experiment was conducted in 2003 and 2004 by 4 pilot-farmers in Djibo. Experiment dealt with the effect of various manure types (cattle manure, mixed sheep and goat manure and compost) applied either alone or in combination of mineral compound fertilizer ($N_{15}P_{15}K_{15}$) on crop yield. Selection criteria for the pilot-villages were herd size and soil fertility management practices. The objective was, on the basis of the results of the PRAS, to promote sustainable cropland forage production for mixed farming systems through a judicious combined use of locally available resources (manure, compost, crop residues) and external inputs (seeds of improved variety, fertilizer etc.). Test crops for both experiments were millet and dual-purpose cowpea, identified by farmers as the most economically promising and suitable for mixed farming systems.

At the onset of the rainy seasons (early July), prior to land preparation and sowing, different manure types including cattle, sheep-goat and compost were distributed in the field by farmers as small heaps, applied by broadcasting and incorporated when hand ploughing. After land preparation, crops were sown in hills, distance between rows and hills within rows were 80 cm x 60 cm for millet and 75 cm x 40 cm for dual-purpose cowpea. Sowing was done in late July and all plots were weeded twice by hoeing at 15 and 45 days after sowing (DAS). To reduce/avoid damage to pods caused by aphids (*Aphids* spp.), cowpea was kept insect-free by application of Dimethoate at a rate 100 ml 20 l^{-1} of water twice at flower bud initiation and at mid-pod (50% of plants with pods).

The experiment design was a randomized complete block with 4 replicates including the application of locally available organic amendments (sheep/goat, cattle and compost) either alone at 2500 kg ha^{-1} (full recommended manure rate in Burkina Faso) or at 1250 kg ha^{-1} dry matter of sheep/goat manure (half of the recommended manure rate in Burkina Faso) in combination with compound fertilizer (N_{15} P_{15} K_{15}) applied at 50 kg ha^{-1} rate (half of recommended rate of 100 kg ha^{-1} $N_{15}P_{15}K_{15}$ in Burkina Faso), resulting in following treatment combinations:

- T_0: no manure and fertilizer applied (control)
- T_1: application of recommended fertilizer rate (100 kg ha^{-1} $N_{15}P_{15}K_{15}$ fertilizer)
- T_2: application of recommended manure rate of 2500 kg ha^{-1} dry matter
- T_3: ½ fertilizer rate + ½ sheep/goat rate.
- T_4 2500 kg ha^{-1} of sheep-goat manure
- T_5 2500 kg ha^{-1} of cattle manure
- T_6 2500 kg ha^{-1} of compost

The cultivars tested were IKMV 8201 for millet and KVX 745-11 P for dual-purpose cowpea, commonly grown in the Soum province. Each combination was repeated on-farm by four (4) farmers. Plot size for each treatment of experimental plots was 10 x 10 m. To confirm the results of the year 2003, treatments were once more imposed to both crops in 2004 in Djibo. But in 2004, unfortunately, most of the experimental plots were damaged by the erratic rainfall and worsen by the presence of desert locust in the sahel zone of Burkina Faso. The number of observation made was therefore not sufficient for accurate data analysis. Nevertheless, the minimum data recorded in 2004 is presented in Table 4.

2.5 Data collection and analysis

In 2003, soils samples were taken prior to land preparation (0-20 cm depth), ground to pass through a 2mm sieve, prior to chemical analysis of organic carbon, total N, pH-H_2O, pH-KCl and exchangeable Ca, K and Mg (IITA, 1982). Animal manure mixed with minor amounts of bedding and wasted feed from different cattle and sheep-goat corrals in the villages was collected in May 2003 and dried separately within feces type in the sun. Upon sample collection, the remaining quantity of manure was cleaned with a brush to remove loosely adhering soil particles. Manure samples were oven-dried to constant weight at 55°C to determine total dry matter. Samples were thereafter ground to 2 mm to analyze macro nutrient (N, P and K) content. The central 36m^2 area of each plot including millet (grain and stover) and cowpea (grain and haulm) were harvested at maturity. Crop sub-samples were oven-dried at 60°C, weighed and ground to pass 0.5mm and then analyzed for total N and P. Dry matter production and grain yield were expressed on a dry-weight basis. All analyses were conducted using standard methodologies in the ICRISAT laboratory in Niamey, Niger.

2.6 Statistical procedures and evaluation of economic performance

Analysis of variance (ANOVA) was carried out on grain yield and stover yield. When a significant treatment effect was found, comparison of treatment means were carried out using the least significant difference (LSD) test at the α=0.05 probability level (Gomez) and Gomez, 1984). Principal component analysis was used to classify production units in different groups, using information from the PRAS. The characteristics used to describe the profile of the smallholder mixed farming in the Soum province include age of farmer, number of compost pit, herd and farm size, and household labor unit.

The economic performance of the test crops was estimated using the value:cost ratio (VCR). VCR is calculated as followed: (yield with inputs – yield without inputs) / (cost for inputs). The inputs include NPK fertilizer, manure and seeds. The background of the millet and cowpea VCRs is as followed: the price including handling and transportation costs of 50 kg/bag of $N_{15} P_{15} K_{15}$ fertilizers (7.5 % N, 3.3% P and 6.2% K) is 11500 F cfa; 1 kg manure is about 2 F cfa; 1 kg millet and cowpea grain is worth 90 and 145 F cfa, and 1 kg millet stover and cowpea haulm is valued at 20 and 40 F cfa, respectively.

3. Results and discussion

3.1 Participatory on-farm survey to characterize crop-livestock production systems

3.1.1 Typology of the smallholder mixed farming in the Soum province

Results of the PRAS conducted in the Soum province were used to distinguish production units (PU) of different groups. Table 1 gives an overview of the profile of households studied. Major criteria for the distinction between PU were age of farmer, number of compost pit, herd and farm size and household labor unit. Results showed three different PU with distinct characteristics that were significant ($P<0.05$). The typology defines farmer groups with specific combinations of land uses and resource endowment, as they are expected to react differently to specific circumstances or proposed technologies (Samaké, 2003). The following differences were observed between the production units.

Characteristics per household	Production unit (PU)		
	PU-A	PU-B	PU-C
Number of households	152	197	11
Age of farmer	48	45	42
Number of compost pits	0.7	0.6	1
Household labor units	13	6	6
Farm size (ha)	9	5	6
Number of cattle	17	4	4
Number of sheep	17	5	3
Number of goats	17	5	5
Number of pigs	0	0	23

Table 1. Typology of 360 mixed farming households, categorized by three production types, in the Soum province, Burkina Faso

The number of active persons ha[-1] decreased from PU-A to PU-C and was 1.4 for type A, 1.2 for B and 1 for C. The average farm size were 9 ha, 5 ha and 6 ha for PU-A, PU-B and PU-C, respectively. Average number of compost pit per PU does not exceed one, indicating there is a pressing need for manure and compost production. Research should therefore promote technologies for compost and manure production. In addition, PU-A has more livestock (51) and should in principle have more manure available per unit area cultivated than PU-B (14) and PU-C (12). But manure from PU-A is scattered over the surrounding outfields and natural pastures as they are mainly pastoralists. They have large herd size and are therefore always in search of forage and drinking water for livestock, whereas those of PU-B and PU-C are characterized by regular seasonal migrations or are sedentary. There is closer integration of livestock with crop production on PU-B and PU-C farms. Thus, to facilitate the adoption of technologies for cropland forage production, the pilot-farmers selected for training came from PU-B and PU-C as livestock husbandry is more settled and feed shortages are a major constraint for livestock production on their farms

3.1.2 Farmers' perception of constraints limiting crop production

Indigenous farmers in the Soum province identified rainfall as a major limiting factor to crop production. However, poor soil fertility was implied in the problem analysis as farmers pointed out low crop yields was a major problem, linking this continuous cultivation of land, lack of soil moisture and nutrient deficiency (yellow leaves). Although it is often assumed that rainfall is the preliminary constraint to crop production in the Sahel, research in SWA demonstrated that the potential production from any given rainfall is limited by nutrient availability (Zougmoré et al., 2003; Fofana et al., 2005). Participatory research should therefore create awareness of the interaction between soil moisture and fertility. Solutions proposed by farmers to overcome declining soil fertility included use of manure (100%), compost (80%), inorganic fertilizers (56%) and crop rotation (15%) (data not presented). Irungu et al. (1996) conducted a survey in the semi-arid zone to assess soil fertility as perceived by farmers and reported that farmers' opinions considered crop vigor, invading plants, presence of earthworms and soil color. Indigenous farmers categorized soils according to their productivity. They define soil fertility by the duration under cultivation, linking continuous cultivation to soil degradation. In our survey, different soil types were classified on the basis of color and texture. White and/or sandy soils were associated with poor crop growth resulting from low soil fertility and poor water-holding capacity. Reddish and black soils were considered fertile but are known to be prone to water-logging and difficult to work. Also, farmers made associations between natural vegetation and soil fertility level. Weed species were also associated with soil fertility status, particularly *Digitaria horizontalis, Eragrostis tremula, Striga hermonthica* and *S. gesneriodes* in infertile fields and *Andropogon gayanus* in fertile outfields or natural pastures.

3.1.3 Farmers' perception of constraints limiting livestock production

Results of the survey showed two main forms of livestock production systems, i.e pastoralism and mixed crop-livestock farming. Pastoralism comprises two different movements including nomadism and transhumance. The former implies continuous movement of the herds in search of forage and drinking water, whereas the latter is characterized by more or less regular seasonal migrations from a permanent homestead

(Ayantunde, 1998). Farmer ascribed pastoralism to seasonal fluctuation in feed and water supply, the main constraint to livestock production. In the Soum province, natural pastures remain the major source of livestock feed. Thus, animals lose weight during the last months of the dry season and the beginning of the wet season (from March to July) as grazing and water resources diminish. Given the lack of alternative dry season feed, crop residues are used as vital livestock feeds rather than soil amendments. To overcome food shortage during this period, most farmers send their livestock to the southern zones of Burkina Faso where there are enough natural pastures and water reservoirs. As a consequence of the fluctuation of feed availability, manure output and distribution also vary. Hiernaux et al. (1999) reported weight losses associated with movement of animals in search of forage and drinking water. For sustainable livestock production, researchers need to investigate sustainable cropland forage production instead of rangeland grazing. This might reduce the need for constant movement, resulting in greater animal weight gains and give farmers access to more manure.

3.1.4 Nutrient sources and agricultural practices in relation to spatial variability of soil fertility

Most livestock in the Soum province are integrated with cropping systems. Traditionally, farmers in the Soum province have enriched soil fertility in the compound fields or 'infields' close to the homestead through regular application of household waste, crop residues and animal manure. Continuous cropping and grazing on fields further away from the homestead ('outfields') with little or no nutrient inputs, and use of crop residues as building materials and fuel resulted in nutrient depletion and decline in soil fertility. Farmers in the Soum province have therefore developed strategies to optimize crop production on these field types using various crops on the basis of their nutrient requirement. Pearl millet (*Pennisetum glaucum*) is grown on both field types, while sorghum (Sorghum bicolor) and rice (*Oryza sativa* or *O. glaberrima*) are grown on dark fertile soil (vertisols) in low lands. Outfields are however used for legumes and communal grazing. Farmers in the Soum province consider millet and cowpea as cash crop and cropland forage. As cowpea (*Vigna unguiculata*) and groundnut (*Arachis hypogaea*) are not manured, most of collected manure is used for pearl millet, sorghum and rice production. There is a weak link between the commonly practiced crop rotations and soil fertility improvement, apparently because of the relationship between the observed spatial variability in soil fertility and crop requirements for nutrients. Thus, farmers continuously crop cereals on infields and legumes on outfields. The observed crop sequences include pearl millet-maize-pearl millet on infields, and continuous cowpea, groundnut or Bambara- groundnut (*Voandzeia subterranean*) on outfields. Limited availability of fertile land is also considered the major cause for the lack of systematic crop rotations.

3.1.5 Soil chemical properties

Results showed large difference in natural fertility of infields and outfields. While soil pH (water and KCl) and potassium content were comparable for both fields, many other soil characteristics such as organic C, total N, total P, CEC, Bray-1 P,exchangeable Ca and Mg concentrations were significantly different (Table 2). In Mali, Samake (2003) reported C org values ranging from 5.4 to 8.5 g kg^{-1} on infields and 1.0 to 5.2 g kg^{-1} on outfields. C-org values reported by Sédogo (1993) in the Guinea Savanna zone of Burkina Faso ranged from 11-22 g kg^{-1} on infields and 2-4 g kg^{-1} on outfields. The soil properties measured clearly indicate that infields are more fertile than outfields, supporting farmers' perception of higher soil fertility status on infields. These could be ascribed to the continuous accumulation of organic amendments including all kinds of manure and household waste applied in the small ring directly surrounding the villages. The spatial variability in soil fertility between infields and those far away (outfields), was attributed to the long-term fertility management strategies of farmers.

Soil characteristic	Outfield	Infield	P value
pH – H$_2$O	5.7	6.1	0 .16 (ns)
pH – KCl	4.7	5.4	0.07 (ns)
K+ (cmol kg^{-1})	0.43	0.49	0.32 (ns)
Ca++ (cmol kg^{-1})	0.80	3.67	0.0001
Mg++ (cmol kg^{-1})	0.41	2.10	0.0045
CEC (cmol kg^{-1})	2.10	7.10	0.0008
C-org (g kg^{-1})	1.5	3.9	0.001
N tot (mg kg^{-1})	140	350	0.0009
P tot (mg kg^{-1})	88.6	130.4	0.0022
Bray-1 P(mg kg^{-1})	2.2	3.3	0.015
C/N ratio	11	11	1 (ns)

Significant at $P<0.05$, ns = not significant; infield is in close proximity to the homesteads while an outfield is located about 500 m and more away from the homesteads

Table 2. Average soil characteristics of eight infields and eight outfields, Soum province, Burkina Faso

3.1.6 Livestock management and soil fertility management

Livestock ownership defines the availability of manure. Results obtained showed positive correlation between access to livestock, manure availability and crop production (data not shown), depending on livestock variety and feed availability. Beside the provision of manure, cattle and donkey provide transport and draught power, giving farmers owning livestock and animal traction better access to manure and opportunity to cultivate more land. Transhumance and communal grazing systems allows for uncontrolled grazing animals on rangeland and crop residues, enabling livestock owner to "harvest" nutrient from natural pastures and fields of farmers who do not own livestock. From the 360 interviewed farmers, 50% are sedentary and integrates livestock with crop production, 34% practice transhumance, which implies seasonal migration during the later part of the dry season and early wet season (March to July), and 16% are pure pastoralists always in search of forage and water for livestock. While pure pastoralism is the major livestock production system in the northern Sahel (Powell et al., 1996), results of our survey indicate that mixed crop-livestock farming system is the dominant production system in the Soum province. There is little doubt that, especially in sub-Saharan Africa, increasing integration of crops and livestock is going to occur over at least the next 30 years (Thornton and Herrero, 2001). In such a system, there is a pressing need for more cropland forage production as crop residues including cereal stover, groundnut and cowpea hays provide vital feeds during the 6-8 month dry season. Hence, the potential of arable land to provide forage throughout the year must be enhanced. Developing ISFM-strategies for sustainable cropland forage production is a major research challenge if the important role of livestock in improving soil fertility and household welfare is to be maintained or further developed.

3.1.7 Animal manure production

The majority of farmers in the Soum province obtained manure either from their own livestock, from the livestock of other farmers, or through exchange relationship with pastoralists. There are three key methods of manure production: gathering manure from corrals, picking up manure from natural pastures and fields (infields and outfields) surrounding households, and composting. But most of interviewed farmers (80%) harvest manure from natural pastures and neighboring fields, as getting enough manure from corralling requires greater herd size than is kept on most farms. Our survey indicates an average herd size per pilot-farmer of about 8 cows, 8 sheep and 9 goats (Table 1). This indicates that most of livestock-owners may not necessary produce adequate manure to replenish nutrients harvested from cropland. Composting is mainly used by farmers who have no livestock, and in most cases crop residues are included in the compost . Manure is continuously dug out from barns and heaped in pits along with wasted feed and crop residues for composting, which is considered by farmers as labor-intensive.

3.1.8 Animal manure quality and use

Manure is the most used organic amendment in the Sahel of Burkina Faso. While the use of manure includes that from cattle, goats and donkeys, sheep manure is the most preferred and widely used. Manure quality has been defined by farmers on the basis of its source (cattle, goat or sheep), structure, color and the crop response to its application. In the Soum province, 75% of interviewed farmers distinguish and strictly separate manure on the basis of the source, 94% consider manure from goat to be the best quality followed in decreasing order by sheep, cattle and donkey. While manure from cattle and donkey are mainly used for millet and sorghum, sheep-goat manures are particularly used for maize and rice cropping. Manure is transported to the fields just prior to the start of the cropping season (May-June). Carts are the usual mode of transport, poor farmers however used wheelbarrows. The most common application method is to broadcast, although some "smart" farmers practice spot application by placing manure into sowing holes. Spot application is considered more efficient than broadcasting (Munguri et al. 1996). Prior to broadcasting, manure is distributed in the field as small heaps for uniform distribution. But exposing manure to the sun may lead to N volatilization, resulting in decreased manure quality (lower N content) (Esse et al., 2001). Broadcasting is usually done immediately prior to sowing, and manure is incorporated through hoeing.

3.1.9 Animal manure availability

The amount of rangeland needed to feed livestock and capture enough manure for subsequent application to cropland needs to take into account the high variability in the productivity of rangelands, which is assessed at various rangeland:cropland ratios (Hiernaux, 1993). Estimates of rangeland:cropland ratios typically range from 15 to 45 ha of rangeland needed to support the number of livestock that will produce sufficient manure for one hectare of cropland (van Keulen and Breman, 1990). Manure availability for cropping is mostly limited by livestock types and numbers, spatial location at manuring time, manure output per animal, efficiency of manure collection, and the amounts of feed and land resources available (Powell et al. 1996). Sedentary farmers practicing mixed crop-livestock systems in the Soum province are poor in livestock assets (Table 1). Fernandez-Rivera et al. (1995) estimated that during the 8-month period when farmers apply manure to cropland, about 300 kg DM of manure could be collected from a 300-kg cow, 60 kg from a 35-sheep and 45 kg from a 25-kg goat. Similarly manure production was reported by Matlon and Fafcahmps (1988) in northwestern Burkina Faso. Based on the above figures, the quantity of collectable manure for sedentary farmers in the Soum province (PU-B and C, Table 1) is on average 1.7 t PU^{-1} year^{-1}, whereas national extension agencies in Burkina Faso recommend an application rate of 5 t ha^{-1} to cropland The collectable manure of 1.7 t PU^{-1} could cover only 0.34 ha of the average 5.5 ha cultivated land area per PU, corresponding to 6% of cultivated land of each PU. These values suggest that the current soil fertility management options cannot adequately sustain the required food and fodder production in the Soum province. Therefore, research should, particularly in SWA, investigate alternative technologies to increased manure production (quantity and quality) and use.

3.1.10 Farmers' perception of good manure

The criteria used by farmers to classify manure include source, color, structure, age and the decomposition rate. Farmers accordingly defined freshly collected sheep-goat manure as being the best quality adapted to their production system because it withstands termite pests and decomposes slowly (effective for 3 years) compared to cattle manure. Thus, sheep-goat manure is directly used on field as organic fertilizer. Yet, farmers in the Soum Province widely believe that sheep-goat manure burns crops, especially when rainfall is low. The N content in fresh cattle manure is shown to be higher than in dry cattle manure, probably because of N volatilization (Powell et al., 1996). However, cattle manure is mainly composted in combination with crop residues (pit-composting) as its water content is higher, contributing to rapid decomposition of organics when pit-composting. Good compost is supposed to be whitish, not sticky, cold and have a crumby structure as it dries. The heat generated during the composting process is believed to kill weed seeds, and this has been confirmed through on-farm research that revealed a 65-70% decrease in weed seed density in manure heaped for 1-5 months (Jonga et al;, 1997).

3.1.11 Chemical manure characteristics

Results of chemical analysis of compost and manure from cattle and sheep-goat are shown in Table 3. Sheep-goat manure has a higher N content and lower C/N ratio than compost and cattle manure. The amount of faecal N that could be collected is dependent of animal diet selection, whichvaries according to the season (Ayatunde et al. 2001). Data from other studies reviewed by Diarra et al. (1995) suggested the average N concentration in cattle's diet varied from 2.0% in the mid-wet-season, 1.2% in the early-dry and 0.6% in the late-dry season. With N content of 6 g kg-1 in compost, 9 g kg-1 in cattle and 14 g kg-1 in sheep-goat manure (Table 3), N concentration in the three manure types were lower than the 20 and 23 g kg-1 respectively, reported by Brouwer and Powell (1998). These differences could be ascribed to the fact that manure collected in this study came from indigenous farmers, left in the sun and unprotected against ammonia volatilization, whereas the manure used by Brouwer and Powell (1998) were mixed with urine right at the trial site. The higher N concentration in sheep-goat manure compared to cattle manure is in line with our observations and confirms also the farmers' perception of good manure. But the observed differences in N concentrations between manure types may be due also to differences in the feed and specific physiological characteristics of both ruminant species. Schlecht et al. (1997) showed that faecal N concentrations were significantly higher for small ruminants than cattle when fed with green forage, whereas concentrations were similar for both types of ruminants with straw feeding. C/N ratio in sheep/goat manure is lower as compared to cattle. High quality of organics has generally been defined as having low C/N ratio and decomposed relatively faster, and are therefore able to release nutrients than can be utilized by growing plants in the short term.

Manure type	pH-H$_2$O (1:2.5)	C Org	N Total	P Total	K+	Ca^{2+}	Mg^{2+}	C/N
		%			cmol+/kg			ratio
Cattle	7.8	17.4	0.9	0.4	5.6	1.5	2.1	19.4
Compost[1]	7.2	10.1	0.6	0.9	3.0	2.4	3.1	16.8
Sheep-goat	7.6	19.5	1.4	0.4	5.4	2.1	3.4	14.0

[1]Compost is a mixture of cattle manure and all kinds of crop residues and organic wastes

Table 3. Chemical characteristics of manure and compost in the Soum province, Burkina Faso

The constraints to manure use in SWA include the lack of adequate quantities recommended, the slow or no release (N immobilization) of nutrient from low quality of organics, and the lack of water and labor for composting (Enyong et al., 1999). To overcome this constraint, on-the-job training was organized for pilot-farmers in the Soum province. The training topics include techniques of production of high quality manure and efficient use, application methods and crop nutrient requirements.

3.1.12 Mineral fertilizer use and availability

The amounts of fertilizers used by farmers in the Soum province largely depend on availability and accessibility of nutrient sources and crop requirements rather than recommendations by extension services. Most farmers do not apply fertilizers to fields (infields) that they perceive to be fertile. Also, they clearly differentiate crops on the basis of their nutrient requirements and perceive maize, rice and sorghum as having high nutrient demands compared to millet and legumes. Poor adoption of mineral fertilizers by farmers is the major highlight of soil fertility management in the Soum province and largely attributed this to the risk associated with the use of expensive fertilizers under unfavorable cropping conditions including erratic rainfall and very eroded and degraded soils (Fofana et al., 2005). Under these conditions, most farmers also believe that fertilizers burn crops. Most of the available fertilizers are used for pearl millet, sorghum and particularly for irrigated rice production. The extension services in Burkina Faso recommend 100 kg ha-1 and 50 kg ha-1 of fertilizer urea (46% N) for millet and sorghum production. While the actual fertilizer use in Burkina Faso is 7 kg ha-1 (Hien et al. 1994), our survey in the Sahel of Burkina Faso revealed an average fertilizer rate of about 5.6 kg ha-1, including compound (mainly NPK_) and single (urea) fertilizers. Farmers traditionally mix seeds and fertilizers in the calabash and apply fertilizer simultaneously in micro-doses (i.e., in the planting hole) when sowing.

Soil fertility management practices in the Soum province indicate that current cropping systems are largely based on extensive management of resources and are not adapted for more intensive farming systems. Maintenance of soil fertility involves the replacement of nutrients removed from the soil through harvested crop products, leaching, erosion and other pathways. This concept has apparently not been built into the indigenous management practices.

3.1.13 Scenario impact assessment of current farmers' practice on soil fertility changes

Farmers in the Soum province rely mainly on manure and crop residues recycling to sustain soil fertility under current management conditions. Simulated data show decreasing soil organic matter (SOM) and total P content over time on infields, in spite of the application of animal manure and regardless of the crop rotation (Figure 2). The SOM content in the Soum province is generally less than 0.5%. Murage et al. (2000) reported that among soil organic carbon pools and fraction, total soil organic C seems to be a suitable indicator of soil quality. Giller and Cadisch (1997) reported that animal manure additions of 60 t ha-1 only increased soil carbon from 0.25 to 0.66 % over a period of 18 years. These studies revealed that the storage capacity for SOM in soil depends on the amount of clay content and silt in the soil. Most of upland soils in the Soum province are very sandy, have a poor aggregation and weak physical stabilization effect on SOM. Sandy soils therefore are not good at preventing SOM from microbial decomposition. This may explain the small C org content in SWA sandy soils, regardless of the quantity of soil organic amendments applied and the land use systems.

Apparently, current soil fertility management practices in the Soum province lack the capacity to maintain or increase the level of soil organic matter and total P content. The current farmer practices which are mainly based on using locally available organic resources and exploiting spatial variability of soil fertility should be complemented with options aimed at encouraging farmers to judiciously combine fertilizers with locally available and accessible soil amendments including all kind of organics (compost, manure, crop residues) and minerals (rock phosphate, dolomite etc.).

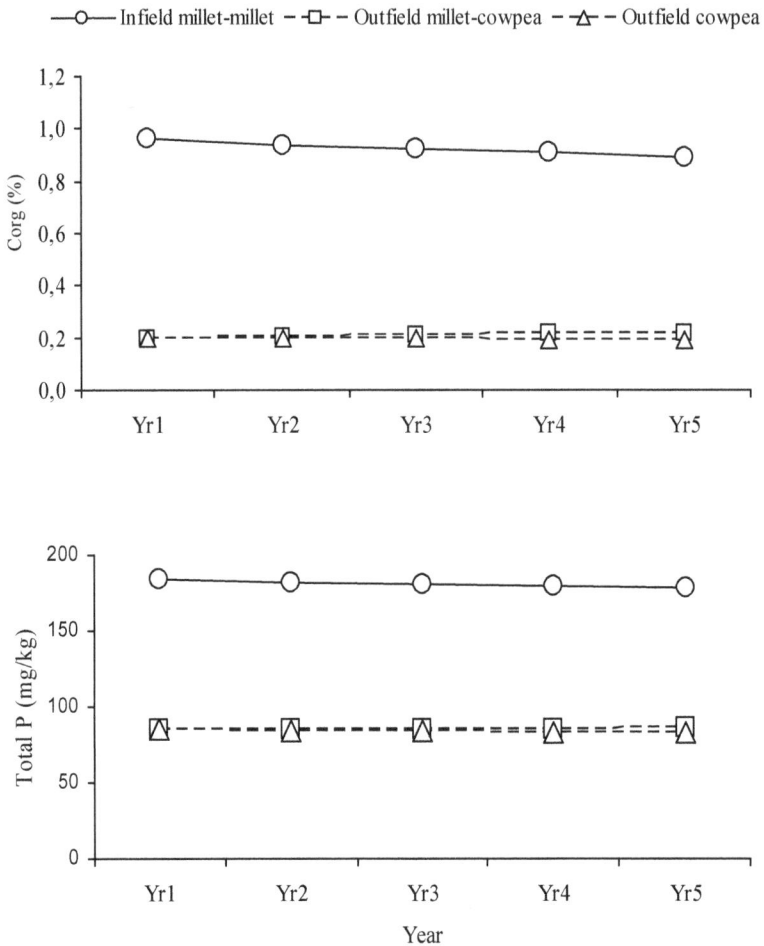

Fig. 2. Simulated evolution of soil organic matter and total P under different crop sequences on infields and outfields using the QUEFTS model (Janssen et al. 1990) (quantitative evaluation of the fertility of tropical soils). Solid lines refer to infields; dotted line to outfields, Soum province, Burkina Faso, 2004

3.2 Combining animal manure and fertilizer for sustainable and productive cropland forage production

3.2.1 Yield performance of tested ISFM technical options

Cowpea yields were significantly affected by manure and fertilizer application (Table 4). Low grain yield were obtained when sheep-goat manure was applied. Application of compost or cattle manure resulted in higher grain yield, probably because of higher P and Ca content, particularly in compost (Table 3). Authors reported higher cowpea yield performance with appropriate P, Ca and Mg fertilization (Giller et al. 1995, Bado et al., 2006). Yields were lowest without manure and fertilizer application (control). Highest yields were obtained either with fertilizer applied at recommended full rate (100 kg ha^{-1}) or with both manure (2500 kg ha^{-1}) and fertilizer (50 kg ha^{-1}) applied at half rates.

Dual-purpose cowpea						Pearl millet		
	Grain		Shoots			Grain		Shoots
Year	2003	2004	2003		2004	2003		2003
Treatment								
Manure (M)								
T$_4$	499 a	356	1989 b		1419	284 b		997 b
T$_5$	370 b	306	1934 b		1599	520 a		1585 a
T$_6$	521 a	492	2894 a		2733	549 a		1640 a
Fertilization (F) + Manure								
T0	366 c	330	1994 bc		970	367 b		1206 a
T1	508 ab	400	2710 a		1095	552 a		1578 a
T2	421 bc	350	1877 c		715	393 b		1349 a
T3$_{1)}$	558 a	550	2509 ab		1355	493 ab		1499 a
Probability Level								
Manure (M)	0.05		0.05			0.00		0.00
Fertilization (F)	0.00		0.01			0.05		ns
Interaction M x F	ns		ns			ns		ns

Values affected by the same letters within the column are not significantly different according to LSD$_{0.05}$;
[1] T$_0$: no manure and fertilizer applied (control); T$_1$: application of recommended fertilizer rate (100 kg ha^{-1} NPK fertilizer); T$_2$: application of recommended manure dry matter rate of 2500 kg ha^{-1} ; T$_3$: ½ fertilizer rate + ½ sheep/goat manure rate, T$_4$: cattle manure at 2500 kg ha^{-1}, T$_5$: sheep/goat manure At 2500 kg ha^{-1}; T$_6$: compost including all kinds of manure and organic wastes

Table 4. Yield performance of dual-purpose cowpea and pearl millet as affected by fertilizer and manure application in the Soum province, Burkina Faso.

Manure and fertilizer application resulted also in significant millet yield increase (Table 4). Application of sheep-goat manure or compost significantly increased millet grain and shoots yields. Low yields were obtained when cattle manure was applied. Lower C/N ratio observed in compost and sheep-goat manure apparently contributed to nutrient release that meets millet nutrient requirement, resulting in increased yields. Lowest yields were recorded when no fertilizer or manure was applied. Highest yields were achieved either with NPK fertilizer applied at recommended full rate or with the combination of both manure and fertilizer at half rates. The highest yields results suggest the efficiency of integrated use of fertilizers and organic soil amendments. The crucial role of ISFM for judicious nutrient management and use has been reported by many authors (Fofana et al., 2005, Wopereis et al. 2006).

3.2.2 Economic performance of tested ISFM technical options

Calculation of the value:cost ratio (VCR) shows that combined application of manure and compound fertilizer (NPK) at half rate result in the highest economic benefits, and this is true for both dual-purpose cowpea and pearl millet production. With a VCR value of 5.9 (2.1 for millet), dual-purpose cowpea production seems to be more profitable than millet (Table 5). Higher fertilizer application rates (100 kg ha-1) also resulted in relatively high VCR values for both crops. Results suggest that fertilizing dual-purpose cowpea with half of the recommended fertilizer rate and manure at recommended full rate to produce cropland forage might be profitable for sedentary farmers from PU-B and C (Table 1). A combined use of locally available phosphate rock (PR) (Kodjari) with manure (with PR-enriched compost), especially for dual-purpose cowpea, might even be more profitable than the soluble fertilizers. The VCR concept provides a rough estimate of economic profitability and could be helpful in formulating recommendation for seasonal soil fertility management in mixed farming systems. Policy instrument such as the VCR are required to institutionalize soil fertility management at the grassroots level and facilitate farmers' access to inputs including soil amendments (e.g. compost, manure etc.) and mineral fertilizers. Farmers urgently need more opportunities to connect with markets for their products, and this will give them income to invest in external inputs for their farms.

	Pearl millet					Dual-purpose cowpea				
	Grain	Shoots	TEB	TIC	VCR	Grain	Shoots	TMV	TIC	VCR
	kg ha-1		F cfa ha-1			kg ha-1		F cfa ha-1		
Tmt										
T0	367	1206				366	1994			
T1	552	1578	24094	14375	1,7	508	2710	49230	14375	3,4
T2	393	1349	5218	5000	1,0	421	1877	3263	5000	0,7
T3	493	1499	17204	8250	2,1	558	2509	48408	8250	5,9

Tmt treatment, TEB total economic benefit, TIC total input costs, VCR value cost ratio, T_0 control, T_1 fertilizer rate of 100 kg ha-1 NPK, T_2: sheep/goat manure at 2500 kg ha-1 year -1, T_3: ½ fertilizer rate + ½ sheep/goat manure at 1250 kg ha-1 year-1.

Table 5. Value-Cost Ratio (VCR) of dual-purpose cowpea and pearl millet production for different combinations of fertilizer and manure, Soum province, Burkina Faso

4. Conclusions

The results indicate that there are close links between the perceptions and indicators of soil fertility used by farmers and researcher, and that farmers use their knowledge of soil types and differences in soil nutrient status to tailor their soil fertility management practices to different situations. The PRAS indicated that farmers' decisions to grow some crops on certain portions of their farm are governed by soil fertility status but the selection of those crops also provides opportunities for crop-livestock integration. Under the current management practices, the only viable nutrient sources are manure and crop residues. However, manure (and NPK fertilizer) application rates by farmers in the Soum province are considerably less than those recommended by agricultural extension. Farmers are more likely to allocate their limited organic resources to higher value crops on more productive infields of the farm than to improve degraded soils on outfields that are becoming progressively infertile due to nutrient depletion. Constraints to soil fertility improvement on outfields include lack of adequate quantity of manure and materials (water, crop residues and manure, wheelbarrows, shovels etc.) for making compost and transporting available organic amendments (compost, manure, household waste, crop residues etc.) to outfields, shortage of labor and water for pit-composting. Results of the study indicate that efficient and sustainable use of nutrients in sandy soils should involve frequent combined application of compost at half of recommended rates (2500 kg ha-1 dry matter) and compound fertilizers (N15P15K15), which is consistent with the ISFM approach. Thus, the way forward may lie in combining the best use of locally available organic amendments with external sources of nutrients. As the quantity of collectable manure, especially for sedentary farmers of PU-B and PU-C, is not sufficient to guarantee sustainable mixed farming systems, there is a pressing need to introduce improved fodder and water harvesting technologies as to produce more manure and compost even in the drought years through sustainable increased livestock and crop residues production using ISFM. Furthermore, there is a need to promote widespread adoption/adaptation of mineral fertilizer in SWA. Flexible approaches to management of fertilizers need to be further explored and promoted, including site- and crop rotation-specific fertilizer recommendation.

Research should therefore, especially for sedentary production units that mainly rely on crop-livestock integration, identify alternative nutrient sources, promote efficient use of existing nutrient resources in combination with fertilizers and industrial byproducts so as to intensify both cropland forage production for insuring forage availability during shortage period, and settled livestock production for increasing manure production. Thanks to farmers' perception on soil fertility management practices, we started investigating the residual effect of a 3-year continuous mono-cropping of pearl millet, dual-purpose cowpea, sorghum (*Sorghum bicolor*) and andropogon, on soil physical and chemical properties and nutrient use efficiency as well. Sustainable intensification of soil fertility management on arable land for cropland forage production will increase livestock and manure production at farm gate, saving natural pastures and sparing these ecosystems from degradation. These will contribute to the maintenance of crop and animal productivity by smallholder farmers in SWA.

5. Acknowledgements

We are indebted to the 360 smallholders in the Soum province for their cooperation during the household survey and on-farm field experiment. We are grateful to L. Konfe and O. Tambora for field work, and to the national project's Coordinator Mr. A.D. Somé for his fruitful peace of advice. Financial support to the investment project was provided by the African Development Bank, the Burkina Government and USAID. Additional funding was provided through Desert Margins Program funded by the Global Environment Facility (GEF) from United Nations Environment Program.

6. References

Ayantunde, A.A., 1998. Influence of grazing regimes on cattle nutrition and performance and vegetation dynamics in Sahelian rangelands. Ph.D thesis, Wageningen Agricultural University, Wageningen.

Ayatunde, A.A., Fernandez-Rivera, S., Hiernaux, P.H.Y., van Keulen H. and Udo H.M.J., 2001. Effect of timing and duration of grazing of growing cattle in the West African Sahel on diet selection, faecal output, eating time, forage intake and live-weight changes. *Animal Science* 72: 117-128.

Bado, B.V., Bationo, A. and Cescas M.P., 2006. Assessement of cowpea and groundnut contribution to soil fertility and succeeding sorghum yields in the Guinea savannah zone of Burkina Faso (West Africa). Biol. Fertil. Soils *(in press)*.

Breman, H. and Debrah, S.K. 2003. Improving African food security. SAIS Review vol. XXIII no. 1, pp. 153–170.

Brouwer, J. and Powell, J.M., 1998. Increasing nutrient use efficiency in West-African agriculture: the impact of micro-topography on nutrient leaching from cattle and sheep manure. *Agriculture Ecosystems and Environment* 45: 229-239.

Chambers R. Pacey A., Thrupp LA, 1989. Farmer First: Farmer Innovation and Agricultural Research. Intermediate Technology Publications: London; 219.

Diarra, L., Hiernaux, P. and de Leeuw, P.N., 1995. Preferential grazing of herded cattle in a semi-arid rangeland in central Mali at three spatial levels: vegetation type, biomass on offer and species composition. In Livestock and Sustainable Nutrient Cycling in Mixed-Farming Systems of Sub-Saharan Africa. Vol.II: Technical Paper, eds. J.M. Powell, S. Ferandez-Rivera, T.O. Williams and C. Renard. Proceedings of an International Conference held in Addis Ababa, Ethiopia, 22-26 November 1993. International Livestock Centre for Africa (ILCA), Addis Ababa, Ethiopia, pp. 99-114.

Enyong, L.A, Debrah S.K. and Bationo A., 1999. Farmers' perception and attitudes towards introduced soil-fertility enhancing technologies in western Africa. *Nutrient Cycling in Agroecosystems* 53: 177-187.

Esse, P.C., Buerkert, A., Hiernaux, P. and Assa, A., 2001. Decomposition of and nutrient release from ruminant manure on acid sandy soils in the Sahelian zone of Niger, West Africa. *Agriculture, Ecosystems and Environment* 83, 55-63.

FAO-Unesco, 1973. Base Map VI-1. FAO, Rome, Italy.

Fernandez-Rivera, S., Williams, T.O., Hiernaux, P., Powell, J.M., 1995. Livestock, feed, and manure availability for crop production in semi-arid west Africa. In: Powell, L. M., Fernandez-Rivera, S., Williams, T.O. and Renard, C. (Eds.), Livestock and Sustainable Nutrient Cycles in Mixed-Farming Systems of Sub-Saharan Africa. Vol. II: Technical Papers. Proceedings of an International Conference held in Addis Ababa, Ethiopia, 22-26 November, 1993. International Livestock Centre for Africa (ILCA), Addis Ababa, Ethiopia, pp. 149-170.

Fofana, B, A Tamélokpo, MCS Wopereis, H Breman, K Dzotsi, RJ Carsky (2005) Nitrogen use efficiency by maize as affected by a mucuna short fallow and P application in the coastal savanna of West Africa. *Nutrient Cycling in Agroecosystems* 71:227-237.

Giller, K.E. and Cadisch, G., 1997. Driven by nature: a sense of arrival or departure? 393-399 p. in Cadisch G. and Giller K.E. (Eds.): Driven by nature: plant litter quality and decomposition, CAB International, Wallingford, Oxon, UK.

Giller, K.E., Mcdonagh J.F., Toomsan B., Limpinuntana V., Cook, H.F. and Lee, H.C., 1995. Legumes in the cropping systems of North-East Thailand. Proceedings of the Third

International Conference on Sustainable Agriculture, University of London, UK.
Wye College Press, Ashford.

Gomez, K.A., Gomez, A.A., 1984. Statistical Procedures for Agricultural Research. An
International Rice Research Institute Book, John Wiley, Brisbane, Australia, pp. 187-
233.

Hien, V., Sédogo, P.M. & Lompo, F., 1994. Gestion de la fertilité des sols au Burkina Faso. Bi-
lan et perspectives pour la promotion de systèmes agricoles durables dans la zone
Soudano-Sahelienne. In: FAO and CIRAD (Eds.), Compte rendu d'un séminaire
régional sur la promotion des systèmes agricoles durables dans les pays d'Afrique
Soudan-Sahelienne, Dakar, Sénégal, pp. 47-59.

Hiernaux, P., 1993. The Crisis of Sahelian Pastoralism: Ecological or Economic? ILCA. Addis
Ababa, Ethiopia, 15pp.

Hiernaux, P, Ayatude A., de Leeuw P, Fernandez-Rivera, S, Sangaré M. and Schlecht E.,
1999. Foraging efficiency, stocking rates, grazing pressure and livestock weight
changes in the sahel. In Eldridge, D and Freudenberger D. (Eds.), People and
Rangelands Building the Future. Proceeding of the VI International Rangeland
Congress, Vol 1, Townsville, Queensland, Australia.

Irungu, J.W., Warren, G.P. and Sutherland, A., 1996. Soil fertility status in smallholder farms
in the semi-arid areas of Tharaka Nithi District: farmers' assessment compared to
laboratory analysis. Kenya Agric. Res. Inst. 5th Scientific Conf., Nairobi, 30 October-
1 November 1996.

International Institute of Tropical Agriculture (IITA), 1982. Automated and Semi-automated
Methods for Soil and Plant Analysis. Manual series 7. IITA, Ibadan, Nigeria.

Janssen, B.H., Guiking, F.C.T., van der Eijk, D., Smaling, E.M.A., Wolf, J. and H. van Reuler,
1990. A system for quantitative evaluation of the fertility of tropical soils
(QUEFTS). Geoderma, 46, 299-318.

Jonga, M., Mariga, I.K., Chivinge, O.A., Munguri, M.W. and Rupende, E., 1997. Towards the
improved management of organic and inorganic fertilizer in dry land maize
production in the smallholder sector of Zimbabwe. Pages 202-206 in Maize
productivity gains through research and technology dissemination: proceedings of
the fifth eastern and southern Africa regional maize conference, 3-6 Jun 1996,
Arusha, Tanzania. (Ransom, J.K., Palmer, A.F.E., Zambezi, B.T., Mduruma, Z.O.,
Waddington, S.R., Pixley, K.V., and Jewell, D.C., eds.). Ethiopia: Centro
International de Mejoramiento de Maiz y Tigro (CIMMYT).

Legoupil J.C, Lidon B., 1993. Quelle maitrise d'eau pour quelle intensification agricoledes
zones de bas-fonds en Afrique de l'Ouest ? Les leçons des expériences passées. Les
perspectives pour une nouvelle approche. In Jamin J.Y., Andriesse W., Thiombiano
L., Windmeijer P.N. (eds). La recherche sur les bas-fonds en Afrique sub-
Saharienne. Priorités pour un consortium régional. Actes du 1er atelier annuel du
Consortium bas-fonds, ADRAO, Bouaké, 8-10 Juin 1993.

Matlon, P.J. and Fafchamps, M., 1988. Crop Budgets for three Agroclimatic zones for the West
African semi-arid tropics. Progress Report No 85, ICRISAT, Patancheru, India.

Munguri, M.W., Mariga, I.K. and Chivinge, O.A., 1996. The potential of optimizing cattle
manure use with maize in Chinyika resettlement Area, Zimbabwe, 46-50p. In
Research results and network outputs in 1994 and 1995: proceedings of the second
meeting of the network working group, 18-21 Jul. 1995, Kadoma, Zimbabwe.
(Waddington, S.R., ed.). Harare, Zimbabwe: Soil Fertility Network for Maize-Based

Cropping Systems in Countries of Southern Africa, Centro International de Mejoramiento de Maiz y Trigo (CIMMYT).

Murage, E.W., Karnja, N.K., Smithson P.C and Woomer, P.L., 2000. Diagnostic indicator of soil quality in productive and non-productive smallholders'fields of Kenya's Central Highlands. *Agriculture, Ecosystems and Environment* 79, 1-8.

Penning de Vries, F.W.T., Ditèye, M.A., 1991. La Productivité des Pâturages Saheliens. Wageningen, The Nertherlands : Pudoc.

Powell, J.M., Fernandez-Rivera, S., Hiernaux, P. and Turner, M.D., 1996. Nutrient Cycling in Integrated Rangeland/Crop Systems of the Sahel. *Agric. Sys.* 52: 143-170.

Powell, J.M., Williams, T.O., 1993. Livestock, Nutrient Cycling and Sustainable Agriculture in the West African Sahel. Gatekeeper Series N0. 37. International Institute for Environment and Development, Sustainable Agriculture Programme, London, 15pp.

Quansah C. Drechsel P., Yirenkyi BB, Asante-Mensah S. 2001. Farmers' perceptions and management of soil organic matter – a case study from West Africa. Nutrient Cycling in Agroecosystems 61:205-213.

Samaké, O., 2003. Integrated crop management strategies in Sahelian land use systems to improve agricultural productivity and sustainability: A case study in Mali. Ph.D. thesis, University of Wageningen, Netherlands.

Savadogo, M., 2000. Crop Residue Management in Relation to Sustainable Land Use. Case study in Burkina Faso. Ph.D. thesis, Wageningen, The Nertherlands.

Schlecht, E., Ferandez-Rivera, S. and Hiernaux, P., 1997. Timing, size and N-concentration of faecal and urinary excretions in cattle, sheep and goats: can they be exploited for better manuring of cropland? In: Renard, G., Neef, A., Becker, K., von Oppen, M., (Eds). Soil Fertility Management in West African Land Use Systems. Margraf Verlag, Weikersheim, Germany, ICRISAT, INRAN, University of Hohenheim and Niamey, Niger, pp. 361-367.

Sédogo, P.M., 1993. Evolution des sols ferrugineux lessivés sous culture: Incidence des modes de gestion sur la fertilité. Ph.D. thesis, University of Abidjan, Ivory Coast, 343 pp.

Struif Bontkes, T.E., 1999. Modeling the dynamics of agricultural development: a process approach. The case of Koutiala (Mali). PhD Thesis, Wageningen Agricultural University, pp. 233.

Struif Bontkes, T.E., 2006. SImulating Mixed Farming In the Sahel (SIMFIS). Description and Tutrial, Technical Bulletin, IFDC, An International Center for Soil Fertility and Agricultural Development, Muscle Shoals, Alabama, USA, pp. 69. (*in press*)

Thornton P.K. and Herrero M., 2001. Integrated crop-livestock simulation models for scenario analysis and impact assessment. *Agricultural Systems* 70: 581-602.

Van Keulen, H. and Breman, H., 1990. Agricultural development in the West African Sahelian region: A cure against hunger? *Agriculture Ecosystems and Environment* 32: 177-197.

Winrock International, 1992. Assessment of Animal Agriculture in Sub-Saharan Africa. Winrock International Institute for Agricultural Development. Morrilton, Arkansas, USA, 162 pp.

Wopereis MCS, A Tamélokpo, D Gnakpénou, G Ezui, B Fofana, H Breman (2006) Mineral fertilizer management strategies for maize on farmer fields with differing in organic input history in northern Togo. *Field Crops Research* 96: 355-362.

Zougmore R., Zida Z and Kambou N.F., 2003. Role of nutrient amendments in the successs of half-moon soil and water conservation practice in semiarid Burkina Faso. Soil & Tillage Research 71: 143-149.

14

Integrated Soil Fertility Management in Bean-Based Cropping Systems of Eastern, Central and Southern Africa

Lubanga Lunze et al.*

Institut National pour l'Etude et la Recherche Agronomiques (INERA), Kinshasa
Democratic Republic of Congo

1. Introduction

Common bean (*Phaseolus vulgaris* L.) is an important grain legume in Eastern, Central and Southern Africa (ECSA), where it is grown on over 3.7 million of hectares every year. In this region, bean consumption per capita exceeding 50 kg a year is perhaps the highest in the world, reaching over 66 kg in densely populated western Kenya (Wortmann et al., 1998). Bean provides the essential dietary protein, fiber and income to at least 100 million people in Africa (Kimani et al., 2001). Beans are grown primarily as a food crop and to generate income by smallholder, resource-poor farmers, in holdings that rarely exceed 1.5 hectares. They are grown in pure stands or in association with maize, bananas, or root or tuber crops, and in recent years, between rows of fruit crops, banana and coffee, especially in the early establishment phase of these traditional cash crops. About 22% of the production area is sole-cropped, 43% is in association with maize, 15% with bananas, 13% with root and tuber crops, and 7% with other crops (Wortmann et al., 1998). In southern Africa, beans are either grown in pure stands (42%) or in association with maize (47%), or to a lesser extent, with root crops (6%) or other crops (5%). Production is mainly rain-fed, except in Mauritius and the Nile Valley of Sudan where beans are grown as an irrigated crop. In lowland areas of Madagascar, Malawi, Mozambique, and DR Congo, beans are sown after another crop in order to use residual moisture and to take advantage of lower temperatures during the winter months.

Bean is produced essentially in the Eastern and Central African highlands, where the population density is the highest (Wortmann et al., 1998). It is produced by smallholder farmers with few resources to allocate to soil improvement. In the Great Lakes region for example, Dreschsel et al. (1996) report extremely low use of mineral fertilizer, only about 0.4 kg ha^{-1}. Moreover, because of the high population density, farmers are faced with rapid soil

* Mathew M. Abang[2], Robin Buruchara[2], Michael A. Ugen[3], Nsharwasi Léon Nabahungu[4], Gideon O. Rachier[5], Mulangwa Ngongo[1] and Idupulapati Rao[6]
[1]*Institut National pour l'Etudes et la Recherche Agronomique (INERA), Kinshasa, R.D. Congo*
[2]*International Center of Tropical Agriculture (CIAT), Kampala, Uganda*
[3]*National Crop Resources Research Institute (NaCRRI), Namulonge, Uganda*
[4]*Institut des Sciences Agronomiques du Rwanda-Rubona, Butare, Rwanda*
[5]*Kenyan Agricultural Research Institute (KARI), Kakamega, Kenya*
[6]*International Center of Tropical Agriculture (CIAT), Cali, Colombia*

fertility decline as a result of continuous cropping and inappropriate cropping systems with very little or no external nutrient input to replenish soil fertility. Therefore, bean yield is generally low in most regions and is most likely to decline because of the ever increasing population density. In fact, under current farming systems in small holders' fields, soil nutrient balances are negative (Bationo et al. 2006), except in banana based systems (Wortmann and Kaizzi, 1998).Although bean grain yields are variable across countries and regions, they generally vary from 200 kgha[-1] in less favorable environments to 700 kg ha[-1] in more favorable environments when grown in pure stands, and about half of this when intercropped (Kimani et al., 2001).

In ECSA, low soil fertility is the most important yield-limiting factor in most of the bean-producing regions. The major soil fertility related problems are found to be low available phosphorus (P) and nitrogen (N), and soil acidity, which is associated aluminum (Al) and manganese (Mn) toxicity. According to the Atlas of common bean production in Africa (Wortmann et al., 1998), P is deficient in 65 to 80% of soils and N in 60% of soils in bean production areas of Eastern and Southern Africa, while about 45 to 50% of soils are acidic with a pH less than 5.2, containing high levels of either Al or Mn. The details on the importance of the edaphic stresses for common bean production are presented in table 1. Low soil fertility causes considerable losses in productivity as the estimated bean production losses due to edaphic stresses in the ECSA are about 1,128 million tons per year (Table 1).

Bean grows well in deep well drained, sandy loam, sandy clay loam or clay loam with clay content of between 15 and 35% with no nutrient deficiencies (Thung and Rao, 1999). The optimum soil pH range is 5.8 to 6.5 and Al saturation is below 10% (Lunze, 1994). It will not grow well in soils that are compacted, too alkaline or poorly drained.

Constraints	Eastern and Central Africa	Southern Africa	Eastern and Central Africa	Southern Africa	Sub Saharan Africa
	% Total bean area affected		Annual losses (t)		
N deficiency	50	60	263,600	125,200	389,900
P deficiency	65	80	243,200	120,400	355,900
Acidity	45- 50	45 - 50	152,700	65,800	220,000
Al/Mn toxicity	52	42	97,500	60,300	163,900
Total			757,000	371,700	1129,700

Table 1. Major soil related production constraints and bean yield losses in Africa (Wortmann et al. 1998)

In light of these constraints, the Pan African Bean Research Alliance (PABRA) has undertaken regional efforts to develop integrated soil fertility management (ISFM) technologies that improve sustainability, productivity and quality of the bean crop in various environments across ECSA. Over the past two decades, there have been research initiatives from both national programs and bean research networks Eastern and Central Africa Bean Research Network (ECABREN) and the Southern Africa Bean Research Network (SABRN), both of which belong to PABRA, to develop strategies and technologies to improve sustainably bean crop productivity and production in various production environments in ECSA. A range of technologies have been evaluated and developed to address the regional constraints, and efforts were made to promote the promising technologies widely.

Several technologies have been developed through collaborative research efforts within PABRA. The objective has been developing strategies and technologies that enhance resilience to environmental stresses and improve bean productivity and product quality. These include: (i) development of diagnostic tools for soil fertility assessment that are adapted to local conditions; (ii) replenishing soil nutrient pools, maximizing on-farm recycling of nutrients, and reducing nutrient losses to the environment; and (iii) improving the efficiency of external inputs. As common bean can derive part of its N from the atmosphere under low input agriculture (Giller et al., 1998), improving biological nitrogen fixation using seeds inoculation with appropriate rhizobacteria and soil management was considered. Currently recommended ISFM options in bean based cropping systems include farmyard manure, compost, biomass transfer, green manure and cover crops, liming, phosphate rock (PR) and mineral fertilizers in different combinations with organic resources. The soil management options are complemented by utilization of resilient bean germplasm that perform well under low soil fertility conditions.

This paper reviews ISFM options developed by the ECABREN and SBRN and the approaches for effective and efficient delivery of these technologies to farmers.

2. Integrated soil fertility management (ISFM) options

2.1 Concept of ISFM

Integrated soil fertility management (ISFM) is an approach that stresses sustainable and cost-effective management of soil fertility (Sanginga and Woolmer, 2009; Vanlauwe et al., 2010). This soil fertility strategy relies on a holistic approach that embraces the full range of driving factors and consequences of soil degradation- biological, chemical, physical, social, economic, health, nutrition and political (Bationo et al., 2006). ISFM attempts to make the best use of inherent soil nutrient stocks, locally available soil amendment resources and mineral fertilizers to increase land productivity while maintaining or enhancing soil fertility. Vanlauwe et al. (2010) define ISFM as 'A set of soil fertility management practices that necessarily include the use of fertilizer, organic inputs, and improved germplasm combined with the knowledge on how to adapt these practices to local conditions, aiming at maximizing agronomic use efficiency of the applied nutrients and improving crop productivity. A conceptual diagram is shown in Figure 1.

2.2 Local soil fertility diagnostic tools

Soil fertility can vary drastically from one end of field to the other (Vanlauwe, 2006). Therefore, response to applied soil management and external inputs can vary accordingly and no single recommendation can be made for a whole farm. As ISFM technologies are generally complex, labour intensive and costly, a more accurate intervention and recommendation system is crucial to improve the chance of adoption of these technologies by farmers. To assure farmers get the maximum return from the investment in inputs for

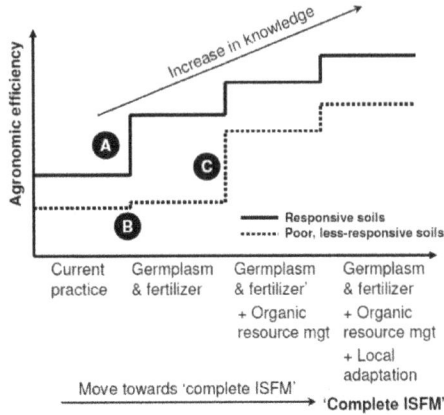

Fig. 1. Conceptual relationship between the agronomic efficiency of fertilizers and organic resource and the implementation of various components of ISFM, (Vanlauwe et al., 2010)

soil improvement, it is important that recommendations well target specific local conditions. Hence the indicator of soil quality is important for local communities to better manage their soil resources though better decision making (Barrios et al., 2001). These have to be simple enough for use by farmers and extension personnel. Dominant plant species on a farmland have the potential to integrate changes in soil quality, reflecting changes in the physical, chemical and biological characteristics of the soil (Pankhurst et al., 1997).

There have been research initiatives from both national programs and PABRA to integrate farmers' perceptions of soil fertility in simple soil fertility assessment. In South Kivu province of D.R. Congo, Ngongo and Lunze (2000) tested bean response to fertilizer application on soils with varying levels of weed infestation. Field trials were conducted for two seasons to test the effect of compost applied at 20 t ha^{-1} rate. Where *Gallisonga parviflora* is the dominant weed species, soil nutrient levels were high, bean (*Phaseolus vulgaris L.*) yield was high and did not increase further compost application. Where *Pennisetum polystachia* is the dominant species, soil fertility and bean productivity were low. Bean yield on these fields was increased considerably with the application of compost. The results (Table 2) confirm the farmer's perception that *Pennisetum polystachia* is an indicator of low soil fertility. *Conyza sumatrensis* and *Bidens pilosa* indicate intermediate level of soil fertility and response to compost was also intermediate.

Similar results were reported by Ugen and Wortmann (2006) in Uganda. Relative densities of *Digitaria scalarum* (blue couch), *Eleusine indica (L.) Gaertn.* (goosegrass), *Euphorbia hirta* (garden spurge), *Cyperus spp., Oxalis latifolia H.B.K.* and *Sorghum halepense (L.) Pers. (johnsongrass)* varied more with nutrient supply than did other species. Soil properties had less effect on the distribution of *Ageratum conyzoides* L. (*tropic ageratum*), *Bidens pilosa* (*hairy beggarticks*), *Commelina benghalensis* L. (*tropical spiderwort*), and *Galinsoga parviflora Cav.* (*smallflower galinsonga*). High relative densities of *Digitaria scalarum* and *Euphorbia hirta* were generally associated with low soil nutrient levels. *Eleusine indica (L.) Gaertn.* (*Goosegrass*), *Sorghum halepense (L.) Pers. (johnsongrass)*, and *Oxalis latifolia* were associated with higher nutrient levels in soil.

Dominant weeds	Bean yield (kg ha⁻¹)			
	1994B		1995A	
	+C	-C	+C	-C
Gallinsoga parviflora	995.3a	1000.0a	1713.3a	1700.0a
Pennisetum polystachia	427.7cd	170.7cd	696.7cd	41.7e
Conyza sumatrensis	720.1b	627.8b	827.7c	419.7d
Bidens pilosa	1012.0a	950.0a	933.3c	873.3c
Digitaria vestida var scalarum	892.2a	457.7c	1335.3b	1251.0b
Mean (kg ha⁻¹)	809.5	641.2	1101.3	857.1
C.V. (%)	25.9	25.9	21.2	21.2

+C: with compost and –C: without compost

Table 2. Bean response to compost on fields with different weed species in South-Kivu, D.R. Congo (Ngongo and Lunze, 2000)

A more detailed assessment of soil chemical properties based on weed species density was done in Uganda by Ugen, et al. (1999). In this study, they made observations on 39 fields with annual crops over 4 locations in Eastern and Central Uganda. Densities of weed species relative to the total weed population were determined and surface soil samples were collected and analyzed for organic carbon (OC), soil pH, available P, K, Ca and Mg and total N, P and K levels. The results are presented in table 3.

Species	OC	pH	P	K	Ca	Mg	N total	P total	K total
Eleusine			+	+	+	+		+	+
Euphorbia			-		-	-		-	
Sorghum	+	+		+	+		+		
Oxalis	+	+	+	+					
Nutsedge	-					-			

Source: Ugen, M., Wien A. D., Wortmann C.. CIAT Annual Report 1999

Table 3. Positive (+) and negative (-) relationships of the relative densities of weeds species with soil properties which may be useful in diagnosis of soil fertility status

Oxalis latifolia was correlated with OC content, as this weed was unlikely to be present when percentage C was less than 2.3, and increased in density as soil C increased. Soil pH was well correlated with both Sorghum halepense and Oxalis latifolia. Sorghum halepense was found to be a good indicator for soil pH, being generally absent when soil pH was less than 5.8, while Oxalis latifolia increased in importance when soil pH increased. Eleusine indica was not likely to occur when soil P availability was less than 13 mg kg⁻¹ and increased with increased P availability, while the total P was positively correlated with

Eleusine indica, but negatively correlated with *Euphorbia hirta* densities. *Sorghum halepense* occurred infrequently when exchangeable K was less than 0.6 cmol(+)kg^{-1}. *Eleusine indica* and oxalis increased as exchangeable K; *Eleusine indica* and *Bidens pilosa* tended to increase as total K increased. Relative densities of *Digitaria scalarum*, *Euphorbia hirta* and *Cyperus esculentus L.* (yellow nutsedge) were negatively related to exchangeable Mg and accounted for 62% of variation in exchangeable Mg. Soil fertility assessment using natural weeds is the most convenient tool available to smallholder farmers and extension workers because it requires minimal training. Therefore, weed flora was used to develop a decision guide to aid farmers in the identification of areas in their fields with severe nutrient deficiencies in Uganda and Eastern DR Congo. Thus, farmers and extension workers have at their reach a quick and inexpensive way to assess the soil fertility levels and make decisions concerning soil management.

3. Genetic approaches

Genetic variation for abiotic stress tolerance exists within common bean germplasms in 15 national and regional bean programs (Singh, 1991; Aggarwaal, 1994; Lynch and Beebe, 1995; Rachier et al., 1999; Rao 2001; Beebe et al., 2009), and significant number of new lines bearing these traits were developed at CIAT headquarters in Cali, at CIAT Africa regional breeding program hosted by the University of Nairobi, Kenya and the national bean programs. Singh (2001) stated that development of high yielding cultivars adapted to low soil fertility and low input sustainable systems is essential to maximize yield of common bean to enhance food security, reduce production costs and generate income. PABRA considers development of low soil fertility-tolerant bean varieties as an option to increase bean yield at no additional cost (Kankwatsa et al., 2008). Thus, bean improvement for low soil fertility adaptation has become an important component of ISFM strategy for optimum bean production in low input systems of smallholder African farmers.

Screening common bean cultivars for low soil fertility tolerance is done under field conditions following a harmonized protocol (CIAT, 1994). The methodology consists of screening the same sets of bean genotypes at several locations under a single stress and combining results across sites with different stresses (Wortmann et al., 1995). Screening is done at two stress levels: moderate stress and no stress. The moderate stress corresponds to the stress level at which a well-adapted control variety under stress performs at 40 to 50% of its normal unstressed performance. The criterion for selection is essentially bean yield, but farmers' preferences and market preference were also considered. Five most popular market classes in ECSA are red kidney, red mottled, small red, white navy and pintos (Kimani et al., 2001). Such intentional market class choice was intended to link production to well-established markets and favor adoption by farmers. The potential genotypes identified are evaluated with farmers following the participatory varietal selection (PVS), a strategy developed within PABRA for heterogeneous environments where farmers have a range of preferences (Sperling et al., 1993). The selected varieties are then promoted through other participatory methods in on-farm trials. The sites selected in different countries represent a range of soils characteristics and agro-ecologies. Soil chemical characteristics at the experimental sites in different countries are presented in Table 4.

Sites and stress	Country	Altitude (m)	pH H₂O	Organic C (%)	Bray-I P (mg kg⁻¹)	Exch. Ca (cmol(+) kg⁻¹)	Exch. Mg (cmol(+) kg⁻¹)	Exch. K (cmol(+) kg⁻¹)	Exch. Al (cmol(+) kg⁻¹)	Al saturation (%)
Nyamuny-unye (Al)	DRC	1730	4.7	2.3	0.5	2.6	1.1	0.07	2.6	41
Gikongoro (Al)	Rwanda	1900	4.8	0.5	2.2	1.4	0.3	0.01	2.6	60
Antsirabe (P &Al)	Madaga-scar		4.6	3.7	1.2	0.45	0.65	0.28	2.43	59
Kakamega (low P)	Kenya	1550	4.9	2.6	7.3	1.6		2.3	na	
Mulungu 2 (low N)	Uganda	1690	5.8	2.4	9.6	5.5	3.1	0.56	na	
Selian (low N)	Tanzania		6.5	1.8	18.2	9.2	1.6	6.2	na	
Kawanda (low N)	Uganda	1190	5.5	2.1	20.0	2		0.5	na	
Kawanda (low K)	Uganda	1190	5.3	3.3	10.0	2		0.2	na	
Ikulwe (Mn)	Uganda	1200	5.2	2.8	3.0	4		0.8	na	

Table 4. Characteristics of surface soils (0-20 cm) at test sites in several African countries

To date, a total of 1,400 beans lines have been evaluated through BILA (Bean Improvement for Low fertility soils in Africa) for their relative tolerance to the stresses under consideration, particularly low N, low P and soil acidity, which is associated Al and/or Mn toxicity. Considerable genetic variability in germplasm was detected and several varieties identified with specific single or multiple edaphic stress tolerance. The initial screening did not consider market factors, but allowed selection of tolerant cultivars from national and regional bean programs, already well adapted to local environments. The following cultivars were identified as tolerant lines to different stresses from the evaluation at different BILFA sites (Lunze, 2002).

- Low N: RWR 382, RAO 55, ACC 433, UBR(92)25 and BAT 85
- Low P: Carioca, BAT 25, RAO 55, XAN 76 and MMS 224, ACC 433 and Ikinimba
- Low K: EMP 84, ICA Pijao, RAO 52 and BAT 1220
- Al toxicity : ACC 7/4, Ubusosera
- Mn : MCM 5001, XAN 76 and Urugezi

These cultivars were integrated into national bean breeding programs, and several tolerant varieties have been adopted by farmers and are among the released varieties (Ikinimba, XAN 76, Ubusosera, ACC 7/4, RAO 55, UBR(92)25, MwaMafutala, Ntekerabasilimu) in some countries including Madagascar, DR Congo, Malawi and Uganda, MLB 4089A , RWR 1092 in Kenya

3.1 Low soil fertility adapted bean lines or varieties of various market classes

The most popular bean types grouped in major market classes were screened and lines tolerant to edaphic stresses were selected to allow potential users to choose for their own market. Bean lines or varieties belonging to various market classes and types, grouped by their tolerance to low N, low P and low pH conditions are presented in table 5. Bean genotypes were evaluated for low N adaptation at two sites, Mulungu in DR Congo and Selian in Tanzania. The genotypes varied significantly in their grain yield under N deficient conditions and in their response to applied N. Without applied N, the yield varied from 695 to 1,789 kg ha[-1] while with added N at the rate of 30 kg ha[-1], the grain yield varied from 1,258 to 3,139 kg ha[-1] at Mulungu. Without N, most lines gave significantly higher yield than the local sensitive check Kirundo, and previously selected tolerant variety, MwaMafutala.

Low N			Low P			Low pH		
Line code	Seed size (100 seed weight, g)	Seed color	Line code	Seed size (100 seed weight, g)	Seed color	Line code	Seed size (100 seed weight, g)	Seed color
A 286	17	Carioca	AFR 619	34	Red	37/66/6	23	Tan
AFR 675	24	Navy	AFR 675	24	Navy	A 286	17	Carioca
AFR 699	40	Red	AFR 708	44	Calima	A 344	27	Cream
AFR 714	23	Navy	AFR 714	23	Navy	AFR 708	44	Calima
AND 871	35	Calima	AND 871	35	Calima	AFR 714	23	Navy
CAL 143	50	Calima	ARA 4	21	Cream	ARA 4	21	Cream
CAL 150	50	Calima	CIM 9314-36	41	Calima	BRB 119	31	Calima
CIM 9314-33	42	Red	CIM 9314-37	34	Calima	DB 201/77/1	19	Navy
CIM 9314-36	41	Calima	CIM 9331-1	25	Red	CIM 9314-3	37	Calima
CIM 9315-1	24	Pink	CIM 9331-2	29	Pink	CIM 9331-1	31	Red
CIM 9315-3	27	Calima	CIM 9331-3	23	Red	CIM 9415	38	Calima
CIM 9318-4	27	Calima	DB 196	20	Navy	CNF 5520	44	Calima
CIM 9331-3	23	Red	DOR 663	17	Black	DFA 53	28	White
DB 196	20	Navy	FEB 192	19	Cream	FEB 197	22	Black
DOR 715	18	Red	FEB 196	20	Carioca	G 12489	44	Calima
FEB 192	19	Cream	G 2858	21	Tan	G 2910	21	Calima
FEB 196	20	Carioca	G 5889	15	Cream	G 3480	15	Black
G 5889	15	Cream	LSA 32	32	Carioca	G 5889	15	Cream
LSA 32	32	Carioca	MORE 92018	49	Tan	HM 21-7	45	Red
MORE 92018	40	Tan	PAN 150	24	Carioca	LRK 34	45	Pink
PAN 150	24	Carioca	RWR 1873	35	Calima	LSA 144	22	Red
PRELON	20	Navy	RWR 2075	44	Red	PAN 150	24	Carioca
RAB 482	17	Red	RWR 2091	37	Red	RAB 482	17	Red
REN 22	21	Navy	SDDT 49	20	Carioca	RWR 1742	22	Red
RWK 10	40	White speckled	SDDT 54-C5	31	Pink	RWR 1873	35	Calima
SDDT 55-C4	42	Calima	UBR(92)24/11	15	Navy	UBR(92)11	18	Carioca
UBR(92)25	18	Navy	VEF 88(40)L1PYT 6	24	Red	VEF88(40)L1 PYT6	24	Red
VEF88(4O)L1 PYT6	24	Red	XAN 76	18	Calima	XAN 76	18	Cream
			ZAA 5/2	33	Cream	ZAA 5/2	33	Calima

Table 5. Promising bean lines in different market classes and types, grouped according to their tolerance to stressful soil conditions- low N, low and pH (Lunze et al., 2002)

The performance of the best lines selected for their tolerance to low P was evaluated at Kakamega and Antsirabe in Madagascar. Several lines were outstanding compared with the local released bean variety GLP 585. Considerable yield advantage of up to 80% was observed compared with the local check. The outstanding genotypes tolerant to low P are AFR 619, XAN 96, ARA 4, AFR 708 and RWR 1873.The screening of genotypes for low pH was done at Mulungu, DR Congo; Gikongoro, Rwanda; and Antsirabe, Madagascar (in acidic soils that demonstrate positive response to liming) revealed that the genotypes selected as Al resistant at three low pH sites were generally the same and performed consistently across sites. However, the slight variability noticed across sites could be attributed to the difference in adaptation to local environments (Wortmann et al., 1995). At Mulungu, with no lime, most test lines gave significantly higher yield than the sensitive check variety Kirundo, while only two lines VTTT 923-6-1, HM 21-7 and AFR 593-1 outperformed the tolerant check MwaSole (Table 6). The yield advantage of Al-resistant genotypes could be as high as 300% compared with the check.

Line	Bean grain yield		Loss due toxicity (%)	% Yield advantage over check
	Stress (kg ha-1)	Non-stress (kg ha-1)		
VTTT 923-6-1	1494	1588	5.9	310.6
HM 21-7	1317	1455	9.5	273.8
AFR 593-1	1040	1801	42.2	216.2
MwaSole	999	1264	20.9	207.7
ARA-8-5-1	957	1394	31.3	199.2
AND 932-A-1	932	1206	22.7	193.8
BZ 12984-C-1	866	874	0.9	180.0
Mwamafutala	625	1510	58.6	129.9
Kirundo (control)	481	988	51.3	100

Adapted from Lunze et al., 2007

Table 6. Bean lines and varieties resistant to Al toxicity in soil with a yield advantage over the check

At Antsirabe, Madagascar, several lines outperformed the local check Soafianarana and improved check Goiano Precoce. Most genotypes selected as resistant to Al toxicity at Antsirabe showed the same at Mulungu. The following lines gave consistently better yield under acid Al-toxic soil at the two low pH sites: AND 1056-1, AND 932-A-1, ARA 8-1B, BZ 12984-C-1 and VTTT 920-26. At Gikongoro, Rwanda many lines gave higher yield than the Al-resistant checks 7/4, Acc and RAB 478. However AND 93-A-1, BZ 12894-C-1 and AFR 593-1 were also identified as resistant to Al at the other two low pH sites, Mulungu and Antsirabe (Lunze et al., 2007).

The overall evaluation across sites and stresses allowed selection of bean lines that had consistently high yield under different stresses and across sites, with identified tolerance to one, two or even all three soil constraints considered. BZ 12894-C-1, AND 932-A-1, DRK 137-1,

Nm 12806-2A are outstanding across sites and in response to all stresses - high aluminum toxicity, low N and low P availability. Several other lines have manifested tolerance to two stresses: ARA 8-B-1, AFR 709-1, AFR 703-1 and AND 1055-1 are tolerant to low P and low pH: RWK 10, ARA 8-5-1, and T 842-6F11-6A-1 tolerant to low N and low pH. These appear to have multiple tolerances to edaphic stresses and good adaptation at all environments. The results show existence of tolerance to low soil fertility in market class bean types. The promising varieties identified provide opportunities for higher bean productivity on acid soils and those with limited N and P supply (Lunze et al., 2007).

3.2 Dissemination and impact

Efforts have been made to promote all the promising and potential cultivars in all regions with problem soils as their yield advantage in similar environments has been confirmed. The benefit of low soil fertility adapted bean varieties has been demonstrated in Eastern DR Congo (Njingulula, 2003; Mastaki, 2006), Malawi, Kenya and Uganda (Kankwatsa et al., 2008). In Eastern DR Congo, with very low fertility soils, Ubusosera and RWR 382 have become the main varieties grown by farmers, replacing previously grown local or improved varieties. Kimani (2005) reported that different countries have selected and widely promoted bean genotypes that are adapted to their own environment (table 7). More than 10 years later, most farmers still kept BILFA varieties because of their adaptation to marginal soil conditions (Musungayi et al., 2008). Njingulula (2003) conducted a study to assess the impact of two low soil fertility adapted varieties in Eastern Congo and has indicated that farmers who have adopted those two varieties (Ubusosera named MwaSole, resistant to Al and RWR 382, locally named MwaMafutala adapted to low N) have noticeably improved their socio-economic conditions. In this study, (1) 32% of respondents noted that bean quantity for consumption and sale was increased, (2) 28.4% of respondents said that bean become more permanent throughout the year, (3) 30% of respondents mentioned that nutrition was improved as they could eat twice a day compared to only once previously, (4) and other farmers responded they purchased livestock and various household items such as bicycles, radio, etc and paid medical care and school fees.

Lines	Countries where adopted
RWR 1873	Uganda, Kenya, DRC
RWR 1946	Uganda, Kenya, DRC
RWR 2075	Uganda, Kenya, DRC
UBR(92)25	DRC, Malawi, Tanzania, Uganda
RWR 719	Ethiopia, Rwanda and Kenya
ACC 7/4, Ubusosera	Rwanda, DRC
DFA 54, DFA 53, DOR 633, DOR 715, AFR 708	Uganda, Kenya

Source: Kimani, 2005

Table 7. Bean lines/varieties adapted to low soil fertility in different countries

4. Use of organic resources

Organic matter based soil nutrient management is a traditional practice that continues on smallholder farms. Among the organic resources used are animal manure, compost, crop residues for soil incorporation, natural fallowing, improved fallows, relay or intercropping of legumes, and biomass transfer (Place *et al.*, 2003). Organic manure, compost and farmyard manure are the most common inputs used to improve soil fertility by small scale farmers (Musungayi et al. 1990; Kankwatsa et al., 2008). The need for both organic and mineral inputs to sustain soil health and crop production through ISFM has been highlighted due to their positive interactions (Vanlauwe et al. 2010).

4.1 Compost and farmyard manure

FYM use is the only possible practice for many resource poor farmers to improve crop and soil productivity. FYM and compost application are the most common practices used to improve soil fertility on smallholder farms although the quantities available are inadequate to meet crop nutrient demand. Organic manure available on farm estimates on dry matter basis varied from to 3.1 to 18.9 t ha^{-1} in Central Kenya (Opala, 2011) and 0.6 t ha^{-1} in Eastern DR Congo, sufficient to cover only 25% of their land under crops every season (Musungayi et al, 1990).Thus, scarcity of organic materials on farm limits their use at recommended rates. Because of scarcity of these resources, most farmers have developed localized application rather than broadcasting. This involves applying FYM or compost in the furrow or planting hole and covering it with soil before placing bean seeds, to avoid direct contact of manure with seeds. Considerable yield increase is achieved using compost alone, and the increase is dependent on initial fertility level in soil (Table 8).

Soil dominated by	Bean yield (kg ha^{-1})		
	Without compost	With compost	Yield increase
Pennisetum polystachia	41.7	697	655
Conyza sumatrensis	418	828	408
Bidens pilosa	876	933	60
Digitaria sestida	1251	1335	84
Mean	857	1101	244
CV (%)	24.2	21.2	

Adapted from Ngongo, 2001

Table 8. Effect of compost application to fields dominated with different weeds on bean yield

The potential of organic resources, such as FYM and compost to improve bean productivity, either as source of nutrients or by improving mineral fertilizer efficiency is well established in the region and various fertilization recommendations have been formulated. However, any substantial effect of organic amendments requires very large quantities that are not readily accessible to the majority of bean growers who are smallholder farmers (Thung and Rao, 1999).

4.2 Biomass transfer

Search for alternative sources of organics and more economical resources have always been a concern of the national bean programs. Many shrubs and trees, such as *Lantana camara*, *Tephrosia vogelii* and *Tithonia divesrifolia* which are common on smallholder's farms in ECSA have been studied in different countries.. Among all shrubs, *Tithonia* is the most common shrub with substantial biomass production in most countries of the region. It is grown for land stabilization along the road and for erosion control on cropland, and as plot or compound boundary, and as an ornamental plant. *Tithonia* use and popularization as soil improving resource has gained interest with the increasing need of intensifying bean-based production system with the recognition of the potential of this species to accumulate nutrients. Jama et al., (2000) estimated *Tithonia* biomass available after nine months for transfer to fields at 2 t ha^{-1} kg of dry matter in Western Kenya. *Tithonia* has very high shoot vigour with relatively high nutrient concentrations in its biomass. Concentration of N ranges from 3.0 to 4.1%, P from 0.24 to 0.56% and K from 2.7 to 4.0% (Jama et al. 2000). Besides this advantage in nutrient concentration, the biomass of *Tithonia* is also known to be rapidly decomposing (Buresh and Tian, 1998) due to proper balance in lignins, polyphenols and N (Palm et al., 1997). In fact, it has the ability to extract relatively high amounts of nutrients from the soil, a property which may make the practice not sustainable in the long run due to nutrient depletion. On soils with high natural nutrient stock however, such as soils of volcanic origin of the Great Lakes region, *Tithonia* use has shown considerable potential in improving crop production. In most cases, as in the productions systems of a growing number of small scale farmers without livestock, *Tithonia* remains the only and essential source of organic material available on the farms (Rabary, 2001).

Tithonia biomass transfer is a practice extensively studied in Rwanda, Kenya, Tanzania and DR Congo for its integration into bean-based production systems. In Rwanda and DR Congo, considerable bean yield increase is obtained, 227% in Rwanda (Ruganzu and Nabahungu, 2002) and 68% in DR Congo (Ngongo, 2002).

4.2.1 Rate application of *Tithonia* biomass

Determining the appropriate rate of biomass transfer is essential for integration of *Tithonia* biomass use in farmers' cropping systems, either to beans grown as monocrop or as a mixed crop. The effect of the rate of application on bean yield of monocrop was evaluated in DR Congo and Rwanda (Ngongo, 2002; Ruganzu and Nabahungu, 2002; Nabahungu and Ruganzu, 2001). In Rwanda, the study was conducted on soils with pH varying from 3.8 to 4.9. The rate of application of *Tithonia* biomass varied from 0 to 12 t ha^{-1} on dry matter basis (Table 9). The optimum rate of application was determined to be 6 t ha^{-1} in Rwanda, and 4 to 6 t ha^{-1} depending on the initial soil fertility in DR Congo (Table 10). This slight difference is due to the difference in soil fertility at experimental sites, Rubona in Rwanda and Mulungu in DR Congo. In light of these results, the appropriate rate of application of *Tithonia* biomass on climbing bean may vary between 4 and 6 t ha^{-1}, which is slightly lower than the recommendation on maize of 7.5 t ha^{-1} in Tanzania (Ikerra et al. 2007). In the later study, they found that the optimum rate of application on maize could be set at 5 t ha^{-1} in the long rainy season while in short rainy season, the recommendation could be higher at 7.5 t ha^{-1}.

Tithonia rate of application (t ha-1)	Bean yield (kg ha-1)
0	917
2	1314
4	1714
6	2086
8	1486
10	942.9
12	1271

Table 9. Climbing bean yield response to applied *Tithonia* biomass in Rwanda (Ruganzu and Nabahungu, 2002)

Rate of application *Tithonia* (t ha-1)	Bean yield (kg ha-1)		
	Soil fertility level		
	Low	Medium	High
0	230 d	949 c	1777 b
2	533 c	1358 b	1841 ab
4	838 b	1788 a	1901 b
6	1565 a	1788 a	2034 a
CV %	39.2	39.2	39.2
Mean yield (kg ha-1)	791.5	1470.7	1887.7

Source: Ngongo, 2002

Table 10. Climbing bean yield response to applied *Tithonia* biomass in DRCongo

The rate of application in bean- maize mixed cropping, a more common cropping system in farmers' field, was evaluated by Ngongo (2002) in DR Congo (Table 11). In this system, the best yield of both crop either sole crop or intercropped was obtained at the rate of 8 t.ha-1 of biomass.

Tithonia Biomass (t/ha)	Bean yield (kg. ha-1)		Maize yield (kg. ha-1)	
	Bean sole crop	Bean- Maize intercrop	Maize sole crop	Maize- bean intercrop
0 t ha-1	703.6 c	748.0 c	1250 c	2250 b
4 t ha-1	887.0 c	900.1 c	1930 bc	2450 b
8 t ha-1	1410.0 a	1139.0 b	2220 b	3470 a
C.V. (%)	16.34	16.34	21.72	21.72
Mean (kg ha-1)	1000.2	929.03	1800	2720

Figures followed by the same letter are not significantly different at 0.05 probability. Source: Ngongo, 2002

Table 11. Bean and maize yield response to *Tithonia* biomass in sole cropping and intercropping systems in DR Congo.

4.2.2 Method of application of *Tithonia* biomass

The technique of *Tithonia* use was fine-tuned by Ngongo (2002), who studied the appropriate time for biomass application. The time of application prior to bean sowing varied from 0 to 7 days. The best bean yield was obtained when biomass application was made at least 4 days prior to sowing. It was concluded that there was no reason to apply *Tithonia* biomass 7 days before planting (Table 12). The mode of application involved applying the fresh biomass on soil surface and incorporating it into soil 24 hours later, leaving it to decompose for 4 to 7 days before sowing bean.

Biomass application	Bean yield (kg ha-1)	
6 t ha-1	2000A	2000B
Control	563 fg	375 g
At sowing	1,253 bc	985 cde
At 7 days before sowing	1,327 bc	1,174 bcd
At 4 days before sowing	1,703 a	1,273 bc
At 2 days before sowing	1,423 ab	1,105 bcde
Surface applied and worked in at sowing	776 ef	854 def
C.V. (%)	19.4	
Mean yield (kg ha-1)	1,174	

Figures followed by the same letters are not significantly different at 0.05 %

Table 12. Effect of different application of *Tithonia diversifolia* on bean yield

4.3 Green manure and cover crops

Research and extension efforts were made to identify legumes that present farmers with as many options as possible so that they can choose cropping systems that are best suited to their needs. Many plant species have been selected and their potential benefit demonstrated using both on-station and on-farm experiments. More detailed field investigations have been done in Uganda and DRC. The green manure and cover crops are primarily used to enhance land productivity by improving N economy in the production systems. Several legume species have been evaluated and found to be well adapted to the region and produce substantial amounts of biomass. Among the potential green manure species identified were *Crotalaria ochroleuca*, *Mucuna pruriens* (L.) DC, *Lablab purpureus* (L), and *Tephrosia vogelii* Hook, which were tested extensively in on-farm experiments for their potential to improve productivity and to obtain feedback based on farmers' perception (Wortmann et al., 1994; Esilaba et al., 2001). High N concentration in *Crotalaria* biomass grown under N-limiting conditions indicated that large quantities of N were biologically fixed. Mean maize grain yields following *Crotalaria* sole crop were 180% and 240% of maize grain yields following maize in two on-station trials and nine on-farm trials, respectively. In the study done by Fishler and Wortmann (2001), grain yields of maize and bean following one season of *Crotalaria* fallow were 41% and 43%, respectively, and the yields were more than following a two-season weedy fallow. Grain yields of maize following a one-season fallow with *Mucuna* and *Lablab* were 60% and 50% higher, respectively, compared to maize following maize. Maize and bean yield were more, although effects were small, during the second and third subsequent seasons, indicating additional benefits from the residual effects of the green manures.

In spite of the demonstrated advantages of green manure and cover crops, the adoption by farmers has always been very poor, hampered by several factors such as availability of seeds, difficulties in planting and establishing the crop, and labor intense practice such as mowing and incorporating the cover crop (Esilaba et al. 2005). Alternative methods of green manure integration on farm were studied, particularly simultaneous cultivation of green manure with main crop, bean or another crop. In fact, in densely populated regions, leaving land under green manure for the whole season is not always acceptable to farmers. Therefore, intercropping with main crop has been developed. In Eastern Uganda, *Mucuna* and *Lablab* were successfully produced by intersowing into maize at three weeks after sowing maize, although the yields of the associated maize crop were reduced by 24% to 28%, and farmers estimated the labor requirements for *Mucuna* and *Lablab* to be less than for *Crotalaria* (Fishler and Wortmann, 2001). In Uganda, the yields of the green manure species were reduced by 40–70% when intercropped with a food crop as compared to sole crop production and that yields of food crops were reduced by 61–87% when intercropped with *Crotalaria ochroleuca* G. Don. In contrast, maize grain yield response in the first season following sole-crop of green manure production ranged from 0 to 240% (Fishler and Wortmann, 2001). Production of the green manure by intercropping *Crotalaria* with either maize or beans was found to be feasible with little reduction in food crop yield with a mean Land Equivalent Ratio of 1.3.

Additional benefits of cover crops reported in Uganda are weed suppression, soil erosion control and control of mole rats (Fishler and Wortmann, 2001). Results from these trials also indicated that *Mucuna pruriens* (L.) and *Lablab purpureus* (L) were best for weed suppression and to control soil erosion. Also the requirements for tillage and weeding could be lower following green manure and cover crops. In addition, maize can often be planted directly in the holes left from uprooting *Mucuna* and *Lablab*, thus reducing labor requirements during the following season. *Tephrosia vogelii* Hook. f. was effective in controlling mole rats and there was significant adoption by neighboring farmers.

4.4 Climbing bean rotation effects

Climbing bean is by far the bean type with high biomass production and probably with high BNF, and therefore considerable green manure effect is expected. Its integration in the production system can improve productivity and sustainability of bean-based cropping systems. In fact, climbing bean develops extensive nodulation, up to three times more than bush bean (Van Schoonhoven and Pastor-Corrales, 1994). This is an indication of higher capacity for N fixation. It has been suggested that the contribution from N fixing legume in rotation could be responsible for most of the beneficial rotation effect observed in the subsequent maize crop (Sanginga, 2003; Bado, 2002; Balldock et al, 1981). Evidence of net positive soil N balance by climbing bean has been reported (Kumarasinghe et al. 1992). In this study, they reported that at the late pod-filling stage the climbing bean had accumulated 119 kg N ha[-1], 84% being derived from fixation, 16% from soil, and only 0.2% from the ^{15}N fertilizer. In a long-term experiment, Wortmann (2001) reported improved sorghum yield in the rotation with climbing bean and estimated that N derived from the atmosphere was 40% to 57% of plant N, depending on the calculation used. It is clear that evidence of benefit of climbing bean cultivation exists, either as rotational effects or improved N nutrition.

On-station and on-farm farmer participatory trials were conducted by Lunze and Ngongo (2011) to assess the beneficial effects of the climbing bean on the subsequent maize crop in rotation, compared to bush bean and continuous maize cropping systems. The estimate of N contribution from the climbing bean to the system estimated as N fertilizer replacement values (Bado, 2002), and these values varied from 15 kg N ha^{-1} to 42 kg N ha^{-1} in the first season of rotation. Note that this method is believed to overestimate the amount of N supplied by the legumes in the systems (Bullock, 1992). Without applied N fertilizer, the average maize grain yield increase over three cropping seasons in response to the preceding climbing bean effect were 489 kg ha^{-1} and 812 kg ha^{-1} compared to bush bean and maize as preceding crops, respectively, which is 17.5% and 33.8% increase. However better yield advantage of climbing bean over continuous maize was obtained in the long rains cropping season (Season A), 43.2% compared to 24.2% in short rains season (Season B) (Table 13).

Preceding crop	N Rate (kg/ha)	Maize grain Yield (kg/ha)		
		Season 2002A	Season 2002 B	Season 2003 A
Bush bean	0	2911	2735	2625
	33	4542	3738	3656
	66	5297	4206	4003
Climbing bean	0	3821	2847	3072
	33	4869	4115	4000
	66	5019	4375	4278
Maize	0	2724	2477	2102
	33	4141	3556	2872
	66	4836	3931	3897
Mean Yield		4540	3558	3336
LSD (0.05)		334.9	707.1	379.2
CV (%)		6.5	9.8	10.0

Table 13. Effects of preceding crop and fertilizer N on maize grain yield over 3 seasons at Mulungu, DR Congo (Lunze and Ngongo, 2011)

These studies in Uganda and DR Congo provide evidence for beneficial effects of climbing bean to improve N nutrition of the following maize crop. It is presumed that climbing bean promotion is an appropriate strategy for higher productivity and sustainability on smallholder farms, as substantial gain of N will be achieved by proper integration of climbing bean in the bean-based production systems in ECSA. Additional effects of growing climbing bean are expected, such as reducing erosion because it provides more extensive soil cover through its canopy compared to bush bean. These beneficial effects are considerable on relatively medium fertility soils where climbing bean effects were highest, because of the large quantity of biomass produced.

5. Biological nitrogen fixation in bean

The enhancement of the capacity of beans for biological fixation through symbiosis with *Rhizobium phaseoli* was recognized by biological nitrogen fixation(BNF) working group within PABRA as an important option to improve the productivity of bean crop on farms (Nyabienda, 1988). This was particularly true for regions where farmers have limited access to N fertilizers. However, improvement of bean BNF requires a multiplidisciplinary approach that will require the plant breeder to increase the host capacity to fix N_2 as bean is considered as a poor N_2 fixing legume (Giller, 2001), and selection of effective *Rhizobium* strains that can compete for nodulation with native populations of bacteria present in most soils.

5.1 N_2 fixation variability in bean germplasm

BNF is among the strategies extensively investigated as one of the avenues to improve bean productivity. Matheson (1997) measured N_2 fixation at Kawanda and Namulonge, Uganda using two common methods: ^{15}N abundance method and N-difference method using non-nodulating beans lines INIAP 404 and EXRICO. N2 fixed was lower at Namulonge than at Kawanda. It ranged from 10.6 to 35.1 kg Nha-1 and from 0.7 to 20.4 kg N/ha, respectively, at Kawanda and Namulonge. This study found that the genotypes identified as tolerant to low soil N had greater N fixing capacity. Table 14 summarizes the genotypic variation in N_2 fixation of common bean at Kawanda, Rwanda.

The same author calculated the net N balance and found a negative net N balance for all varieties, ranging from - 23.6 kg ha-1 to - 5.3 kg ha-1, indicating that more N had been exported in grain at harvest than had been fixed and retained in soil. The yield was negatively related to net N balance: the higher the yield, the more negative the net N balances. However, unlike bush bean, climbing beans exhibit greater N_2 fixation capacities (Wortmann, 2001; Lunze and Ngongo, 2011). Lunze and Ngongo (2011) estimated N_2 fixation by climbing bean r from 16 to 42 kg ha-1 per season. N fixation is further enhanced by several cultural practices and agronomic management, i.e. inoculation, P fertilization and liming, that are discussed below.

5.2 Inoculation

BNF is extensively investigated as one of the avenues to improve bean productivity. Rhizobial inoculation is a common practice to promote BNF in bean, given that the *Rhizobium* strain is well adapted to local environments. Strain CIAT 899 is currently used for bean seed inoculation in all regions of ECSA. Inoculation with this strain in Burundi showed a yield advantage up to 59% in Gitega (Ruraduma, 2002). Evaluation by the same author of local *Rhizobium* strains so far has failed to identify a more efficient strain.

5.3 Liming and P fertilization effects

The efficiency of BNF is further improved by liming and P fertilization. This is not always affordable for smallholder farmers. Considering these factors, bean seed pelleting with lime and phosphate was evaluated in Burundi and Malawi. On-farm trials at Ikulwe, Iganga District of Uganda, the %N_2 fixation was estimated for a well-adapted high yielding variety

Variety	Biomass kg/ha R9	Yield (kg/ha)	Total N (kg/ha)	% N fixed	N2 Fixed (kg/ha) 15N method	N2 fixed (kg/ha) Difference Method
BAT 1297	218,4	1033	63,9	43	26,8	28,6
BAT 308	217,9	969	51,5	34,4	17,6	16,2
INIAP 404	136	601	31,1	-12,5	-3,7	-4,2
CNF 5513	220,6	1413	61,5	35,6	22,1	26,2
H2 MULATHINO	230	1061	61,5	39,7	23,1	26,5
IBHBN 69	241,9	1161	57	45,4	25,2	21,7
IBR (92B) 43	184	890	52,6	31,2	16,1	17,3
MCM 1015	201,7	1090	52,9	41,4	22	17,6
MCM 1016	223,8	1015	60	35,1	22,2	24,7
MLB645689A	202,4	1279	55,8	42,1	21,7	20,5
MMS 243	230,5	1000	69,4	36,2	23,3	34,1
MMS 250	207	1175	56,3	38,7	21,9	21
MMS 253	176,4	1080	50,2	51	25,9	14,9
MORE 90040	230	1318	62,1	55,1	33,9	26,8
MUS 97	217,7	1465	58,8	44,4	25,9	23,5
EXRICO	199,1	545	39,5	12,5	5,1	4,2
RWK 5	186,9	1326	51,5	39,7	19,9	16,2
RWR 109	209,8	1025	59,7	44,8	27,4	24,2
RWR 382	267,9	956	68,7	50,9	35,1	33,4
UBR (92) 09	198,3	1304	57,8	58,4	33,7	22,5
UBR (92) 10	233,2	941	66,2	33,1	21,3	30,9
UBR (92) 11	197,5	1155	53,9	49,5	26,1	18,6
UBR (92) 12	209,7	1209	58,7	29,4	10,6	23,4
UBR (92) 17	180,4	888	52,3	49,4	25,4	17
UBR (92) 20	258,2	889	69,9	42	28,8	34,6
UBR (92) 25	202	1132	61,9	47,6	29,4	26,6
UBR (92) 38	208,8	788	65,3	38,9	52,1	30
XAN 76	185,3	1416	54	50,6	26,1	18,7
CAL 96	201,4	962	59,9	39,2	23,4	24,6
MCM 5001	171,7	1004	52,8	49,6	26,1	17,5
Mean	208,3	1070	57,2	39,9	22,9	21,9
s.e.d.	27,37	129	8,77	10,93	6,129	8,77

INIAP 404 and EXRICO: non-nodulating reference lines; CAL and MCM 5001: check varieties

Table 14. Nitrogen fixation, biomass and yield of 30 varieties grown under low N conditions at Kawanda (1996)

MCM 5001, considering the effects of inoculation and P application on N_2 fixation. Neither inoculation nor 100 kg ha^{-1} of TSP (triple super phosphate) application increased bean yield, but 100 kg ha^{-1} TSP with inoculation yielded significantly higher yields than these treatments individually. Inoculation had therefore greater effects on bean yield at higher P levels. The estimate of % N_2 fixation without addition of P fertilizer and inoculation ranged from 5% to 32%, while the addition of 50 kg ha-1of TSP increased N_2 fixation by 50%. However, they did not observe any significant differences (P<0.05) in %N_2 fixation between treatments and farms.

In Malawi, Chilimba and Kapapa (2002) evaluated bean seed pelleting with dolomitic lime. On-farm trials conducted in Dedza where soils are strongly acidic showed significant responses of four bean varieties to inoculation and seed pelleting with lime on BNF and grain yield. Seed pelleting and inoculation showed beneficial effects in increasing plant N content, nodule numbers and even grain yield. In acidic soils, both inoculation plus seed pelleting significantly improved BNF and grain yield whereas in a normal non-acidic soil, inoculation only or pelleting only significantly improved BNF and grain yield. The on-farm evaluation was conducted in Bembeke Extension Planning Area where soil pH is 4.5 and the soil is fine, kaolinitic, thermic, Kandiudafic Eutrudox while at Chitedze research station the soil pH is 5.9 and the soil is classified as fine, kaolinitic, thermic, Udic Kandhaplustalfs.The results on the evaluation on soils with varying soil pH are presented in Table 15.

Treatments	Marieta			Ester			Chitedze		
	% N	Nodule No.	Grain yield (kg ha-1)	% N	Nodule No.	Grain yield (kg ha-1)	% N	Nodule	Grain yield (kg ha-1)
Control	3.12	15.5	1056	2.05	21.6	389	2.30	10,8	1438
Inoculation	3.64	82.7	1599	3.48	85.5	1119	3.56	78,8	1937
Pelleting	2.18	18.3	1043	2.68	28.0	911	2.01	12,3	2410
Inoc & Pell	4.52	115.1	1981	4.37	117.6	1612	4.40	94,4	2184
Mean	3.35	57.9	1419,75	3.12	63.2	1007,7	3.07	49.0	1992.25
SE	0.483	4.839	231.2	0.428	5.07	230.4	0.4677	2,66	156.1
CV%	21.3	28.97	28.3	12.8	27.83	39.7	58.9	18,8	26.13

Adapted from Chilimba and Kapapa, 2002

Table 15. Effect of inoculation and seed pelleting on % N, nodule number and the yield of beans at three locations (Marieta, Ester and and Chitedze) in Malawi

The results of the above studies confirm that both lime and P fertilizer application rate can be considerably reduced by coating bean seeds to achieve relatively higher yield. This technology is considered a good ISFM strategy.

6. Mineral fertilizers

Most smallholder farmers in the ECSA regions are well aware of the value of mineral fertilizers, but the rate of application remains low, below the recommended rates, except for commercial farmers. The essential reason for this is the high cost of mineral fertilizers, and low profitability. In most regions however, bean comes after cereals, which are commonly fertilized so that bean benefits from residual fertilizer effects. Nonetheless, fertilizer recommendations have been developed in several countries for bean, particularly in Kenya, Tanzania, Burundi and South Africa.

On Ferrarsols (Oxisols) of Burundi, mineral fertilizer applied alone on bean was found to be non-economical, and under certain conditions bean yield was depressed. The recommended rate (kg ha^{-1}) for Burundi was 15 – 30 for N, 50 – 60 for P_2O_5 and 30 for K_2O, with 2 to 5 t organic manure (Ruraduma, 2002). The ISFM strategy suggests options that combine organic resources and mineral fertilizers to achieve higher yield, economically. Fertilizer application to bean crop is developed for those grown as a sole crop or intercropped with maize. In addition to determining the rate of application, the proper time of application is important to maximize N_2 fixation, as early application of N can inhibit nodulation of bean. In Uganda, Wortmann, (1998) recommended applying a small amount of N at the planting time and the major part of fertilizer N at the second weeding time, to favor bean productivity and N2 fixation. That is an application of 5 kg N ha^{-1} and 20 kg P ha^{-1} at sowing time and applying 35 kg N ha^{-1} at second weeding time for both maize and bean that are intercropped. In on-farm trials with the objective of developing appropriate fertilization practice on snap bean, Ugen et al. (2009) evaluated rates and timing of the mineral fertilizers CAN, NPK 17-17-17 and DAP (diammonium phosphate), which are the common fertilizers in the marketplace. In this study, application of DAP at planting and topdressing with NPK-17:17:17 at 21 days after emergence appeared to be appropriate for bean production. More complete mineral fertilizer recommendations used in South Africa are presented in Tables 16, 17 and 18 (Liebenberg, 2002).

6.1 Nitrogen

According to Liebenberg (2002), inoculation of dry bean seed is regarded as ineffective in South Africa. Consequently, dry beans should be considered as incapable of satisfying all of their N requirements through N_2-fixation. The application of all the N fertiliser at planting time is recommended, particularly where non-decomposed material has been ploughed in before planting.

Yield potential (t ha^{-1})	1,5	2,0	2,5
N fertilisation (kg ha^{-1})	15,0	30,0	45,0

Source: Liebenberg, A.J., 2002.

Table 16. Guidelines for nitrogen application

6.2 Phosphorus

Commercial production in South Africa showed modest yield responses to P fertiliser application in dry beans and P is not normally a yield-restrictive factor. Under subsistence production, where small quantities of fertilizer are applied, P can be a yield-limiting factor. Where the P content of the soil is lower than 20 µg g^{-1} (Bray 1), it is recommended that TSP be broadcasted and ploughed into the soil to a depth of 15 to 20 cm before planting. P fertiliser must still be band-placed at the time of planting. In low pH soils, P can be utilised efficiently by band-placing 3.5 cm to the side of the row and 5 cm below the seed.

Soil analysis Bray 1 (mg kg⁻¹)	P application for potential (t ha⁻¹)		
	1,5	2,0	2,5
	P fertilization (kg/ha)		
13	16	22	28
20	12	16	20
27	10	13	16
34	9	12	15
> 55	5	5	5

Source: Liebenberg, A.J., 2002.

Table 17. Guidelines for phosphorus fertilisation

6.3 Potassium (K)

When dry beans are grown on soils with high clay content, K is not normally a limiting factor. Deficiencies are most likely to occur on sandy soils with an analysis of less than 50 mg kg⁻¹ K. The optimum leaf K content is 2% potassium.

Soil analysis NH₄OAc, pH 7 K (mg kg⁻¹)	K application for potential (t ha⁻¹)		
	1,5	2,0	2,5
	P fertilization (kg ha⁻¹)		
40	22	27	32
59	19	24	29
78	17	21	26
98	15	19	24
> 98	0	0	0

Source: Liebenberg, A.J., 2002. Dry bean production

Table 18. Guidelines for potassium fertilisation

6.4 Effect of cropping systems

Wortmann et al. (1996) studied the relationship between maize monoculture to fertilizer N and P and the intercrop response from 62 fertilizer response trials conducted in Kenya. Their results indicate that intercrop was more productive than the monoculture with no fertilizers applied, but overall responses of the systems to applied nutrients did not differ. Maize both in monoculture and intercropped responded more frequently to applied N than did the intercrop bean. Frequency of response to applied P was similar for both crops and both production systems.

6.5 Combination with organic manure

It is well established that combining mineral fertilizer with organic resources improves fertilizer use efficiency. In Western Kenya's Vihiga District, the recommendation is 50 N kg ha[-1] and 50 P kg ha[-1] with 5 t ha[-1] of farmyard manure (Rachier et al., 2001). The N and P rates are reduced by 50% in the following season. If money or credit is not available, in addition to 5 t ha[-1] FYM, only 25 kg ha[-1] N and P are recommended. The following season, the N rate is reduced by 25% and no P is applied. If FYM is available on farm, 5 t ha[-1] FYM together with 25 kg ha[-1] N and P are recommended. This rate is reduced by 50% in the second season. If the green manure is grown, then N, P and FYM rates are reduced by 50% in the first season, and in the second season only 25 kg P ha[-1] is needed. In the study conducted by Ngongo (2002) in DRC, the aim was to reduce the quantities of both mineral fertilizer and *Tithonia* biomass recommended. The author evaluated bean response to application of 4 t ha[-1] *Tithonia* biomass with varying rates of mineral fertilizer (Table 19).

Treatments	Bean grain yield (kg ha[-1])	
	1998B	1999 A
Control	551 d	1066 cd
4 t/ha *Tithonia*	1132 cd	1744 a
4 t/ha *Tithonia* + 20 kg NPKha[-1]	1341 bc	2201 ab
4 t/ha *Tithonia* + 40 kg NPKha[-1]	1313 bc	2296 a
4 t/ha *Tithonia* + 60 kg NPKha[-1]	1402 bc	2390 a
60 kg/ha NPKha[-1]	1405 bc	
C.V. (%)	23,41	23.41
Mean (kg. ha[-1])	1190,67	2004

Table 19. Effects of *Tithonia* biomass in combination with mineral fertilizers on bean yield in DR Congo.

The results confirmed that less fertilizer could be recommended when *Tithonia* is applied. The application of 4 t ha[-1] *Tithonia* with 20 kg ha[-1] of NPK fertilizer was as efficient as the recommended mineral fertilizer rate of 60 kg ha[-1]. Thus a 3-fold reduction of inorganic fertilizer is possible when *Tithonia is* applied.

7. Mijingu phosphate rock

Bean growing environments are well described in all ECSA countries and P deficiency is widely reported as the most important bean production constraint (Wortmann et al., 1998). The low P availability to plants is explained by the nature of the soils that are highly weathered with low total P and/or high P fixing capacity (Rao et al., 1999). Rock phosphate (RP) which is a resource relatively common throughout ECSA countries is recognized as a fertilizer with high potential to improve bean crop productivity having low cost, compared to conventional mineral P fertilizers. The deposits of the RP are reported in many countries and exploited to some extent in Tanzania (Mijingu), Malawi (Tundulu), Zambia (Isoka), South Africa, DR Congo (Kanzi) and Burundi (Matongo). Minjingu Phosphate Rock (MPR), a sedimentary biogenic deposit which contains about 13% total P and 3% neutral ammonium citrate soluble P is reported to be highly reactive (Jama and Van Straaten, 2006).

On acidic soils of Tonga in Rwanda, Nabahungu et al. (2002) studied the effects MPR associated with limestone and green manures (GM) (*Tithonia* and *Tephrosia* biomass) on P uptake and on maize yield. This study showed that MRP significantly contributed to P increase in the soil and resulted in increased P uptake and maize yield. Green manures in combination with MPR increased P uptake significantly. The results indicated that using a combination of limestone, MPR and GM is the best strategy in improving maize productivity on acid soils (Table 20). The effect of RP alone or GM alone is low, whereas the combination of MPR with organic resources improved its effect.

Treatment	Maize Yield (kg/ha)
Control	148
MPR	3267
Tithonia	3349
Tephrosia	1452
Tithonia + MPR	3993
Tephrosia + MPR	4554
Tithonia + Lime + MRP	5594
Tephrosia + Lime + MRP	5907
CV %	11.1

Table 20. Maize yield response to Mijingu rock phosphate and fertilizers (adapted from Nabahungu, 2007)

A combination of RP and GM produces a demonstrated yield advantage over control or each of these treatments alone. The recommendation for acidic soils of Rwanda is to combine organic resources with RP and lime (ground limestone or burnt lime) allow maize yields to increase up to approximately 6 t ha[-1]. The same recommendation is applied for bean. In Northern Tanzania, the recommended rate of application is 250 kg ha[-1] MRP combined with FYM at 5 to 10 t ha[-1] or a GM grown in the previous season (*Mucuna pruriens or Vernonia subligera*).

Other practices for MRP utilization have been developed in Kenya and Tanzania. The RP-fortified compost technology is well accepted by farmers in Western Kenya (Odera and Okalebo, 2009). Another option which combines PR, liming, *Rhizobium* inoculation and bean seeds, developed by a Phosphate rock evaluation project (PREP), known as PREP-PAC. This product comprises PR (2 kg), urea (0.2 kg), legume seed, rhizobial seed inoculants, seed adhesive and lime pellet, packed to fertilize 25 m[2] of land. In Western Kenya the use of PREP-PAC and climbing bean package increased maize and bean yield by 0.72 and 0.25 t ha[-1] respectively, resulting in a 161% return in investment (Nekesa et al., 1999). PREP-PAC use on bean for the two seasons in seven districts in Easter Uganda showed significant increases of 881 kg ha[-1] in bean yield, increasing from 1,316 to 2,197 kg ha[-1] (Esilaba et al., 2005). This product is commercialized, and makes the PR more accessible to smallholder farmers, thus offers opportunity for easier handling and use for increased bean productivity. Another PR from Uganda, Busumbu RP (BRP) was studied by Nabahungu et al. (2007) who reported that composting increased P availability and P recovery from BRP to the extent similar to that of TSP, as well as the availability of Ca and Mg.

8. Liming

Lime application is a common soil improvement practice generally recommended on acidic soils. The rate of application is based on exchangeable soil Al concentration. For tropical soils, lime recommendation is 1.5 times the exchangeable Al in t ha-1, the rate which is sufficient to neutralize toxic Al. One of the first studies to determine the optimum rate of lime for bean was done in Malawi by Aggarwal et al. (1994). In this study, bean response to applied calcitic lime at the rate of 0, 25, 50, 75 and 100% of the exchangeable Al was evaluated in 1992-93 cropping season using 15 varieties and in 1993 -94 season using 8 varieties. Additional treatment was lime applied at 100% of exchangeable Al plus P since the soil was P- deficient. The results indicated that although the performance of the varieties was poor due to low fertility, liming caused a linear increase in nodule number, nodule weight, and grain weight up to 75% level of Al neutralization (Table 21). The yield declined at higher rate of lime application due to nutrient imbalance, which might have been induced by lime. Lime recommendation is therefore soundly based on neutralizing the exchangeable soil Al concentration, rather than increasing soil pH

Lime level to neutralize (%) exch. Al	Seed Yield (kgha-1)	Nodule/plant	Nodule weight (mgplant-1)	Root weight (gplant-1)	Shoot weight (gplant-1)
0	147	0.85	7.59	0.41	1.59
25	161	1.52	13.98	0.42	1.63
50	153	2.43	19.11	0.40	1.41
75	266	5.01	37.29	0.43	1.78
100	235	3.95	32.46	0.43	1.64
100 + P	253	4.84	44.93	0.44	1.71
SE	194	0.92	6.88	0.04	0.16

Table 21. Effect of lime application on seed yield, nodule number, root weight and shoot weight of common bean varieties grown during 2 seasons at Bembeke, Malawi

Finely powdered ground limestone, cheaper than lime (neutralizing value of 90 to 100) evaluated in Rwanda, was comparable to agricultural lime found in the marketplace. Thus lime recommendation was devised for ground limestone. Table 22 presents the results of bean response to lime (agricultural lime) and limestone in two seasons (Nabahungu and Ruganzu, 2001).

Treatments	Bean yield (kg.ha-1)	
	2000 A	2000 B
Control	733	456
Lime 1 t ha-1	1283	295
Limestone 0.5 t ha-1	1300	209
limestone 1 t ha-1	1817	247
Limestone 1.5 t ha-1	1467	333
LSD (0.05)	831	241

Adapted from Nabahungu and Ruganzu, 2001

Table 22. Bean response to agricultural lime and ground limestone at Rubona, Rwanda

Bean yield was very low without lime and doubled with lime application, but was still considered low. For better yield, lime application should be associated with organic resource application, which according to Palm et al. (1997) improves productivity on acid soils by its interactions with the mineral soil in complexing toxic cations and reducing the P sorption capacity of the soil. Examples from DR Congo, Rwanda and Burundi indicate that combining lime with organic manure results in considerable bean yield increase. Rutunga et al. (1998) studied crop rotation system of maize and beans, established at Rubona (Rwanda) from 1984 to 1992. They evaluated the effects of different types and rates of fertilizers in improving the productivity of acidic Oxisols. Continuous cropping of maize followed by beans for a period of 8 years gave no yield in control plots. A single application of 2 t ha^{-1} of lime increased significantly (p = 0.01) the soil pH, Ca^{2+} content, cationic exchange capacity, and decreased the level of the exchangeable Al. However, this quantity of lime, when applied every two years for a period of eight years, led to overliming. The application of more than 8 t ha^{-1} of FYM annually, combined with 300 kg ha^{-1} of NPK (17:17:17) fertilizer (every six months) significantly improved soil organic C and crop production at Rubona, Rwanda. The high rate 35 t ha^{-1} of FYM or the combination of "lime, FYM and NPK fertilizers" gave the best crop performance.

9. Decision support tools

Several technologies have been developed and widely tested with successful results, resulting in development of soil fertility management decision guide, particularly in Uganda (Farley, 1998;Esilaba et al. 2001), Kenya (Rachier et al., 2001) and Eastern Congo (Ngongo, 2001). Soil fertility management research results have been translated into recommendations that take into account the soil physical conditions as well as farmers' socio-economic status. Decision support systems (DSS) to the use of inorganic and organic sources have been developed for different environments and socio-economic conditions in bean based production systems in the region at different locations. These guides are efficient tools for realizing better bean crop yield in areas where they are developed. However, the extrapolation to other sites and location with different soil conditions, climate and population density and socio-economic conditions is not guaranteed with acceptable results as they are site specific (Esilaba et al., 2001). A model decision tool for Uganda is presented in Table 23 while for Eastern DR Congo the tool was developed using soil fertility assessment based on farmers' perception (Table 24). Other DSS are similar, but the local conditions and available resources vary.

10. Promotion and use of ISFM options

PABRA has developed and disseminated different technologies to address farmers' production constraints and to increase bean productivity in sustainable ways (Kimani et al., 2001). These comprise simple technologies such as new bean varieties and more complex, knowledge intensive ones such as integrated pest and disease management (IPDM) or ISFM. The technologies and their dissemination methods were based on low cost options that were in many cases developed through collaboration with multiple partners, and engaged and empowered end users in participation and adoption (Kankwatsa et al., 2008). The spread and adoption of new bean varieties have been very impressive; reaching several million farming families, while the spread and adoption of ISFM technologies have been slow. Scaling up and disseminating information on recommended technologies was through Farmer Research Groups using a modified farmer field school approach. Because farmers were driving the experimentation and dissemination processes, all aspects of local culture were taken into consideration (Kankwatsa et al., 2008) and they often made recommendation for improvement to suit their production circumstances, both in the social and the technological aspects: this led to the development of the ownership by the farming communities and a high rate of adoption among participating farmers.

Conditions	Maize, sole crop	Bean, sole crop	Maize-bean intercrop
Adequate money or credit available	Apply 50 kg ha^{-1} TSP and 25 kg ha^{-1} urea at sowing; apply 50 kg ha^{-1} urea at second weeding	Apply 100 kg ha^{-1} TSP and 20 kg ha^{-1} urea at sowing	Apply 100 kg ha^{-1} and 20 kg ha^{-1} urea at sowing; apply 50 kg ha^{-1} urea at 2nd weeding
Money or credit is inadequate	Apply 50 kg ha^{-1} urea at first weeding	Apply 50 kg ha^{-1} TSP and 20 kg ha^{-1} urea at sowing	Apply 50 kg ha^{-1} TSP and 20 kg ha^{-1} urea at sowing; apply 50 kg ha^{-1} urea at 2nd weeding
Green manure was produced the previous season	Do not apply inorganic fertilizer	Do not apply inorganic fertilizer	Do not apply inorganic fertilizer
Lantana, etc is available	Reduce application of urea at 2nd weeding by 30% for each ton of fresh leafy material applied	Do not reduce fertilizer rate	Reduce application of urea at 2nd weeding by 30% for each ton of fresh leafy material applied
Sowing is delayed until after 15 march or 15 September	Reduce fertilizer rate by 50%	Do not reduce fertilizer rate	Reduce fertilizer rate by 50%
Sowing is delayed until after 30 march or 30 September	Do not use fertilizer at sowing; top-dress urea at 50% rate if conditions are promising	Do not reduce fertilizer rate	Apply 50% of TSP at sowing; top dress urea at 50% rate if conditions are promising
Farm yard manure in available	Reduce fertilizer by 25% for each ton/ha of dry FYM applied	Reduce fertilizer by 40% for each ton of dry FYM applied	Reduce fertilizer by 20% for each ton/ha dry FYM applied
FYM was applied last season	Reduce fertilizer by 15% for each ton/ha of dry FYM applied	Reduce fertilizer by 30% for each ton of dry FYM applied	Reduce fertilizer by 10% for each ton/ha dry FYM applied
Land was rotated from banana or fallow within last one season	Apply N at 2nd weeding, but only if maize is yellowing	Do not apply fertilizer	Apply N at 2nd weeding, but only if maize is yellowing

Top-dress with urea if the crop is well established, the season appears promising, and especially if the lower leaves are yellowish-green in color.

Table 23. Tentative guide to fertilizer use for maize and bean in Uganda

Farmer Conditions		Soil type according to farmer criteria*	Bean type	Cropping system	Soil fertility management Recommendations
Finances	Organic resources				
No money, no Credit	Compost not available	Plot dominated by *Gallinsoga parviflora*	Bush	Bush beans sole crop	No fertilizers
				Bush beans-maize intercropped	Apply 20 t ha⁻¹ Kitchen ash
			Climbing	Climbing beans sole crop	Apply 4 t ha⁻¹ *Tithonia* fresh biomass + 20 t ha⁻¹ ash
				Climbing beans- maize intercropped	Apply 6 t ha⁻¹ *Tithonia* fresh biomass + 20 t ha⁻¹ ash
		Plot dominated by *Pennisetum polystachia* or *Bidens pilosa* (poor soils)	Bush	Bush beans sole crop	6 t ha⁻¹ *Tithonia* + 20 t ha⁻¹ ash
				Bush beans-maize intercropped	8 t ha⁻¹ *Tithonia* + 20 t ha⁻¹ ash
			Climbing	Climbing beans sole crop	10 t ha⁻¹ *Tithonia* + 20 t ha⁻¹ ash
				Climbing beans- maize intercropped	10 t ha⁻¹ *Tithonia* + 30 t ha⁻¹ ash
	Compost available	Plot dominated by *Gallinsoga parviflora*	Bush	Bush beans sole crop	No of fertilizers
				Bush beans-maize intercropped	20 t ha⁻¹ ash
			Climbing	Climbing beans sole crop	6 t ha⁻¹ *Tithonia* or 20 t ha⁻¹ compost
				Climbing beans- maize intercropped	8 t ha⁻¹ *Tithonia* or 20 t ha⁻¹ Compost + 20 t ha⁻¹ ash
		Plot dominated by *Pennisetum polystachia* (poor soils)	Bush	Bush beans sole crop	4 t ha⁻¹ *Tithonia* or 10 t ha⁻¹ compost
				Bush beans-maize intercropped	(6 t ha⁻¹ *Tithonia* or 20 t ha⁻¹ compost)
			Climbing	Climbing beans sole crop	6 t/ha *Tithonia* or 20 t ha⁻¹ compost
				Climbing beans- maize intercropped	(8 t ha⁻¹ *Tithonia* or 30 t ha⁻¹ compost) + 30 t ha⁻¹ ash
Money or credit Available	Compost not available	Plot dominated by *Gallinsoga parviflora*	Bush	Bush beans sole crop	30 kg/ha (P,K)
				Bush beans-maize intercropped	4 t ha⁻¹ *Tithonia* + 50 kg ha⁻¹ (P,K)
			Climbing	Climbing beans sole crop	6 t ha⁻¹ *Tithonia* + 50 kg ha⁻¹ (P, K)
				Climbing beans- maize intercropped	6 t ha⁻¹ *Tithonia* + 50 kg/ha (N.P,K)
		Plot dominated by *Pennisetum polystachia* or *Bidens pilosa* (poor soils)	Bush	Bush beans sole crop	6 t/ha *Tithonia* + 50 kg/ha (N,P,K)
				Bush beans-maize intercropped	6 t/ha *Tithonia* + 75 kg/ha (N,P,K)
			Climbing	Climbing beans sole crop	8 t ha⁻¹ *Tithonia* + 50 kg ha⁻¹ (N,P,K)
				Climbing beans- maize intercropped	8 t ha⁻¹ *Tithonia* + 75 kg ha⁻¹ (N,P,K)

Farmer Conditions		Soil type according to farmer criteria*	Bean type	Cropping system	Soil fertility management Recommendations
Finances	Organic resources				
	Compost available	Plot dominated by *Gallinsoga parviflora*	Bush	Bush beans sole crop	4-6 t ha⁻¹ *Tithonia* or 20 t ha⁻¹ compost
				Bush beans-maize intercropped	(8 t ha⁻¹ *Tithonia* or 30 t ha⁻¹ compost) + 50 kg/ha (N,P,K)
			Climbing	Climbing beans sole crop	(8 t ha⁻¹ *Tithonia* or 20 t/ha compost) + 50 kg ha⁻¹ (P,K)
				Climbing beans- maize intercropped	(8 T t ha⁻¹ *Tithonia* or 20 t ha⁻¹ compost) + 75 kg ha⁻¹ (N,P, K)
		Plot dominated by *Pennisetum polystachia* or *Bidens pilosa* (poor soils)	Bush	Bush beans sole crop	(4 t ha⁻¹ *Tithonia* or 10 t ha⁻¹ compost) + 50 kg ha⁻¹(P,K)
				Bush beans-maize intercropped	(6 t ha⁻¹ *Tithonia* or 20 t ha⁻¹ compost) + 50 kg ha⁻¹ (N,P,K)
			Climbing	Climbing beans sole crop	(8-10 t ha⁻¹ *Tithonia* or 20 t ha⁻¹ compost) + 50 kg ha⁻¹(N,P,K)
				Climbing beans- maize intercropped	(10 t ha⁻¹ *Tithonia* or 30 t ha⁻¹ compost) + 75 kg ha⁻¹ (N,P,K)

Table 24. Decision guide for ISFM in a Bean-based cropping system in Eastern DR Congo

Awareness and adoption of ISFM technologies by farmers is found to be essential factor to their eventual widespread adoption in ECSA. Kankwatsa et al. (2008) found that farmer's awareness varied significantly with their participation to the technology promotional activities of farmers. They reported that participating farmers were more rational in their choice of technologies to solve specific constraints and more participating farmers adopted the IPDM and ISFM technologies than the non-participating farmers. To favor farmers' capacity to experiment and eventually to adapt technological options to their biophysical and socioeconomic conditions, they were exposed to a whole range of new soil fertility management options through farmers' field schools and dialogue with researchers in the process of farmers' participatory research. They have continued to experiment and ended to suggest modifications to the technologies. In Bushumba, Estern DRC, Njingulula and Ngongo (2007) reported the results of participatory evaluation of the rate of *Tithonia* application. They found that although the rate recommended by research was not much modified, farmers found the better manner to express the rate of application using a common volume measure. Farmers' recommendation of *Tithonia* biomass application was the rate of 3 to 4 basins per 4 m x 4 m portion of land, which is equivalent to 5 to 7.5 t ha⁻¹, in perfect agreement with the recommendation by researchers. Going through this process increased their confidence in their own ability to find solutions to different problems, and improved the rate of adoption of suitable technology. The impact studies were conducted by Kankwatsa et al (2008) in PABRA member countries to evaluate the fate of promoted ISFM technologies.

11. Conclusions

African smallholder farming conditions are worsened by declining soil fertility as a consequence of population pressure on a limited landbase. PABRA, as an African research for development program has made outstanding efforts in establishing partnership with numerous stakeholders, farmers, rural communities, non-governmental and governmental organizations, private sector, traders and research organizations (national, regional and international).   In spite of complexity of ISFM technologies developed, bean productivity has been improved under low soil fertility conditions by developing strategies and technologies that enhance resilience to environmental stresses and by enhancing farmers' access to adapt and use cost effective integrated environmental stress management options (PABRA, 2008). The strategy of PABRA on ISFM has two broad avenues for achieving the objective. These include soil fertility management and deployment of resilient bean germplasm.

Genotypes with improved performance on low fertility soils is valued by farmers and readily adopted, and have made it possible to improve bean production in regions where they are adopted at no additional cost. Identification and use of cultivars tolerant to mineral deficiencies and toxicities are essential for reducing production costs and dependence of farmers on soil amendment inputs. The resilient bean cultivars have been shown superior to existing popular varieties developed without consideration of tolerance to soil infertility tolerance. Many bean genotypes have been selected with tolerance to single or multiple edaphic stresses (Wortmann et al., 1995; Rao, 2001). Therefore, as stated by Lynch et Beebe (1995), efficient cultivars would have several benefits, including increased food production, increased farm income, increased bean availability and consumption, and thereby improved nutrition for low income producers and consumers. Indeed, improved variety is considered as a fast adoption technology (Kankwantsa et al., 2008). The nurseries of these genotypes are maintained with the PABRA regional networks with characteristics of market preferences, besides tolerance to low soil fertility.

Bean productivity is further enhanced by various soil management options developed in participatory manner.   The use of farmer participatory approach to technology dissemination enabled farmers to be familiar with, understand, experiment and adapt ISFM technologies. Through capacity building through farmers' groups and farmer research groups, farmers have been empowered to obtain various technologies and services from appropriate partners. Various locally available resources are used, as limestone and phosphate rock, alone or in combination with mineral fertilizers, and the effect of organic resources in improving nutrient use efficiency is well demonstrated. Considerable and beneficial bean yield increase is achieved. However, the low production of organic resources on farms calls for biomass transfer. *Tithonia* biomass is readily available and adopted by farmers and contributes considerably to bean productivity enhancement as an essential source of nutrients and organic amendment. Maximal use of locally available nutrients through low-external input technologies, techniques, combined with optimal use of external nutrients appears to be the most appropriate strategy in the existing economic environment (de Jager, et al., 2004).

Based on research data on soil fertility management in major bean growing areas, a comprehensive decision support system (DSS) was developed for use by farmers and extension workers in different agro-ecological zones and under diverse socio-economic scenarios. In contrast to standard guidelines, the DSS provides farmers with ISFM options under different scenarios, thereby allowing farmers the ability to choose the option(s) that best suit their needs and socio-economic conditions. Soil fertility conditions are assessed using criteria and indicators that are easy to measure, such as dominant weed species. The desired outcome is that the tool will be used by researchers, extension workers, and farmers for assessing and implementing options of using scarce resources for maintaining soil fertility and improving crop yields in bean-based cropping systems.

PABRA places special emphasis on increased access to cost effective and environmentally friendly integrated bean production, especially by female bean growers. Currently, two approaches are used to facilitate and increase farmer's access to ISFM technologies: a deliberate promotion and delivery of improved varieties and ISFM technologies as a single package; and policy advocacy and harnessing of enabling policies (including input support system) to deliver ISFM technologies to bean farmers. The Alliance (PABRA) feels confident that these approaches will greatly enhance the adoption of ISFM technologies by bean growers and overall bean production in eastern, central and southern Africa.

12. References

Aggarwal, V. D., Mughogho, S. K., Chirwa, R. & Mbvundula, A. D. (1994). Results of testing for tolerance to a low pH complex in Malawi. *Proceedings of a Working group meeting. Bean improvement for a low fertility soils in Africa.* CIAT Africa Workshop Series No. 26, Kampala, Uganda, January 23 – 26, 1994

Bado, B.V. (2002). Rôle des légumineuses sur la fertilité des sols ferrugineux tropicaux des zones guinéenne et soudanienne du Burkina Faso. *Thèse de Philosophiae Doctor (Ph. D.)*, Département des Sols et de Génie Agroalimentaire, Faculté des Sciences del'Agriculture et de l'Alimentation Université Laval, Québec, Décembre 2002

Barrios, E., Bekunda, M., Delve, R., Esilaba, A. & Mowo, J. (2001). Identifying and classifying local indicators of soil quality: Methodology for decision making in natural resource management. Eastern Africa version. *CIAT, SWNM, TSBF, AHI.* 2001

Bationo, A., Hartemink, O., Lungu, M., Naimi, P., Okoth, E., Smaling& Thiombiano L.(2006). African soils: Their productivity and profitability of fertilizer use. *Background paper prepared for the African Fertilizer Summit,* Abuja June 9 – 13, 2006

Beebe, S., Rao, I. M., Blair, M. W. & Butare L. (2009). Breeding for abiotic stress tolerance in common bean: present and future challenges. In: *Proc 14th Australian Plant Breeding & 11th SABRAO Conference*, 10 to 14 August, 2009, Brisbane, Australia.

Bullock DG (1992) Crop rotation. Crit Rev Plant Sci 11(4):309–326

Chilimba, A.D.C. & Kapapa, C. (2002). On-Farm Evaluation of Rhizobium inoculation and seed pelleting and Inoculation of Field Beans (*Phaseolus vulgaris*) in acid soils to enhance nodulation and grain yield.*Soil fertility management and bean improvement for low fertility soils in Africa "BILFA" Working Group Meeting,* Awassa, Ethiopia, November 12 – 16, 2002

Dreschel, P., Gyiele, L., Kunze, D., & Coffie, O. (2001). Population density, soil nutrient depletion and economic growth in Sub-Saharan Africa. *Ecological Economics,* Volume 23, pp. 152 – 258

Esilaba, A.O., Byalebeka, J.B., Delve, R.J., Okalebo, J.R., Ssenyange, D., Mbalule, M., & Ssali H. (2005). On farm testing of integrated nutrient management strategies in eastern Uganda. *Agricultural Systems*, Volume 86, pp. 144–165

Esilaba, A.O., Byalebeka, J.B., Nakiganda, A., Mubiru, S., Delve, R.J., Ssali, H. & Mbalule M. (2001). Integrated Nutrient Management strategies in Eastern Uganda. CIAT, Kampala. *CIAT Africa Occasional Publication Series*, No. 35, pp. 71

Fishler, M. & Wortmann, C.S. (2008). Green manures for maize-bean systems in Eastern Uganda: Agronomic performance and farmers' perceptions. *Agroforestry Systems,* Volume 47, Number 1 – 3, pp. 123 – 138

Giller, K. E. (2001). Nitrogen fixation in tropical cropping systems. 2 ed. CABI, Wallingford, Oxon UK.

Giller K.E., Amijee F., Brodrick S.J.& Edje O.T. 1998. Environmental constraints to nodulation and nitrogen fixation of Phaseolus vulgaris L. in Tanzania. II. Response to N and P fertilizers and inoculation with Rhizobium. African Crop Science Journal 6(2):171-178

Ikerra, S. T., Semu, E. & Mrema, J. P. (2007). Combining tithonia diversifolia and Mijingu phosphate rock for improvement of P availability and maize grain yield on a Chromic Acrisol in Morogoro, Tanzania, In: *Advances in integrated soil fertility management in sub_Saharan Africa: Challenges and Opportunities,* Bationo, A., Waswa, B., Kihara, J. & Kimetu J. pp. 333 – 344, Springer, Dordrecht, The Netherlands

Jama, B. & Van Straaten, P. (2006). Potential of East African phosphate rock deposits in integrated nutrient management strategies. *Anais da Academia Brasileira de Ciencias,* Volume 78, Number 4, pp. 781-790

Kumarasinghe KS, Danso SKA,& Zapata F (1992) Field evaluation of N-2 fixation and N-partitioning in climbing bean (Phaseolus vulgaris L.) using N-15. *Biol Fertil Soils* 13:142–146

Kankwatsa P., Ampofo K., Kasambala C. & Mukankusi C. (2008). Analysis of the socio-economic validity of new bean based integrated soil fertility management and integrated pest and disease management technologies: Uganda, Malawi, Swaziland and Eastern DRC. CIAT, Kampala

Kimani, P. M., Buruchara, R., Ampofo, K., Pyndji, M., Chirwa, R. M., & Kirkby, R. (2001). Breeding Beans for Smallholder Farmers in Eastern, Central, and Southern Africa: Constraints, Achievements, and Potential. *PABRA Millennium Workshop,* Arusha, Tanzania, 28 May – 1 June 2001

Liebenberg, A.J. (2002). Dry bean production. Directorate Agricultural Information Services, Department of Agriculture in cooperation with ARC-Grain Crops Institute, South Africa

Lunze, L. (1994). Screening for tolerance to aluminum toxicity in bean. In. *Bean Improvement for Low Fertility in Soils Africa: Proceedings of a Working Group Meeting.* Kampala, Uganda, May, 23-26, 1994

Lunze, L. & Ngongo, M. (2011). Potential Nitrogen Contribution of Climbing Bean to Subsequent Maize Crop in Rotation in South Kivu Province of Democratic Republic of Congo. In: *Innovations as Key to the Green Revolution in Africa,* A. Bationo et al. (eds.), 677 – 681, Springer Science, Dordrecht, The Netherlands . In press

Lunze, L., Kimani, P.M., Ndakidemi, P., Rabary, B., Rachier, G.O., Ugen, M.M. & Nabahungu, L. (2002). Selection of bean lines tolerant to low soil fertility conditions in Africa. *Bean Improvement Cooperative, BIC* Volume 45, pp. 182–183

Lunze L., Kimani P.M., Ngatoluwa R., Rabary B., Rachier G.O., Ugen M.M., Ruganza V. & Awad elkarim E.E. 2007. Bean improvement for low soil fertility in adaptation in Eastern and Central Africa. In: *Advances in integrated soil fertility management in sub_Saharan Africa: Challenges and Opportunities,*Bationo, A., Waswa, B., Kihara, J. & Kimetu J. pp. 325 – 332. Springer, Dordrecht, The Netherlands

Lynch, J.P., & Beebe, S.E. (1995). Adaptation of beans (Phaseolus vulgaris L.) to low phosphorous availability. *HortScience*, Volume 30, pp. 1165-1171

Mastaki, N. J. L. 2006 Le rôle des goulots d'étranglement de la commercialisation dans l'adoption des innovations agricoles chez les producteurs vivriers du Sud-Kivu (Est de la R.D.Congo). *Dissertation originale présentée en vue de l'obtention du grade de docteur en sciences agronomiques et ingénierie biologique. Filière : Economie et Développement Rural.* Faculté Universitaire des Sciences Agronomiques de GEMBLOUX

Mauyo, L.M., Okalebo, J.R., Kirkby, R.A., Buruchara, R., Ugen, M. & Maritim, H.K. (2007). Spacial pricing efficiency and regional market integration of cross-border bean (Phaseolus vulgaris L.) marketing in Rest Africa: The case of Western Kenya and Eastern Uganda. In: *Advances in Intergrated Soil Fertility Management in Sub-Saharan Africa: Challenges and Opportunities*, A. Bationo et al. (eds), pp. 1027 – 1033, Springer

Musungayi, T., Njingulula, M., Mbikayi, N., & Lunze, L. 2006. Expansion des variétés biofortifiées de haricot et de patate douce à chair orange dans les provinces du Sud Kivu et Nord Kivu en R. D. Congo. Rapport d'Enquête. INERA (Unpublished)

Musungayi, T. Sperling, L. Graf, W. & Lunze, L. 1990. Enquêtes diagnostiques de la zone de Walungu, zone d'action de la Femme Solidaire pour le développement du Bushi, Rapport d'enquêtes (Inédit).

Nabahungu L. et Ruganzu, V. 2001. Effet du Travertin et de Tithonia diversifolia sur la Productivité du Haricot Volubile en Sols Acides du Rwanda. *PABRA Millenium Symposium.* Arusha, Tanzania, 28 May – 1 June, 2001

Nabahungu, N.L, Semoka, J.M.R &. Zaongo, C. (2007). Limestone, Minjingu Phosphate Rock and Green Manure Application on Improvement of Acid Soils in Rwanda. *Advances in Integrated Soil Fertility Management in sub-Saharan Africa: Challenges and Opportunities.*Springer, pp 703-712

Nekesa P., Maritim H.K., Okalebo J.R. & Woomer P.L. 1999. Economic analysis of maize-bean production using a soil fertility replenishment product (PREP-PAC) in Western Kenya. *African Crop Science. J.* Vol 7. No. 4, pp 585-590

Ngongo, M. & Lunze, L. (2000). Espèces d'herbe dominante comme indice de la productivité de sol et de la réponse du haricot commun à l'application du compost. *African Crop Science J.* Volume 8, Number 3 pp. 251–261

Ngongo, M. (2001). Management of soil fertility in South Kivu with green manure of tithonia. In: PABRA Millenium Symposium..Arusha, Tanzania, 28 May – 1 June, 2001

Njingulula, M. (2003). Etude de l'impact socio-économique des variétés améliorées du haricot dans le système de culture paysan au Congo, République Démocratique. Proceedings of a Regional Workshop to Develop a Work Plan for Impact Assessment Studies and Monitoring and Evaluation.Kampala, Uganda. PABRA, CIAT. pp 83–87.

Njingulula, P. M. & Ngongo, M. (2007). Initiating rural farmers to participatory research: case of soil fertilization in Bushumba, East of DR Congo. In. *Advances in integrated soil fertility management in Sub_Saharan Africa: Challenges and Opportunities.* Bationo, A., B. Waswa, Kihara, J. & Kimetu J. (Eds) pp. 1051-1059, Springer

Nyabienda, P. & Hakizimana, A. (1988). Atelier sur la fixation biologique de l'azote du haricot en Afrique, CIAT African Workshop Series, No. 8, Rubona, Rwanda, 27 – 29 Octobre 1988

Opala PA. 2001. Management of organic inputs in East Africa: A review of current knowledge and future challenges. Archives of Applied Science Research, 2011, 3 (1): 65-76. Available from: www.scholarsresearchlibrary.com

PABRA. 2008. Supporting nutrition and health, food security, environmental stresses and market challenges that will contribute to improve the livelihood and create income of resource poor small holder families in Sub-Saharan Africa. PABRA, CIAT, Kampala, Uganda

Palm, C.A., Myers R. J. K. & Nandwa, S. M. (1997) Combined use of organic and inorganic nutrient sources for soil fertility maintenance and replenishment. In: *Replenishing Soil Fertility in Africa*, Buresh RJ, Sanchez PA & Calhoun F (eds), pp 193–217. SSSA Special Publication 51. SSSA, ASA, Wadington

Pankhurst, C., Doube, B.M. & Gupta, V.V.S.R. (Eds.) (1997). Biological Indicators of SoilHealth. CAB International, Wallingford.

Rabary, B. 2001. Participatory research for improved agroecosystemmanagement: a mean for rural community development. PABRA Millenium Symposium. Arusha, Tanzania, 28 May – 1 June 2001

Rachier G.O., Wortmann C.S., Tenywa J.S. and Osiru D.S.O. 1999. Phosphorus acquisition and utilization in common beans: mechanisms and genotype differences. African Crop Science Journal Vol. 4: 187-193

Rachier, G.O., Salaya, B.D. & Wortmann, C.S. (2001). Verification of recommended rates of inorganic fertilizers and farmyard manures in maize intercrop under contrasting soil types. In: *PABRA Millenium Symposium*. Arusha, Tanzania, 28 May – 1 June 2001

Rao, I. M. (2001). Role of physiology in improving crop adaptation to abiotic stresses in the tropics: The case of common bean and tropical forages. In: *Handbook of Plant and Crop Physiology*, Pessarakli M, (ed). Marcel Dekker, Inc, New York, USA, p. 583–613.

Rao, I. M., Friesen, D. K. & Osaki, M. (1999). Plant adaptation to phosphorus-limited tropical soils. In: M. Pessarakli (ed.) Handbook of Plant and Crop Stress. Marcel Dekker, Inc., New York, USA, pp. 61-96

Ruraduma, C. (2002). Improving biological nitrogen fixation of beans. *Soil fertility management and bean improvement for low fertility soils in Africa "BILFA" Working Group Meeting*. November 12 – 16, Awassa, Ethiopia.

Rutunga, V, Steiner, K.G., Karanja, N.K., Gachene, C.K.K. & Nzabonihankuye, G. (1998). Continuous fertilization on non-humiferous acid Oxisols in Rwanda "Plateau Central": Soil chemical changes and plant production. *Biotechnol. Agron. Soc. Environ*. Volume 2, Number 2, pp. 135–142

Rutunga, V. (1997). Sols acides de la région d'altitude de la Crête Zaïre-Nil (Rwanda): Potentialités agricoles et forestières. Lengo Publishers, Nairobi, 68pp.

Sanginga, N. (2003). Role of biological nitrogen fixation in legume based cropping systems; a case study of West Africa farming systems. *Plant and Soil*, Volume 252, Number 1, pp. 25-39 (15)

Sanginga, N., & Woolmer, P. L, eds. (2009). Integrated Soil Fertility Management in Africa:*Principles, Practices and Developmental Process. Tropical Soil Biology and Fertility Institute of the International Centre for Tropical Agriculture*. Nairobi. 263 p.

Singh, S.P. (2001). Broadening the genetic base of common bean cultivars: a review. *Crop Science* Volume 41, pp. 1659–1675

Sperling, L., Scheidegger, U., Buruchara, R., Nyabienda, P. & Munyanesa. (1993). Intensifying production among smallholder farmers: The impact of improved climbing bean in Rwanda. *CIAT African Occasional Publication Series* No. 12. CIAT/RESAPAC, Butare, Rwanda, p13

Thung, M. & Rao, I. M. (1999). Integrated management of abiotic stresses. In: *Common Bean Improvemen in the Twenty-First Century*.S. P. Singh (ed.),. Kluwer Academic Publishers, Dordrecht, The Netherlands, pp. 331-370.

Tumuhaire, J.B., Rwakaikara, S.M.C., Muwnga, S. & Natigo, S. (2007). Screening legume green manure for climatic adaptability and farmer acceptance in the semi-arid agro-ecological zone of Uganda. In. *Advances in Integrated Soil Fertility Management in Sub-Saharan Africa: Challenges and Opportunities*, A. Bationo et al., pp. 255 – 259, Springer,Dordrecht, The Netherlands

Ugen, M. & Wortmann, C. S. (2001). Weed Flora and Soil Properties in Subhumid Tropical Uganda, *Weed Technology*, Vol. 15, Number. 3, pp. 535-543

Ugen, M. A., Ndegwa, A. M. Nderitu, J. H., Musoni, A. & Ngulu, F. (2009). Enhancing competitiveness of snap beans for domestic and export markets. ASARECA CGS Revised Full Proposal Document. Entebbe, Uganda

Vanlauwe, B., Bationo, A., Chianu, J., Giller, K. E., Merckx, R., Mokwunye, U., Ohiokpehai, O., Pypers, P., Tabo, R., Shepherd, K., Smaling, E., Woomer, P. L. & Sanginga, N. (2010) Integratedsoil fertility management: Operational definition and consequencesfor implementation and dissemination. *Outl onAgric* 39, 17–24

Vanlauwe B., Chianu J., Giller K.E., Merckx R , Mokwunye U., Pypers P., Shepherd K., Smaling E, Woomer P.L. & Sanginga N. 2010. Integrated soil fertility management: operational definition and consequences for implementation and dissemination. *19th World Congress of Soil Science, Soil Solutions for a Changing World*. 1 – 6 August 2010, Brisbane, Australia.

Vanlauwe B, Tittonell P, &Mukalama J (2006) Within-farm soil fertility gradients affect response of maize tofertilizer application in western Kenya. *Nutrient Cycling in Agroecosystems* 76,171-182.

van Schoonhoven, A. & Pastor-Corrales, M.A. (1994) . Système standard pour l'évaluation du germoplasme du haricot. Publication CIAT, Cali, Colombia, No. 207

Wortmann, C.S.(2001). Nutrient dynamics in a climbing bean and sorghum crop rotation in the Central Africa Highlands, *Nutrient Cycling in Agroecosystems*, Volume 61, Number 3, pp. 267-272

Wortmann, C. S. & Kaizzi, C. K. (1998). Nutrient balances and expected effects of alternative practices in the farming systems of Uganda. *Agriculture, Ecosystems and Environment* Volume 71, pp. 117-131

Wortmann, C. S. & Ssali, H. (2001) . Integrated nutrient management for resource-poor farming systems: A case study of adaptive research and technology dissemination in Uganda. *American Journal of Alternative Agriculture*, Volume 16, pp. 161-167

Wortmann, C.S., Isabirye, M. & Musa, S. (1994). Crotalariaochroleucaas a Green Manure Crop in Uganda. *African Crop Science J.*, Volume 2, Number 1, pp. 55-61

Wortmann, C.S., Kirkby, R.A., Aledu, C.A. & Allen, J.D. (1998). Atlas of common bean (Phaseolus vulgaris L.) Production in Africa, CIAT Cali, Colombia

Wortmann, C.S., Lunze, L., Ochwoh, V.A. & Lynch, J.P. (1995). Bean Improvement for Low Fertility Soils in Africa. *African Crop Science J*, Volume 3, Number 4, pp. 469-477

Wortmann, C.S., Schnier, H.F. & Muriuki A.W. (1996). Estimation of the fertilize response of maize and bean intercropping using sole crop response equations. *African Crop Science Journal*, Vol. 4. No.1, pp. 51-55, 1996

15

Simple and Blended Organic Fertilizers Improve Fertility of Degraded Nursery Soils for Production of Kolanut (*Cola acuminate*) Seedlings in Nigeria

Emmanuel Ibukunoluwa Moyin-Jesu
Agronomy Department, Federal College of Agriculture, Akure, Ondo State
Nigeria

1. Introduction

Feeding the rapidly growing population in sub-Saharan Africa has become a major development concern to policy makers, agricultural experts and international agencies. Populations are increasing in the region where, ironically, the soils are fragile and highly weathered, with low cation exhcnage capacity and low inherent fertility.

Brader (1993) reported that intensive farming in the western world relies heavily on chemical fertilizers and pesticides. In contrast, traditional African farmers practice shifting cultivation of mixed crops which allow long fallows to restore fertility. However, this system is no longer feasible because of increasing population rate and limited land base.

The practice of using chemical fertilizers input for agricultural production has not been widely adopted by farmers in tropical Africa because they are expensive, scarce and destructs soil properties on continuous use (Moyin-Jesu, 2003, Moyin-Jesu and Ojeniyi, 2006, Moyin-Jesu, 2009). This is because NPK, Ammonium sulphate and urea when hydrolysed in the soil increased soil acidification through their nitrate phosphate and sulphate component that form strong acids (HNO$_3$, H$_2$SO$_4$ and H$_2$PO$_4$.

Moyin-Jesu (2007) reported that many agricultural wastes were available in tropical countries but only about 20% was returned to the soil for fertility maintenance because they are considered bulky and difficult to transport by poor resource farmers, most of them are either burnt on the field after harvest or eating up by termites. There is no much consciousnesses of using them as compost, green manure and farm yard manure.

Regrettably, soil organic matter is declining with cultivation and urgent research attention is needed to promote the use of organic fertilizers for sustainable crop production. Among the organic materials that could be used for crop production are wood ash, cocoa husk amended with goat, duck and turkey manures or sole application of manures. Their potential as organic amendments for raising kola nut seedlings in the nursery has not been tested previously. Kola belongs to the family Stericuliaceae and is noted for its use in religious and social activities throughout West Africa. Industrially, kola is used for preparation of drinks such as Coca Cola, Pepsi-Cola, dyeing purposes and production of pharmaceutical (Adegeye and Ayejuyo 1994).

Kolanut tree (*Cola accuminata*) is a slender tree, that can grow up to 12m high, but usually reaches 6-9 m. The branches are slender, crooked and markedly ascending, the foliage is often sparse and confirmed to the tips of the branches. The hermaphrodite flower may be up to 25cm across. The perianth segments usually joined for nearly half their length while the anthers are borne on a short, but distinct column.

The fruit consists of five follicles borne at right angles to the stalk or slightly bent downwards. The follicles are sessile and have a straight point or tip up to 20cm long. The surface is rough to the touch, russet or olive brown. There are up to 14 seeds in each follicle and the embryo may have three to five or more cotyledons which are pink, red or sometimes white in colour. The fruits mature in the period from April to June.

Kolanut (*Cola accuminata*) grows in an area with rainfall between 1200-1500mm and responds well to fertile soils with high organic matter. Well drained soils are suited for deep tap rooted tree. It is not a mycorrihzal plant. Kola trees are both grown in farms and plantations which are established by individual farmers (2-3 ha). In some government farms settlement, Kola plantations are about 50 – 100 ha but they have been abandoned because of oil boom in Nigeria.

The objective of writing this chapter is to investigate the effectiveness of several organic fertilizers (wood ash, cocoa husk amended with goat, duck and turkey manures, or manure alone) as a source of plant nutrients for kola seedlings in a nursery. Seedling growth, leaf nutrient concentration and soil parameters were evaluated in this study.

2. Materials and methods

The experiment took place at the Teaching and Research Farm Federal College of Agriculture in Akure (7°N, 5°10°E) in the rainforest zone of Nigeria between August 2003 and May 2004 and was repeated between August 2004 and 2005 to validate the results.

The annual rainfall is between 1100 and 1500mm while the average temperature is 24°C. The Soil is sandy clay loam, skeletal, Kaolinitic, isohyperthermic oxic paleustalf (Alfisol) and belongs to Akure soil series. Soil survey staff (1999). The site had been continuously cropped to arable crops for more than 5 years.

2.1 Soil sampling and analysis before planting

Thirty core samples were collected with hand corer (3 cm diameter) from 0-15cm depth of the site, composited, air dried and sieved through a 2mm screen prior to analysis.

The soil pH (1:1 soil/water and 1:2 soil/0.01M $Cacl_2$) solution was determined using a glass calomel electrode system (Crockford and Nowell, 1956) while organic matter was determined by the wet oxidation chromic acid digestion (Walkley and Black, 1934).

The total nitrogen was determined by the microkjedahl method in which the distillate is titrated against the boric acid (AOAC, 1990) while available soil phosphorus was extracted by the Bray P_1 extractant and measured by the Murphy blue colouration and determined on a spectronic 20 at 882um (Murphy and Riley, 1962). Soil K, Ca, Mg and Na were extracted with 1M NH_4OAc, pH 7 solutions. The K, Ca and Na contents were determined with flame photometer while Mg was determined on atomic absorption spectrophotometer (Jackson, 1958). The mechanical analysis of the soil was done by the hydrometer method. (Barycous 1951) and the soil textural name was determined using textural triangle.

2.2 Determination of soil bulk density and porosity

The soil bulk density (mgm^{-3}) was determined by a core method (Ojeniyi, 1980). The soils samples taken were placed in an aluminium can, at 105°C for 48 hours, allowed to cool down and weighed. Thereafter, measured volume of water is put into an eureka can and the mass of the soil is placed inside nylon bag tied to a thread and is suspended by a tripod stand.

Gradually, it is immersed into the eureka can and displaces the water through upthrust displacement. The mass of the soil divided by the volume of displaced water gives the soil bulk density.

The soil porosity was calculated from the values of bulk density using the formulae $ep=\left[I-\frac{eb}{es}\right]\times100$ where eb = bulk density and soil particle density 2.75 mg m $^{-3}$ for tropical soils.

2.3 Source and preparation of organic fertilizers

Cocoa husk and wood ash were obtained from cocoa plantation and cassava processing unit at Federal College of Agriculture, Akure respectively while the poultry; duck and turkey manures were obtained from their pens in the livestock unit of Federal College of Agric, Akure.

The organic materials were processed to allow decomposition. The woodash was sieved to remove pebbles, stones and unburnt shafts while the cocoa husk was partially composted for 6 weeks to reduce C/N ratio. The poultry, duck and turkey manures were air-dried to allow quick mineralization process.

2.4 Chemical analysis of the organic materials

Two grammes from each of the processed forms of the organic materials used, were analysed. The nitrogen content was determined by kjdaht method (Jackson, 1964) while the determination of other nutrients such as P,K, Ca, Mg was done using the wet digestion method based on 25-5-5 ml of HNO_3 – $H2SO_4$ – $HclO_4$ acids AOAC, 1990). The organic Carbon (%) was determined by wet oxidation method through chromic acid digestion. Walkley and Black, (1934).

2.5 Collection of kolanuts feeds for planting

Ripe fruits of kolanut were collected from the kola tree plantation in Federal College of Agric, Akure. The seeds were obtained after extraction of the fruits, its mucilage washed and air dried for 72 hours at room temperature to remove moisture.

2.6 Pre-nursery establishment of kolanut seedlings

Five seed boxes of (90 x 60 x 30cm) size each were filled with top soil and the mature seeds of kolanut were planted. A shed was erected for the pre-nursery to prevent the seeds from dessication and cultural practices such as weeding and watering twice a day were carried out. The planted kolanut seeds germinated after 21days and were transplanted to the nursery.

2.7 Nursery establishment of kolanut seedlings

The nursery site of 70 x 70m (490m^2) was cleared for laying out polybags and a shed was constructed to shade the site from direct evaporation and scorching of plants by the sun. The bulk soil randomly taken from the site (o-15cm) depth was sieved to remove stones and plant debris. Ten (10) kg of the sieved soil was weighed into a poly bag (30x17cm) size. (Manufacturer name is Nigeria Plastic Company, Ibadan, Polyethene type and 2cm in thickness).

Nine organic fertilizer treatments were used for the experiment namely; turkey manure, duck manure, goat manure, goat manure/cocoa husk mix, woodash/ duck manure mix, cocoa husk/turkey manure mix and woodash/turkey manure mix. All mixtures consisted of equal weight on a dry weight basis (50%) of the two components. The treatments were applied at 8t/ha (40g residues per 10kg soil). There was a control treatment (no fertilizer, no manure) and a fertilizer treatment (400kg/ha NPK 15-15-15 fertilizer at 2g per polybag). All the treatments were replicated three times and arranged in a completely randomized design (CRD).

The organic fertilizers and NPK fertilizer were incorporated into the soil filled poly bags using a hand trowel and allowed to decompose for one week by watering twice in a day. Then one pre-germinated kolanut seed from the pre-nursery was planted in each polybag and watered adequately in the morning and evening for the first 12weeks. The periods of the nursery experiments tallied with the commencement of dry season in previous year into the beginning of the rainy season in the following year., also, rains are not evenly distributed throughout the year. Hence, the need for supplementary watering to ensure steady growth of seedlings.

After two weeks, plant height, leaf area and stem girth of the seedlings were measured and these measurements continued every week till 24 weeks after planting.

Weeding of the site was done at 3,6,9 and 15 weeks after planting (WAP). The kola seedlings were sprayed with karate (active ingredient lambda cyhalotrin 720EC) at 10ml/10L of water 2 weeks intervals to control leaf defoliating beetles. At 12 weeks, part of the shed roof was removed to allow more sunlight to thicken the seedlings and permitted watering natural from incoming rainfall (4mm average per day during the study period).

At 10 weeks after planting, representative leaf samples from the top, middle and lower parts of the seedlings (excluding newly emerged leaves) per each treatment were randomly taken, packed into labeled envelopes and oven dried for 24 hours at 70oC.

The dried leaf samples were ground and analysed. The nitrogen content was determined by kjedahl method (Jackson, 1964) while P,K, Ca and Mg contents were determined by wet digestion method using 25-5 - 5ml of HN0$_3$ – H2SO$_4$ and HClO$_4$ acids (AOAC, 1990).

At 24 weeks after planting, seedlings were carefully uprooted from the poly bag and separated into shoots and roots. The shoot weight (g dry mass) and tap root length were measured. Also, soil samples were taken from each polybag at 25 weeks after planting (WAP), air dried and sieved through a 2mm screen prior to soil analyses for total N and extractable N,P,K,Ca and Mg as well as soil pH (H_2O) and soil O.M (AOAC, 1990).

2.8 Statistical analysis

All the data collected on growth parameters, leaf nutrient concentrations and soil analysis after harvesting were evaluated with analysis of variance (ANOVA) and means separation using Duncan Multiple Range Test at P = 0.05 level.

3. Results

3.1 Initial soil analysis before planting the seedlings

The physical and chemical properties of the soils used for raising kola seedlings in the nursery are presented in Table 1.Based on the established critical levels for the soils in South West Nigeria, the soils are acidic and low in organic matter when compared with the 3% critical level (Agboola and Corey, 1973).

Soil parameters	Values
Soil pH (1:1) soil/water	5.35
Soil pH (0.01M Cacl2)	5.10
Organic matter (%)	0.36
Soil nitrogen	0.03
Available P (mg/kg)	5.36
Exchangeable K (mmol/kg)	0.09
Exchangeable Ca (mmol/kg)	0.08
Exchangeable Mg (mmol/kg)	0.13
Exchangeable Na (mmol/kg)	0.11
Soil bulk density (mgm-3)	1.58
Soil porosity (%)	40.80
Textural class	Sandy loam
USDA soil classification	Alfisol (Oxic tropodaulf)

Table 1. Soil chemical composition before planting kola seedlings.

The total % nitrogen was found to be less than 0.15%N, which is considered as the optimum for crops by Sobulo and Osiname (1981). The available soil P was less than 10mg/kg[-1] that is considered as adequate for crop production in this region (Agboola, 1982).

Soil exchangeable bases (K, Ca, Mg and Na) had concentrations lower than the 0.20 mmol/kg[-1] critical level recommended by Folorunso et al (1995). The soil was very sandy and low in clay. The soil bulk density was high (1.58 mg m[-3]) and would adversely affect root penetration and growth.

3.2 Chemical composition of the organic materials used for the experiment

Among the organic residues used, the turkey and duck manure had the highest N, P and lowest C/N ratios. In-addition, wood ash had the highest K, Ca and Mg concentration which was followed by cocoa husk. Goat dung was indicated to be fairly high in N, P, K and Ca. (Table 2).

Treatments	CN Ratio	N (%)	P (mg/kg)	K (%)	Ca (%)	Mg (%)	Fe (mg/kg)	Zn (mg/kg)	Cu (mg/kg)
Cocoa husk	11.0	1.44	100	2.10	0.93	0.71	50.4	1.69	0.16
Wood ash	11.8	1.53	86	2.30	0.94	0.85	65.5	1.83	0.66
Goat manure	7.9	1.82	168	1.00	0.29	0.45	34.5	1.30	0.16
Duck manure	7.2	2.10	260	0.65	0.19	0.15	21.3	1.13	0.16
Turkey manure	7.1	3.86	346	0.79	0.21	0.18	29.7	1.16	0.14

Table 2. Analysis of the organic fertilizers used for the experiment on raising kola seedlings.

The quantity of nutrients (total kg/nutrients) supplied by each of the organic fertilizers for raising kola seedlings is presented in Table 3. More nutrients were supplied by the manure (turkey, duck manure and goat manures), wood ash and cocoa husk than the NPK 15-15-15 and control treatments.

Fertilizers	N	P	K	Ca	Mg
			Total kg/nutrient		
Cocoa husk	144	104	155.9	93.4	71.0
Wood ash	153	90	230.2	94.0	85.2
Goat manure	248	135	49.9	14.5	22.5
Duck manure	286	168	62.3	14.0	19.0
Turkey manure	434	186	48.6	16.0	20.50
NPK 15-15-15	240	240	240	0.2	0.1

+++ - Application of NPK 15-15-15 at 400kg/ha.
++ - Application of manures at 8t/ha.
+ - Application of wood ash and cocoa husk at 4t/ha.

Table 3. Total amount of nutrients supplied kg/nutrient by each of the organic fertilizers used for raising kola seedlings.

The quantities of nutrients supplied by the organic fertilizers were adequate for sustaining growth of kola seedlings in the nursery and later, when seedlings were moved to the field as reflected in the chemical composition of the seedlings and soil chemical composition after this 24 weeks experiment. (Tables 5 and 6)

3.3 The growth parameters of kola seedlings under simple and blended organic fertilizers

There were significant increases (P<0.05) in the plant height, leaf number, leaf area, stem girth, shoot weight and tap root length of kola seedlings under different simple and blended organic fertilizers compared to the control treatment (Table 4).

Treatments	Shoot Weight (g)	Plant height (cm)	Tap root length (cm)	Leaf area (cm^2)	Leaf number	Stem girth (cm)
Duck manure (sole)	180.2f	18.2f	8.5c	26.8d	5.0d	0.83d
Turkey manure (sole)	140.1c	15.8c	7.3b	24.5c	4.0b	0.76c
Goat manure (sole)	130.0b	13.4b	7.0b	22.6b	4.0b	0.50b
Goat manure + Cocoa husk	163.2d	17.3e	10.4d	27.6de	4.4bc	0.92e
Goat manure + Wood ash	175.3e	16.2cd	11.0de	28.2f	5.1de	0.96ef
Duck manure + Cocoa husk	193.2i	28.4j	13.3h	32.4i	7.6h	1.16h
Duck manure + Wood ash	201.4j	31.6k	14.8i	42.3j	8.2i	1.46i
Turkey manure + Cocoa husk	185.1g	26.1h	12.0g	30.5g	7.0g	1.00f
Turkey manure + Wood husk	190.0h	27.2hi	13.8f	31.1h	7.4gh	1.10fg
NPK 15-15-15	188.2h	23.2g	11.8g	27.5de	6.0f	0.92e
Control	32.10a	6.3a	4.1a	9.3a	3.6a	0.26a

Treatment means within each column followed by the same letters are not significantly different, using DMRT at P <0.05.

Table 4. The growth parameters of kola seedlings with simple and blended forms of organic fertilizers treatments between 2 and 24 weeks after planting.

Among the organic fertilizers, the simple forms of duck manure, wood ash blended with duck manure and cocoa husk blended with duck manure had the highest values of plant height for kola seedlings compared to others.

The blended forms of the organic fertilizers were found to increase significantly (P<0.05) the plant height leaf area, leaf number, stem girt, tap root length and shoot weight of kola seedlings compared to the NPK 15-15-15 fertilizer. For instance, the wood ash blended with duck manure increased the shoot weight, plant height, root length, leaf area, leaf number and stem girth of kola seedlings by 6%, 27%, 20%, 35%, 27% and 37% respectively compared to using NPK 15-15-15 fertilizer.

Generally, all the growth parameters of kola seedlings under the blended forms of the organic fertilizers were higher in value than those under the simple forms. For-instance, wood ash blended with duck manure increased the plant height, leaf area, tap root length, leaf number and stem girth by 42%, 37%, 43%, 39% and 43% respectively compared to the simple form of duck manure.

Furthermore, NPK 15-15-15 fertilizer increased the growth parameters of kola seedlings more than the simple forms of duck, goat and turkey manures.

3.4 Leaf chemical composition of kola seedlings with simple and blended organic fertilizers

The leaf analysis of the kola seedlings receiving different organic fertilizer sources are shown in Table 5. There were significant (P<0.05) increases in the leaf N, P, K, Ca and Mg concentrations of seedlings receiving organic fertilizers than the control treatment. The simple and blended forms of the organic fertilizers increased the kola seedling leaf K, Ca and Mg compared to the NPK fertilizer. For-instance, turkey manure + wood ash increased kola leaf K, Ca and Mg concentrations by 73.4%, 84% and 76% respectively, compared to the NPK 15-15-15 treatment. However, the NPK 15-15-15 treatments increased the leaf N and P concentrations more than the organic fertilizers.

Treatments	N	P	K	Ca	Mg
	%				
Duck manure (sole)	1.90f	0.32d	1.63e	0.78de	0.33c
Turkey manure (sole)	1.65c	0.28c	1.53d	0.72d	0.36d
Goat manure (sole)	1.48b	0.25b	1.20c	0.6 3c	0.32b
Goat manure + Cocoa husk	1.78d	0.36e	2.10f	1.56f	0.72e
Goat manure + Wood ash	1.80f	0.42g	2.43g	1.63g	0.75f
Duck manure + Cocoa husk	2.16h	0.43h	3.70j	2.55j	1.26i
Duck manure + Wood ash	1.85fg	0.53i	3.90k	2.76k	1.35g
Turkey manure+ Cocoa husk	1.80f	0.42g	3.20h	2.50h	1.20g
Turkey manure + Wood ash	1.79de	0.41f	3.50i	2.5hi	1.23h
NPK 15-15-15	2.23i	0.56ij	0.93b	0.40ab	0.3
Control	1.25a	0.20a	0.30a	0.2a	0.2a

Treatment means within each column followed by the same letters are not significantly different, using DMRT at P <0.05.

Table 5. The leaf chemical composition of kola seedlings under different simple and blended organic fertilizers.

Among the organic fertilizers, duck manure and blended duck manure with wood ash and cocoa husk increased the kola seedlings leaf N, P, K, Ca and Mg concentrations when compared to other organic fertilizers. In-addition, the simple forms of the turkey manure, duck manure and goat manure had lower values of kola leaf nutrients than the blended forms with wood ash and cocoa husk. Duck manure + cocoa husk increase the Kolanut leaf concentration of N by 12%, P by 74%, K by 56%, Ca by 69%, and Mg by 75% compared to the duck manure compared to duck manure.

The leaf K, Ca, Mg contents of kola leaf in both simple and blended forms of organic
fertilizer treatments were far higher than 1.19% K, 0.8% Ca and 0.25% Mg critical levels
reported by Jones and Eck (1973) while the leaf N and P contents were also higher than 1.5%
N and 0.22% P critical levels recommended by Adepetu et al (1979).

However, the leaf N, P, K, Ca and Mg contents in the control treatments (no fertilizer
applied) were far below the different critical levels mentioned above. Leaves of seedlings in
the control treatment showed deficiency symptoms of N and Mg (loss of chlorophyll and
yellow leaf colouration), P (purple colouration), K (burnt leaf margin and Ca (stunted root
growth). Ojeniyi (1984).

3.5 Soil chemical composition after the experiment on kola seedlings under different simple and blended organic fertilizers

Both simple and blended organic fertilizers and NPK 15-15-15 increased significantly
(P<0.05) the soil N, P, K, Ca and Mg compared to the control treatment (Table 6). However,
NPK fertilizer decreased soil pH and O.M relative to the simple and blended organic
fertilizer treatments.

Treatments	N	P	K	Ca	Mg	Soil pH	O.M
	(%)	(mg/kg)		mmol/kg			%
Duck manure (sole)	0.19d	19.36d	0.83e	0.50e	0.24f	6.50e	1.16e
Turkey manure (sole)	0.18c	17.26c	0.74d	0.48d	0.22e	6.40c	0.98c
Goat manure (sole)	0.15b	15.60b	0.52b	0.36b	0.16c	6.20b	0.70b
Goat manure + Cocoa husk	0.20e	19.10d	0.63ef	0.42c	0.18d	6.60d	0.99cd
Goat manure + Wood ash	0.22g	20.94f	0.94f	0.46c	0.22e	6.90ef	1.20f
Duck manure + Cocoa husk	0.23h	24.4g	1.24i	0.96hi	0.56hi	7.10g	2.10hi
Duck manure + Wood ash	0.33j	26.3h	1.34j	0.92h	0.58j	7.20gh	2.40j
Turkey manure + Cocoa husk	0.21f	22.10f	1.05fh	0.85fg	0.52g	7.00f	1.85g
Turkey manure +Wood ash	0.27i	23.0f	1.19h	0.81f	0.55h	7.00f	1.96h
NPK 15-15-15	0.36k	27.60i	0.66c	0.03a	0.04ab	5.10a	0.25a
Control	0.02a	3.40a	0.04a	0.02a	0.02a	5.10a	0.25a

Treatment means within each column followed by the same letters are not significantly different,
using DMRT at P <0.05.

Table 6. The soil chemical composition of kola seedlings with simple and blended forms of
organic fertilizers.

The cocoa husk and wood ash blended with duck manure treatments gave the highest values of soil N, P, K, Ca, Mg, pH and O.M by 36%, 16%, 22%, 8%, 10%, 3% and 23% respectively, compared to turkey blended with cocoa husk.

In - addition, the simple and blended organic fertilizers had higher values of soil Ca and Mg relative to the NPK 15-15-15 fertilizer treatment. Also, the blended forms of organic fertilizers increased the soil N, P, K, Ca, Mg, pH and O.M compared to their simple forms.

The soil N, P, K, Ca, Mg, pH and O.M values after the experiment were far higher than 0.15% N, P (10mg/kg), K, Ca and Mg (0.2mmol/kg) and 3% recommended by Sobulo and Osiname (1981), Agboola and Corey (1973) and Folorunso et al (1995) as critical soil fertility levels for sustainable crop growth in the field.

The soil K/Ca, K/Mg and P/Mg ratios were 22:1, 17:1 and 690:1 under NPK 15-15-15 fertilizer compared to soil K/Ca (1:1), soil K/Mg (2:1) and P/Mg (43:1) ratios under turkey manure blended with cocoa husk signifying the presence of nutrient imbalance in the NPK fertilized soil.

4. Discussion

The lowest values of Kola seedling growth parameters (plant height, leaf area, stem girth, leaf number, shoot weight and tap root length), leaf nutrient concentration and soil chemical parameters were found in the control treatment consistent with the initial low nutrient status of the soil before application of the organic fertilizers.

Therefore, the kolanut seedlings were having deficiency symptoms of yellow, purple colourations and the marginal burn of leaves consistent with N, P, K and Mg deficiencies. This observation agreed with Adepetu et al (1979), who reported an approximate 55% drop in soil O.M over seven years of continuously cultivating an Iwo soil association in the green house and under field conditions. Hence, this finding corroborates the importance of fertilizer use to enhance crop productivity in the tropics.

The effectiveness of blended forms of wood ash and cocoa husk with duck, turkey and goat manures in improving the growth, soil and leaf chemical composition of kolanut seedlings can be attributed to enhancement of their degradation rate by the manures with lower C/N ratio.

Furthermore, the blending of the organic fertilizers before application to soil also enhanced their decomposition and rate of nutrient release to the soil. Woodash is expected to be a good source of cations and cocoahusk could retain soil moisture also the composting of cocoahusk would reduce its immobilization of N and P. This observation might be responsible for the exceptional difference in the performance of wood ash and cocoa husk blended with manures compared to the work of Adebayo and Olayinka (1984) which used the unprocessed forms of sawdust, wood ash and cocoa husk blended with turkey and poultry manures to grow maize.

The better performance of duck manure blended with cocoa husk and wood ash treatments in increasing the plant height, leaf area, stem girth, leaf number, tap root length and shoot weight compared to the NPK 15-15-15 fertilizer could be traced to their rich nutrient contents (N, P, K, Ca and Mg) which increased the soil nutrients and consequently improved nutrient and water uptake in the plants. This is consistent with the leaf nutrition, based on N, P, K, Ca and Mg concentrations in foliage

The application of NPK 15-15-15 fertilizer at 400kg/ha led to high soil K/Ca, K/Mg and P/Mg ratios, which could have produced imbalances and limited uptake of P, K, Ca and Mg. Results point to lower concentration of leaf K, Ca and Mg with NPK fertilizer than the organic fertilizers.

Unbalanced fertilization with the NPK fertilizer treatment could be responsible for the lower values of soil K, Ca, Mg, concentrations, compared to the wood ash and cocoa husk blended with manures,. Lower soil pH and O.M content in the NPK fertilized poly bags, compared to the wood ash and cocoa husk blended with manures points to soil acidification and O.M loss. These findings are supported by Agboola (1982) who reported that arbitrary use of inorganic fertilizers resulted in signs of toxicities, poor yield responses and deterioration of some soil properties.

The contribution of the organic fertilizers used, in increasing the growth parameters of kolanut, leaf N, P, K, Ca and Mg concentrations, soil nutrients pH and O.M was also consistent with their chemical composition and the total nutrients/kg applied to the soil. The view is also corroborated by Swift and Anderson (1993) who reported that organic manures supplied nutrients which NPK fertilizer could not supply to the crops. This showed the potentials of organic fertilizers in increasing the crop yields.

Furthermore, the exceptional performances of he blended cocoa husk and wood ash with duck, turkey, and goat manures over the simple forms of the manures was due to the fact that duck, turkey and goat manures have high nutrient concentrations and low C/N ratios and their combination with the cocoa husk and wood ash fortified their nutrient supplying power.

This observation explained the superiority in the growth parameters, leaf and soil chemical composition of kolanut seedlings in the wood ash blended with duck manure and turkey manure compared to their simple form of application.

This was in line with Moyin-Jesu (2007) and Moyin-Jesu (2008) who reported nutrient superiority of organically blended fertilizers over their sole forms in coffee seedlings and okra.

The increase in soil pH under duck manure blended with wood ash and cocoa husk compared to other treatments was traced to the high K, Ca and Mg contents of wood ash and cocoa husk and could be effective as liming materials as well as enhancing effective release of nutrients (Gordon, 1998). Unlike the NPK 15-15-15 fertilizer which acidify the soil if used continuously.

Obatolu (1995) reported that oil palm bunch ash, wood ash and cocoa husk improved soil K, Ca and Mg concentrations and corrected acidity in an Alfisol grown to coffee and maize. Therefore, the exceptional increase in soil pH in duck manure blended with wood ash over that turkey manure blended with wood ash could be responsible for better growth performance and soil nutrients due to the importance of soil pH in effective nutrient release.

The balanced nutrient supplying power of the manures blended with wood ash cocoa husk coupled with the simple forms of the manures contributed to the healthy kolanut seedlings in the nursery and for proper establishment on the field when they are transplanted.

4.1 Conclusions and recommendations

The research indicates that the duck manure and turkey manure (simple forms) and their blended forms with cocoa husk and wood ash applied at 8t/ha (40g/polybag) increased the soil nutrient supply, pH and O.M, leaf nutrition and a number of plant parameters (plant height, stem girth, leaf number, leaf area, tap root length and shoot weight) of kolanut seedlings.

Duck manure blended with wood ash and cocoa husk (8t/ha) was the most effective fertilizer materials and is recommended to improve the nutrient availability and ensure sustainable nursery and field production of kolanut seedlings on a commercial basis.

This recommendation agreed with the fact that inorganic fertilizers are becoming very expensive to purchase by the small holder farmers of kolanut. These organic fertilizer materials appear to have beneficial secondary effects on soil properties and could be more favourable to the environment.

5. References

Abulude, F. O. 2004: Composition and properties of *cola nitida* and *cola acuminate* flour in Nigeria. Global Journal of Pure and Applied Sciences, 10(1), 11-16.

Adebayo, A. and A. Olayinka. 1984: The effects of methods of application of sawdust on plant growth, nutrient uptake and soil chemical properties. Ife Jour. of Agric. 2(1): 36-44.

Adepetu, J.A., A.A. Adebayo, E.A. Aduayi and C.O. Alofe. 1978: A preliminary survey of the fertility status of soils in Ondo State under traditional cultivation. Ife Jour. of Agric. 1(2): 134-149.

Adeyeye, E.I. and Ayejuyo, O.O. 1994: Chemical composition of cola acuminate and Garcinia kola seeds grown in Nigeria. Int. J. Food Sci. and Nutri. 45. 223-230.

Agboola, A.A. and R.B. Corey. 1993: Soil testing N, P. K, for maize in the soils derived from metamorphic and igneous rocks of Western State of Nigeria. Nigeria Jour. Of West African Sci. Assoc. 17(2): 93-100.

Agboola, A.A. 1982: Soil testing, soil fertility and fertilizer use in Nigeria. A paper presented at the first National Seminar on Agricultural Land Resources, Kaduna. Pp. 6-8.

AOAC. 1970: Official methods of Analysis 12th ed.; Association of Official Analytical Chemists, Arlington, V.A.

AOAC, 1990. Official methods of Analysis. 15th ed; Association of Analytical Chemists, Arlington, V.A.

Bouycous, H. 1951: Mechanical analysis of soils using hydrometer method. Analytical Chem. Acta. 22:32-34.

Brader L. (1993): Agricultural systems review in Africa. A keynote address delivered at International Symposium on Soil Organic Matter Dynamics and Sustainability of Tropical Agriculture. Jointly organized by the Laboratory of Soil Fertility and Soil Biology Katholieke Universities Leuven (K.U. Leuven) and I.I.T.A., 1993, 1-10.

Crockford, L. and R. Nowell. 1956: Laboratory manual of Physical Chemistry. Exp. 31 and 32. John Wiley and Sons, New York.

Folorunso, O.O., A.A. Agboola and G.O. Adeoye. 1995: Use of Fractional Recovery (FR) to calculate the K needs of maize (Zea mays L). Jour. Tech. Educ. 2(1): 65-75.

Gordon, W. 1998: Coffee: Tropical Agricultural Series. In. H. Murray (Eds.). pp. 1-20. Macmillan Publishing Ltd. London.

Jackson, M.L. 1958: Soil chemical analysis. Englewood Cliffs N.J.Prentice Hall, 57 - 67

Jackson, M.L. 1964: Soil chemical analysis. Englewood Cliffs N.J. Prentice Hall: 86-92.

Jones, B.J. and Eck, H.V. 1973: Plant analysis as an iad in fertilizing corn and grain sorghum. In: L.M. Walsh and J.D. Beaton (eds.) Soil testing and plant analysis rev. ed. Soil Sci. Soc. Amer. Madison, Wisconsin, U.S.A. 349-364.

Melsted, S.W., H.L. Motto and T.R. Peck. 1969: Critical plant nutrient composition values useful in interpreting plant analysis data. Agron. J. 61:1-20.

Moyin-Jesu, E.I. 2003. Incorporation of agro-industrial biomass and their effects on four successive crops of Amanranthus. Pertanika Jour. Trop. Agric. Sci. 26(1): 35-40.

Moyin-Jesu, E.I. and Ojeniyi S.O. 2006: Use of sole and amended plant residues for soil fertility, leaf chemical composition and yield of okra Discovery and Innovation Journal, Africa Academy of Sciences. Vol.4 Dec. 2006. pp. 34-38.

Moyin-Jesu, E.I. 2007: Effects of some organic fertilizer on soil and coffee (Coffee Arabica L), leaf chemical composition and growth. University of Khartoum Jour. of Agric. Sci. 15(1): 52-70.

Moyin-Jesu, E.I. 2008: Determination of soil nutrient levels for okra using sole and amended plant residues. Pertanika Jour. Of Trop. Agric. 3(2): 2008.

Moyin-Jesu, E.I. 2009. Evaluation of sole and amended organic fertilizers on soil fertility and growth of kola seedlings (Cola acuminate). Pertanika Jour. Trop. Agric. Sci. 32 (1): 17-23.

Murphy, J. and J.P. Riley. 1962: A modified single solution method for determination of phosphate in natural waters. Analytical Chem. Acta. 27:31-36.

Obatolu, C.R. 1995: Nutrient balance sheet after coffee and maize cropping on an Alfisol supplied with organic fertilizers in Ibadan, Nigeria. Proc. 3rd All Afric. Soil Sc. Soc. (pp. 250), August 20-23, Ibadan.

Ojeniyi, S. O. 1984: Compound chemical fertilizer and food production. Effect of NPK 15-15-15 fertilizer on pepper, cowpea and maize. Nigeria Jour. Appl. Sci. 2:91-95.

Ojeniyi, S.O. 1988: Soil physical properties under maize and cowpea. Nigeria Journal of Agronomy 3, 65 – 76.

Oladokun, M.A.O. 1990: Vegetative propagation of kolanut trees. Ife Jour. Agric. Forestry 1&2: 25-30.

Sobulo, R. A. and Osiname, O. A. 1981: Soils and fertilizer use in Western Nigeria, Research Bulletin (II), I.A.R. & T., University of Ife, 20-26.

Soil Survey Staff. 1999: Soil Taxonomy. A basic System for soil classification for making and interprinting soil surveys. USDA Hand book No 436, Washington, D.C, USA.

Swift, M.J. and Anderson, J.M. 1993: Biodiversity and ecosystem function in agricultural systems. In Schultz *et al* (Eds.). Biodiversity and ecosystems function (pp.201-203) Berlin, Germany: Springer-Verlag.

Walkley, A. and Black, I. A. 1934: An examination of degtajaroff method for determining soil organic matter and a proposed modification of the chronic acid filtration. Soil Sci. 37:29-38.

Permissions

The contributors of this book come from diverse backgrounds, making this book a truly international effort. This book will bring forth new frontiers with its revolutionizing research information and detailed analysis of the nascent developments around the world.

We would like to thank Dr. Joann K. Whalen, for lending her expertise to make the book truly unique. She has played a crucial role in the development of this book. Without her invaluable contribution this book wouldn't have been possible. She has made vital efforts to compile up to date information on the varied aspects of this subject to make this book a valuable addition to the collection of many professionals and students.

This book was conceptualized with the vision of imparting up-to-date information and advanced data in this field. To ensure the same, a matchless editorial board was set up. Every individual on the board went through rigorous rounds of assessment to prove their worth. After which they invested a large part of their time researching and compiling the most relevant data for our readers. Conferences and sessions were held from time to time between the editorial board and the contributing authors to present the data in the most comprehensible form. The editorial team has worked tirelessly to provide valuable and valid information to help people across the globe.

Every chapter published in this book has been scrutinized by our experts. Their significance has been extensively debated. The topics covered herein carry significant findings which will fuel the growth of the discipline. They may even be implemented as practical applications or may be referred to as a beginning point for another development. Chapters in this book were first published by InTech; hereby published with permission under the Creative Commons Attribution License or equivalent.

The editorial board has been involved in producing this book since its inception. They have spent rigorous hours researching and exploring the diverse topics which have resulted in the successful publishing of this book. They have passed on their knowledge of decades through this book. To expedite this challenging task, the publisher supported the team at every step. A small team of assistant editors was also appointed to further simplify the editing procedure and attain best results for the readers.

Our editorial team has been hand-picked from every corner of the world. Their multi-ethnicity adds dynamic inputs to the discussions which result in innovative outcomes. These outcomes are then further discussed with the researchers and contributors who give their valuable feedback and opinion regarding the same. The feedback is then collaborated with the researches and they are edited in a comprehensive manner to aid the understanding of the subject.

Apart from the editorial board, the designing team has also invested a significant amount of their time in understanding the subject and creating the most relevant covers. They scrutinized every image to scout for the most suitable representation of the subject and create an appropriate cover for the book.

The publishing team has been involved in this book since its early stages. They were actively engaged in every process, be it collecting the data, connecting with the contributors or procuring relevant information. The team has been an ardent support to the editorial, designing and production team. Their endless efforts to recruit the best for this project, has resulted in the accomplishment of this book. They are a veteran in the field of academics and their pool of knowledge is as vast as their experience in printing. Their expertise and guidance has proved useful at every step. Their uncompromising quality standards have made this book an exceptional effort. Their encouragement from time to time has been an inspiration for everyone.

The publisher and the editorial board hope that this book will prove to be a valuable piece of knowledge for researchers, students, practitioners and scholars across the globe.

List of Contributors

Z.M. Zheng
School of Environmental Sciences, University of Guelph, Guelph, Ontario, Canada

T.Q. Zhang
Greenhouse and Processing Crops Research Centre, Agriculture and Agri-Food Canada, Harrow, Ontario, Canada

Alberto C. de Campos Bernardi, Patrícia P. A. Oliveira and Odo Primavesi
Embrapa Pecuária Sudeste, São Carlos – SP, Brazil

I.J. Díaz-Maroto, P. Vila-Lameiro and O. Vizoso-Arribe
Santiago de Compostela University, Spain

E. Alañón and M.C. Díaz-Maroto
Castilla-La Mancha University, Spain

Braulio Valles-de la Mora, Epigmenio Castillo-Gallegos, Jesús Jarillo-Rodríguez and Eliazar Ocaña-Zavaleta
Universidad Nacional Autónoma de México, Facultad de Medicina Veterinaria y Zootecnia, Centro de Enseñanza, Investigación y Extensión en Ganadería Tropical (CEIEGT), México

Ren Wan-Jun, Huang Yun and Yang Wen-Yu
Sichuan Agricultural University, Wenjiang, Sichuan, China

Felix K. Ngetich, Chris A. Shisanya, Jayne Mugwe, Monicah Mucheru-Muna and Daniel Mugendi
Kenyatta University, Kenya

Theodora Matsi
Soil Science Laboratory, School of Agriculture, Aristotle University of Thessaloniki, Thessaloniki, Greece

Roland Nuhu Issaka, Moro Mohammed Buri and Eric Owusu-Adjei
CSIR-Soil Research Institute, Academy Post Office, Kwadaso-Kumasi, Ghana

Satoshi Tobita and Satoshi Nakamura
Japan International Research Center for Agricultural Sciences, Ohwashi, Tsukuba, Japan

Tamas Kismanyoky and Zoltan Toth
University of Pannonia Georgikon Faculty, Hungary

Witold Grzebisz and Jean Diatta
Department of Agricultural Chemistry and Environmental Biogeochemistry, Poznan University of Life Sciences, Poland

Liliana Mahecha and Joaquin Angulo
GRICA Research Group, University of Antioquia, Colombia

Sarita Leonel, Dayana Portes Ramos Bueno and Carlos Renato Alves Ragoso
UNESP. FCA – Depto. de Produção Vegetal, Botucatu, SP, Brasil

Erval Rafael Damatto Junior
APTA – Pólo Regional Vale do Ribeira, Registro, SP, Brasil

Bidjokazo Fofana
IFDC - North and West Africa Division; IFDC Mali - B.P. E 103, Badalabougou-Est Fleuve; Bamako, Mali

Zacharie Zida
IFDC – North and West Africa Division, IFDC-Ouagadougou – CMS 11 B.P. 82, Ouagadougou, Burkina Faso

Guillaume Ezui
IFDC-North and West Africa Division; IFDC-Lomé, BP 4483, Lomé, Togo

Lubanga Lunze and Mulangwa Ngongo
Institut National pour l'Etude et la Recherche Agronomiques (INERA), Kinshasa, Democratic Republic of Congo

Mathew M. Abang and Robin Buruchara
International Center of Tropical Agriculture (CIAT), Kampala, Uganda

Michael A. Ugen
National Crop Resources Research Institute (NaCRRI), Namulonge, Uganda

Nsharwasi Léon Nabahungu
Institut des Sciences Agronomiques du Rwanda-Rubona, Butare, Rwanda

Gideon O. Rachier
Kenyan Agricultural Research Institute (KARI), Kakamega, Kenya

Idupulapati Rao
International Center of Tropical Agriculture (CIAT), Cali, Colombia

Emmanuel Ibukunoluwa Moyin-Jesu
Agronomy Department, Federal College of Agriculture, Akure, Ondo State, Nigeria

www.ingramcontent.com/pod-product-compliance
Lightning Source LLC
Chambersburg PA
CBHW070732190326
41458CB00004B/1137